PROFESSIONAL AND TECHNICAL WRITING STRATEGIES

Communicating in Technology and Science

FOURTH EDITION

Judith S. VanAlstyne

Broward Community College

Prentice Hall, Upper Saddle River, New Jersey 07458

Library of Congress Cataloging-in-Publication Data

VanAlstyne, Judith S., (date)
  Professional and technical writing strategies : communicating in
technology and science / Judith S. VanAlstyne. — 4th ed.
      p.   cm.
  Includes bibliographical references and index.
  ISBN 0-13-954736-3
  1. English language—Technical English. 2. Communication of technical
information—Problems, exercises, etc.   3. Communication in science—
Problems, exercises, etc. 4. Technical writing—Problems, exercises, etc.
5. English language—Rhetoric.
I. Title.
PE1475.V36   1999
808'.066—dc21   98-13106
                                                           CIP

Editorial Director: Charlyce Jones Owen
Editor-in-Chief: Leah Jewell
Sr. Acquisition Editor: Maggie Barbieri
Editorial Assistant: Joan Polk
Director of Production and Manufacturing: Barbara Kittle
Sr. Managing Editor: Bonnie Biller
Production Liaison: Fran Russello
Project Manager: Linda B. Pawelchak
Manufacturing Manager: Nick Sklitsis
Prepress and Manufacturing Buyer: Mary Ann Gloriande
Cover Director: Jayne Conte
Cover Design: Bruce Kenselaer
Electronic Art Creation: Asterisk Group, Inc.
Marketing Manager: Rob Mejia
Copy Editing: Geoffrey Hill
Proofreading: Maine Proofreading Services

This book was set in 10/12 New Century Schoolbook
by Typographics, a division of Black Dot Graphics
and was printed and bound by RR Donnelley & Sons Company.
The cover was printed by Phoenix Color Corp.

© 1999, 1994, 1990, 1986 by Prentice-Hall, Inc.
Upper Saddle River, New Jersey 07458

Printed in the United States of America
10 9 8 7 6 5 4 3

ISBN 0-13-954736-3

Prentice-Hall International (UK) Limited, *London*
Prentice-Hall of Australia Pty. Limited, *Sydney*
Prentice-Hall of Canada, Inc., *Toronto*
Prentice-Hall Hispanoamericana, S. A., *Mexico*
Prentice-Hall of India Private Limited, *New Delhi*
Prentice-Hall of Japan, Inc., *Tokyo*
Pearson Education Asia Pte. Ltd., *Singapore*
Editora Prentice-Hall do Brasil, Ltda., *Rio de Janeiro*

CONTENTS

PART TWO
The Technical Strategies

PART FOUR
The Research Strategies

CHAPTER 15
Writing Professional Papers   **481**

CHAPTER 16
Documenting Reports   **516**

PART FIVE
Presentation Strategies

PREFACE

The fourth edition of *Professional and Technical Writing Strategies: Communicating in Technology and Science* incorporates major revisions and additions. It is designed for students majoring in technical and scientific fields but helpful for all students who will share the writing responsibilities in any field of endeavor. Comprehensive and flexible, the text is suitable for college and university students at any level, technical school students, professional and technical writers, and others in technical/scientific employment seeking a guideline and model text. The materials have been tested in academic classes and in training workshops in a variety of businesses and technical industries. The book has won two professional awards: the Award of Distinction from the Everglades Chapter of the Society for Technical Communication, and the Award of Achievement from the International Society of Technical Communication.

College students of the twenty-first century are a heterogeneous group with a myriad of interests and needs. More often than not, they are computer literate and skillful at word processing, graphics generation, and document design. Their ages range from 17 to over 70 (average age, 28); they major in every field from aviation to zoology. The text includes writing samples that illustrate actual writing demands in a cross-section of career fields: allied health professions, architecture, aviation, computer science, ecology, engineering, fire science, insurance and real estate, landscape technology, medicine and dentistry, nursing and paramedical training, pest control technology, and all of the sciences. The samples in the text have been culled from professionals in all of these fields and more.

The emphasis is on practical writing and its applications. In this new edition, major revisions and additions cover the expanded role of the technical writer, multinational/multicultural communications, ethical considerations, collaborative writing, professional-level graphics and visuals, document design, career strategies on the Internet, new formats for resumes, manuscript submission requirements for publication, library and home computer information retrieval methods, multiple forms of documentation, and sophisticated verbal and visual communications. Each chapter now contains a checklist for the skills covered within the chapter.

In addition to the new aspects, the text covers (1) general communication strategies (the value of technical writing, basic communication theories, ethical considerations, the components of the writing process, graphics, visuals, and documentation design), (2) technical strategies (manual production, definition, description, instructions, and process analysis), (3) professional strategies (correspondence, resumes and cover letters, brief reports, longer reports and proposals), (4) research strategies (accessing information, writing the professional paper, and documentation), and (5) presentation strategies (verbal and visual communications).

Although the organization of the text is intended to offer cumulative skills, the instructor and the student may move about the text as freely as their purposes dictate. Each chapter provides a list of skills to be achieved, strategy guidelines, samples, a new checklist, exercises, and writing projects.

The *Instructor's Manual* offers general notes to the instructor, student preparation guidelines, and sample syllabi for a variety of class structures. In addition, the manual discusses approaches to each chapter, provides exercise solutions, and exhibits models for the writing projects that can serve as transparency masters.

ACKNOWLEDGMENTS

I am indebted to the students in my academic classes and professional workshops and seminars and also the business, technical, and industrial professionals who gave me advice and permission to reproduce their materials.

Without Merrill Tritt and Lawrence Liberman, computer wizards extraordinaire, I could not have mastered the complexities of sophisticated word processing software, computer graphics, document design samples, or Internet information. Their computer skills aided me all along the way. Daniel Barden of the Broward County Main Library offered superb advice about library computer retrieval systems.

I gratefully acknowledge the input received from the following reviewers: Gus Amaya, Florida International University; Tom Lannin, Winthrop University; John E. Moscowitz, Broward Community College, North Campus; and Matilda Delgado Saenz, North Lake College. I have conscientiously tried to incorporate their suggestions, revisions, additions, and deletions whenever feasible.

I particularly thank Linda Pawelchak, production editor, whose professional assistance throughout the editing and production process made this book possible. I am also grateful to Phil Miller, President, Humanities and Social Science, who has encouraged me throughout four editions of the text; and to Maggie Barbieri, Senior Editor/English, and her assistant Joan Polk. A sincere word of appreciation to the marketing department personnel, artists, copy editors, book representatives, and all the good people at Prentice Hall who have had a hand in this edition.

Judith S. VanAlstyne

ABOUT THE AUTHOR

Ms. VanAlstyne taught English at Broward Community College for thirty years and was an adjunct professor at Florida Atlantic University. She has served as an industry technical communications consultant and conducted technical writing workshops and seminars throughout three South Florida counties. Ms. VanAlstyne initiated the Technical Writing Curriculum and Certificate Program at Broward Community College. She has authored numerous graphic design texts, instructional videos for technical writing, and technical workshop/seminar workbooks. In addition to teaching expository and creative writing and literature courses, she introduced the technical writing course curriculum to a private college in Malaysia. She has also published literary criticism, travel articles, and poetry.

PART one

General Communication Strategies

Professional and Technical Communications

DUFFY   by Bruce Hammond

S K I L L S

After studying this chapter, you should be able to:

1. Define technical writing and explain the major difference between technical and scientific writing.
2. Define the *Rhetorical Method* of communication.
3. Define, *sender, receiver, message, channel, noise, encoding/decoding, interpretation,* and *feedback.*
4. Recognize the role of the sender.
5. Recognize the characteristics of an anticipated receiver and tailor the message to a particular audience.
6. List ten questions the sender must consider about the receiver.
7. State the purpose of a message quickly and clearly.
8. Differentiate between mechanical and semantic problems that may occur in the transmission of a message.
9. Name four multinational/multicultural concerns for technical report writers.
10. Name at least five document design elements.
11. Name two ways to elicit feedback on your message.
12. List seven obligations of a collaborative writing team.
13. Name five personal ethical obligations which pertain to writing.
14. Name three motives that may result in unethical organizational or corporate actions.

INTRODUCTION

Changes in communication tools are happening in gigabytes at the speed of milliseconds. The twenty-first century will see more innovations than we can predict or imagine. The information highway, much of it in cyber-space, allows us to spew out messages in mind-boggling quantity and at supersonic speed. Computers with word processing, the Internet, color printers, scanners, CD-ROMs, e-mail and videoconferencing software bombard us with communication materials. Telephone voice mail, cellular phones, beepers, videos, user manuals, formal reports, on-line documents, and graphics demand our attention. So much more is new or on the way, such as the Internet on our television screens, voice-controlled computers, digital cameras, digital video discs (DVD-ROMs), and e-mail capabilities for pagers. There are literally billions of binary impulses converting input into streams of information. Without the scientific and technical writer to provide the explanation and operations of these breakthroughs, however, the twenty-first century technology would overwhelm us.

Effective transactions of all occupations (in business, industry, services, education) depend upon precise communications, which include letters, memos, directions, instructions, reports, and articles as well as graphics and oral presentations. This communication must be concise, orderly, correct, and well-designed in order to elicit results, not confusion. Both the writer and the reader are challenged to participate in collaborative understanding.

THE TECHNICAL WRITER

Industry is pressed for people with both technical and communications training. A new technical product hits the market every 17 seconds. According to the U.S. Department of Labor's *Occupational Outlook Handbook* (1994–1995), the employment of writers and editors is increasing faster than the average for all other occupations. To meet the challenge colleges and universities across the country have developed technical communications courses at all levels. You may take an introductory course such as this one, or continue a degree program, such as a Bachelor of Science in Technical Communications, a Master of Science in Technical Information Design and Management, or even a Doctor of Philosophy in Rhetoric and Scientific and Professional Communications, to name a few.

Not only the professional technical writer, the engineer, or the scientist needs to master the skills of clear technical communication, but also the entry-level employee. The clerk, the firefighter, the pest control technician, the health care worker, or anyone in the technical, scientific, and industrial disciplines will contribute to the written records of an organization.

Countless surveys reveal that employees at all levels, not just the professional technical writers, spend from 25 percent to 80 percent of their time on the job writing. That represents slightly more than two to six hours a day or ten to thirty hours each week. Further, another significant percentage of time is spent dealing with the writing of others. Consider the time spent reading letters, e-mail, memos, reports, operation manuals, professional journals, and so forth. Finally, despite the diversity of careers in the world, the writing skills they require are more similar than dissimilar.

THE VALUE OF TECHNICAL WRITING

The employee who writes well is the one who is noticed by management and marked for promotion. Your ability to write well reveals not only your command of language and grammar but also your organizational skills, your attention to detail, your persuasiveness, and your logic. Your ability to recognize and execute various strategies of technical writing plus your ability to enhance your documents with print variety and graphics will be in great demand. Your writing may even suggest a solid commitment to your employer and profession.

Not every technical writer is an employee of a company. Some people establish themselves as freelance writers, marketing themselves to technological and manufacturing companies and charging very respectable fees. These writers need to study contracts and be aware of ownership issues.

DEFINITION

Technical writing is that which informs, explains, instructs, or persuades a specific audience through mastery of specific strategies and sometimes a special language, so that the readers gain new knowledge or the ability to perform their own complex and specialized jobs more effectively. The writer or collaborative writing team may range from entry-level employees to highly trained technicians and communicators. Increasingly, professional technical writers not only provide written manual and product support materials. They also assume the tasks of other information channels, such as the development of Web instruction, Internet instructional movies for downloading, development of materials to be imbedded into the product similar to the "Help" sections on Windows and computer software, CD-ROM production, and video and audiotape materials. The technical writer also requires skilled interviewing techniques to question product engineers, marketing specialists, and others involved in the development of a product.

Technical and scientific writing differ somewhat in that technical writing serves technology whereas scientific writing serves science investigating natural phenomena. Science writing, however, is taught in journalism courses and sometimes in technical writing courses. Science writing is more like technical writing in that it informs and explains discoveries, theories, research, and systems to lay people. Technical and scientific writers need command of communication strategies. They will routinely be asked to

- Write clear memos or letters seeking action or responding to requests
- Write progress reports on ongoing work
- Develop proposals for new projects, procedures, funding, or personnel increases
- Define and/or describe a product or procedure to employees or customers
- Instruct employees or clients on how to use a product or conduct a procedure

- Explain or analyze a product or procedure for better understanding or improvement
- Persuade management through proposals for new projects, procedures, funding, personnel increases or duty changes
- Contribute articles for the employee newsletter or other publications
- Edit and review documents written by other colleagues
- Make oral and/or visual presentations

As writers improve their communication skills, they will move from writing and editing strategies to the more sophisticated skills of team-producing documents, critical and ethical thinking and problem solving, conducting research, and project management.

Besides textbooks on technical and scientific writing, a number of journals provide up-to-date studies in such professional writing. They include

The ATTW Bulletin

The Technical Communication Quarterly

Studies in Technical Communication

Journal of Business and Technical Communication

Written Communication

IEEE Transactions on Professional Communication

Science, Technology, and Human Values

Journal of Computer Documentation

The serious student of technical communication will examine these journals carefully and subscribe to the one or two that best supplement one's studies.

BASIC COMMUNICATION THEORIES

Aristotle, the famous Greek philosopher who lived 300 years before Christ, analyzed a system of communication that orators used to persuade their audiences to adopt their point of view. He called this the *Rhetorical Method* and emphasized that the orator should analyze his audience carefully in order to adjust the message to the particulars of that audience. More recent research has added new considerations for effective communication in our time.

In the 1940s researchers at Bell Telephone Laboratories devised a model of the process of human communication with attention not only to audience analysis, but to the message itself, the method of transmitting the message, barriers to clarity, and the strategies used to underscore the message. Modern communicators also pay a great deal of attention to the receiver's ability to interpret the message and to react to it.

Increasingly, communication theory recognizes that writing is a collaborative effort between same-level colleagues, others highly trained in computer and graphic capabilities, and those with technical know-how in engineering, marketing, or other specialties. The group or writing team must coordinate skills to produce the end product without causing undue burdens on any one member, jealousies, or hurt feelings.

Each of these theories, the Rhetorical Method, the Bell Telephone Laboratories Model, and the Group Writing Strategy, will be investigated.

Rhetorical Method

Aristotle was concerned with the structure of language, the formation of sentences, and the methods by which one could communicate clearly. The word *rhetoric* is from the Greek for the art of oratory. Aristotle's rhetorical method taught the would-be politician, lawyer, and teacher the art of persuasion through words in order to sway the listeners (or readers) to his or her opinion. Much of the system involved questioning to evoke the desired response. He believed that if one investigates a subject by inquiry and looks at it from every angle, the general meaning will shine through.

Although the rhetorical method still has relevance for persuasive writing, such as proposals, it does not address the all-important aspects of audience perceptions, message particulars, and manuscript design which face modern technical writers. It does, however, address the need for careful self-questioning at every step of document production.

Bell Telephone Laboratories Model

The Bell Telephone Laboratories model of human communications consists of the following elements:

- **Sender:** the person(s) who originates the message
- **Receiver:** the audience to the message
- **Message:** the details and language of the communique
- **Channel:** the method of transmitting the message (print, speech, videotape, and so forth)
- **Noise:** mechanical or semantic barriers
- **Encoding/decoding:** the terms, symbols, and graphics which the sender (the encoder) chooses to carry the message and which the receiver (the decoder) accepts

Two more elements have been added by later researchers:

- **Interpretation:** the receiver's understanding of the message
- **Feedback:** the receiver's reaction to the message

Figure 1.1 shows a sender-receiver communications model:

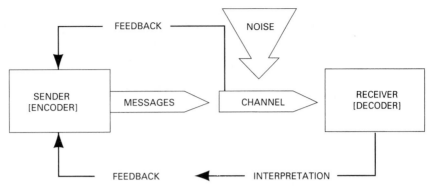

FIGURE 1.1   *Sender-receiver communications model*

APPLICATION OF THEORY

In order to apply this theory, the first step in tackling a professional, technical, or scientific writing project is to sort out answers to the following questions:

- What is my (our) credibility as the originator? (Sender)
- Who *exactly* is the audience? (Receiver)
- What stated purpose, information, and language are necessary for my (our) communication? (Message)
- What means (print, speech, videos, disks, speech, and so forth) are necessary to transmit the message clearly? (Channel)
- What special considerations (mechanical details and semantics) are needed? (Noise)
- What devices and supplements will make my (our) message more understandable? (Encoding/decoding)

Add to these two new and extremely important considerations that involve even more collaboration with the reader:

- What else must be considered, included, or eliminated to make the meaning crystal clear? (Interpretation)
- What reaction is desired? (Feedback)

One more question is vital:

- What considerations, such as overstatement, distortion, omissions, and so forth, must be considered? (Ethics)

Sorting out the answers to these questions is the key to effective professional, technical, and scientific writing. We will examine each separately, but, in fact, as you work out your project, you will be working on all of the areas concurrently.

SENDER

Capabilities.   Analyze yourself or your writing team in terms of the communication project at hand.

- How much do I (we) know about the subject?
- Do I (we) need to do research before proceeding?
- How credible am (are) I (we) in terms of my (our) background?
- What do I (we) need to do to be absolutely authoritative in the presentation?

Recognizing and rectifying your deficiencies will make your communication work.

Purpose.   By keeping one clear-cut purpose in mind, you will write a better document. Purpose is your intention, aim, or plan. Is your purpose to query? To explain? To analyze? To define or instruct? To persuade? Not only should your purpose be clear to you, but you should also establish that purpose as quickly as possible in your writing. Begin your document with such statements as:

> I am seeking additional information on your Corel Clipart software.

> This pamphlet has been prepared to instruct you in the operations of your BellSouth 25-Channel Autoscan Cordless Telephone, Model 33112.

> The feasibility of purchasing ten Mustek computer scanners for use in our ten advertising agencies is presented in this study.

> The purpose of this proposal, regarding a salary index scale, is to provide incentive to our employees to remain with the company and to eliminate the charge of "favoritism" that has contributed to low employee morale.

Knowing and establishing your purpose will keep your message "on track" and will give your reader a clear sense of how to handle your information.

RECEIVER/AUDIENCE

Your receiver is your audience. Anticipating your audience is a major key to your communication project. Your receiver/audience is the reader or readers to whom your written message is directed.

Just as you would make natural adjustments in vocabulary, sentence structure, and overall message in explaining the operation of a toaster to a child or to an adult, so will you make subtle changes in written instructions of any sort to a potential customer or to a group of factory assemblers. Consider the differences in the following two excerpts—one written for an advanced audience and the other for a novice audience:

> The buccal cavity is abounded laterally by the cheeks, anteriorly by the lips, superiorly by the hard and soft palates, inferiorly by the tongue and associated muscles, and posteriorly by the pharynx and uvula. It contains the teeth, which masticate food; the tongue, which aids in mastication and deglutition; and the salivary glands, which manufacture and excrete salivary amylase, an enzyme that attacks starch and hydrolyzes it into maltose and mucin, a very sticky substance that mixes with the food to form the bolus.

The vocabulary and sentence structure are, perhaps, too sophisticated for a novice reading audience. The following rewrite is more appropriate for an uninformed audience.

> The mouth cavity is bounded on the sides by the cheeks, in front by the lips, on top by the roof of the mouth, on the bottom by the tongue, and in the back by the throat. It contains teeth to chew food, and a tongue which helps to chew food and to swallow. There are also salivary glands which produce a chemical that attacks starch and breaks it down into a sugar, forming a sticky substance that holds food in a ball.

You must ask yourself still more questions to determine the characteristics of your receiver:

- Is my audience informed or uninformed on the subject?
- Does my audience have a grasp of the specialized vocabulary of my field?
- Is my audience singular or a group?
- Is my audience a subordinate, a peer, or a superior?

The Audience's Knowledge.   Carefully consider the degree of **knowledge** your reader(s) is already holding. Sometimes a supervisor or colleague is less informed about the writing project than you assume. This lack of knowledge may be why you were asked to prepare a communication in the first place. Perhaps the audience is of a different nationality or different culture, which may require careful investigation of language and multicultural perceptions. These differences may be crucial for understanding even a simple document. Ask yourself these questions:

- How would preliminary knowledge differ for the technician, the marketing member, research and development personnel, an interested layperson, or a potential customer?
- For the uninformed, what detailed statements are necessary to clarify purpose, background data, technical information, and other special information before proceeding?
- What tone, style, and sentence structure will best enable the receiver to understand the message upon completion? (See Chapter 2.)
- What, if any, multicultural and multinational differences should be considered both inside and outside the United States?

The Audience's Technical Level.   In this technological age you must consider the **specialized training** and **vocabulary** of your reading audience and make the necessary adjustments. Your audience may have familiarity with your content or be completely new to the subject. These considerations call for more analysis:

- Is the audience's professional background similar or dissimilar to mine?
- What training in the field can be accurately expected?
- What technical terms, words, abbreviations, and graphics are appropriate?
- What definitions, synonyms, descriptions, explanations, and other helpful details are needed?

Irrespective of the technical ability of your audience, plain English is always necessary, but in international communications other considerations, such as more flowery openings, literary allusions, and complimentary honorifics (titles and terms of respect) may be appropriate.

Whether your audience consists of **subordinates, peers, superiors,** or **laypeople** (everyone outside of your field) should have a bearing on your document or presentation. More questions are in order:

- What are the common denominators among the readers? What are their disparities?
- In the organization hierarchy, are my readers considerably or only slightly above me or my team, about the same, or slightly or considerably below?
- If the audience is within the organization, should the tone be deferential, gracious, familiar, or supportive?
- If the audience is outside my field and/or personal organization, such as readers of software manuals, instructions, and the like, what is the *lowest* level of understanding within that group to which I must address the message?

The Audience's International/Multicultural Aspects.   Other special considerations arise if your audience comprises members of **different nations** or **dissimilar cultures** even within your country. Today these differences simply cannot be overlooked. Tens of thousands of U.S. citizens are employed by overseas companies. In addition, we export many thousands of products abroad with heavy implications for the manuals and other written support materials explaining the installation, operations, and maintenance procedures for these products. There is a wide range of other types of intercommunication with foreign consumers. Our export sales account for billions of the U.S. gross national product. The European Economic Community is an enormous market as is Japan and other parts of Asia.

Even within our country about 30 percent of our work force in the year 2000 will be composed of rather recent immigrants to our country. Among our established citizenry are vast differences in educational systems, ethnic groups, and regional practices. In 1996 the school board of Oakland, California, passed a resolution to teach Ebonics, an amalgam of *ebony* and *phonics,* the prevalent black dialect spoken in America, which was proclaimed by the school board as the "primary language" of African Americans. Further research reveals, despite some definable differences, that the dialect may or may not have West African roots and/or may or may not derive from old Southern white speech. The point is that some African Americans have difficulty speaking and writing in Standard English, a fact communicators must recognize.

Even English-speaking people in different parts of the world attach variant meanings to words, vary spellings, and associate different values to colors. For instance, *toilet* usually suggests a flushing waste apparatus to an urban American but may suggest anything from an outhouse, a hole in the ground, or a ditch to a rural American or foreigner. To an American a *jumper* is a one-piece outfit, and a *torch* is a burning flame. To an Australian or a Brit a *jumper* is a sweater, and a *torch* is a flashlight. Americans spell *color* and *curb* while Brits spell the same words *colour* and *kerb.* The word *American* itself must be used cautiously because any member of a South American country considers himself or herself to be an American. Blue indicates first prize in the United States, but red indicates first prize in Great Britain. In the United States the color red indicates dangers or warnings, but such usage may confuse other people.

Translations into other languages get mangled. In Taiwan, the Pepsi slogan, "Come alive with the Pepsi Generation," translated as "Pepsi will bring your ancestors back from the dead." In Chinese, the Kentucky Fried Chicken slogan, "finger-lickin' good," translated as "eat your fingers off." When General Motors introduced the Chevy Nova to South America, the company was unaware that *no-va* means "it won't go." After GM figured out why it wasn't selling any cars, it renamed the car for its Spanish-speaking markets. One last example is from Italy where a campaign for Schweppes Tonic Water translated the name as Schweppes Toilet Water.

Rhetorical strategies differ from country to country also. Even though the international language of technology is English, the strategies of communication among nations may vary. The U.S. technical writer asserts a main idea and supports it with facts and evidence. The writer most often uses a "you" perspective, which does not mean the use of the second person pronoun but that the writer puts himself or herself in the reader's place and writes naturally, concisely, and objectively. Not so in all other nations. For instance, Arabs may not include a great deal of facts of evidence but use repetition to provide emphasis and persuasion. Some Hispanics do not include straight expository evidence but rely heavily on appeals to emotions to get across an idea. The French include a great deal of literary language and literary allusion in even a straightforward explanation. Asians may begin letters and other material with a great deal of attention to honorifics and flowery generalizations about friendship, the joy of doing business together, and so on. Just knowing a foreign language is not enough to discern multinational and multicultural differences.

Even **graphics layout** bears consideration. We read left to right, but other language groups read right to left. Our culture scans multiple graphics circularly clockwise. Other cultures may scan graphics circularly counterclockwise. Multiple graphics in a document may prove disconcerting to others. For instance, we would scan the following multiple graphics in the direction of the arrows:

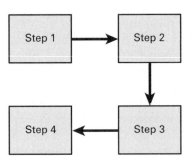

but an Arab-speaking or Chinese-speaking reader would possibly scan the graphics like this:

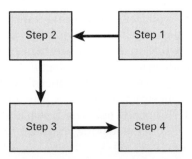

The typical icons used in computers do not always "travel" well. In Asia it is rude to point; therefore, the typical finger that "grabs" material from the toolbars or pointing finger icons indicating direction or continuation may be offensive. The mailbox and trash can icons for sending and receiving mail and deleting a document do not necessarily send universal messages. After all, the mailbox is a largely rural picture, whereas trash cans do not look the same the world over; in Indian and African cultures the trash can is probably a woven basket of some sort. The technical writer who writes for international and multicultural audiences must develop a sharp awareness of such differences.

MESSAGE

Your message is the explanation or analysis, response, request, set of instructions, descriptions, recommendations, or questions—the "meat" of your communication—that will accomplish your transaction. Developing the message involves the processes of preliminary notes, fact lists, organization, outlines, voice and tone, style, draft writing, revision and editing, and the like. These writing processes are discussed at length in Chapter 2.

CHANNEL

It is important to determine the best means of transmitting your message. Print alone may not be sufficient. The use of space as a visual indication of content can enhance your document. Besides space variations, text design elements may include column layout, margins, justifications, indentations, headings, font and type sizes, emphatic feathers, graphics, and other art. Chapters 3 and 4 discuss graphics, document design, and other art. Further, the communication project may suggest oral presentations, computer-enhanced presentations, a slide program, videotapes, and so on. The strategies of verbal and visual communications are covered in Chapters 17 and 18.

NOISE

Noise includes **mechanical** interference, such as illegible print copy due to poor typing, an outdated printer, faded print, or poorly reproduced duplicates. In addition, missing sentences or paragraphs, inappropriate graphics or other art, confusion of fonts and type sizes, and the like can be considered noise. Mechanical noise in verbal messages includes static or background noise in messages transmitted by phone and missing sentences or paragraphs.

Noise also includes **semantic** problems, such as the use of ill-defined terms, incorrect or inappropriate language, poor grammar, questionable usage, incorrect spelling or the omission of essential background information. Other noise or barriers to communication can arise from divergent backgrounds of the participants (e.g., differences in education, sex, race, nationality, intelligence) or the mental and physical stress of the sender or receiver at the time of the communication. Consider all of the possible barriers that might hinder effective communication.

ENCODING/DECODING

Encoding/decoding calls for you to determine the symbols (letters, numbers, words, graphics) and layout (space, headings, fonts, print size, and other design elements) that are most potent for conveying your message. In other words, you encode the message. Your receiver (the decoder) probably has piles of paperwork to handle—letters requiring responses, memos giving instructions for action, bulletins and reports imparting information to retain or react to, and more. That audience must be able to decode your intent exactly.

Bear in mind, too, that the last thing a beleaguered receiver wants is to read a "gray" message, a page or pages of uninterrupted type. To avoid the "gray" look, use the following encoding devices:

- Be as brief as possible.
- Divide your message into an introduction, body, and closing.
- Write in short paragraphs (four to five typed lines).
- Consider spacing on the page.
- Use headings (Completed Work, Work in Progress, Work to Be Completed).
- Vary the font and type size for divisions and headings with discretion.
- Use **boldface,** *italics,* or underlining for emphasis.
- Use numbers or bullets (1, 2, 3, ★, ◆, •, ✔) for lists of items.
- Insert graphics (tables, bar charts, line graphics, circle graphs, organization and flow charts), clip art, icons, and photographs for emphatic understanding.

All of the encoders that pertain to text design and layout are discussed in Chapter 4. Although a message may be as brief as two sentences, it may also be many pages. Careful attention to visuals will make your message more readable by dividing the text into logical units, adding emphasis to essential points, and helping the reader to relocate quickly those passages requiring action.

INTERPRETATION

Language Clarity.   Review all of your work thus far with an eye to interpretation. Eliminate all multiple meanings, possibilities of double interpretations, idioms, or other barriers to understanding.

Make clear if directions are being given from left to right or top to bottom. Indicate time zones if necessary. Clarify your authority (years of experience and/or research documentation). Consider if translations of parts or the whole should be offered in other languages. Remember that your clarity depends on a collaboration between the sender and the receiver.

FEEDBACK

Feedback from your audience constitutes an entirely new message. To determine if all of your communication is interpreted correctly, be alert to letters and phone calls that seek additional information. Be sure to include addresses, phone and fax numbers, and e-mail addresses in your message. Encourage feedback by asking for questions, being available for comment, or enclosing a questionnaire concerning your communication for return to you, the sender. The writers of operations manuals for electronic equipment, such as computers and their software, microwave ovens, VCR players/recorders, and the like often include a questionnaire/comment sheet. Figure 1.2 (see page 18) shows a typical Evaluation and Comments Sheet designed to elicit feedback.

Web pages on products and manufacturers also include questions and answers, ongoing discussion, and other means for interaction between user and manufacturer.

Group Writing Strategy

Theorists are now emphasizing the Group Writing or Collaborative Strategy. One person or writer seldom produces a technical document alone. The writer consults with management, colleagues, engineers, market managers, and others to determine the intentions and the particulars of a document.

Often a writing team will be appointed to co-produce a document. Group Writing calls for careful consideration so that each member understands his or her exact role.

A successful collaborative writing project requires at least seven considerations:

1. The **selection of group members** who can collaborate with common goals and recognize the strengths of others in the group. The document should be "owned" by all members of the group at its completion and not be more the product of one than another. Ideally, each member will have an equal say in the selection of the group leader, the content particulars, the activity plan, the document design, revision/editing decisions, and supporting multimedia.

EVALUATION AND COMMENTS SHEET

FROM:

NAME: _____　　TITLE: _____

BUSINESS

ADDRESS: _____

_____

PHONE: _____ _____

TITLE, PUBLICATION NO., AND DATE _____

1. BASED ON YOUR EXPERIENCE, RATE THIS MANUAL

	Good	Average	Poor
Manual Organization	____	____	____
Manual Readability	____	____	____
Supporting Illustrations / Diagrams	____	____	____
Installation Procedures	____	____	____
Operator's Procedures	____	____	____
Maintenance / Troubleshooting Procedures	____	____	____

2. PLEASE WRITE COMMENTS IN THE SPACE PROVIDED BELOW INDICATING ERRORS YOU HAVE FOUND. INCLUDE PARAGRAPH OR FIGURE NUMBER AND THE PAGE NUMBER.

_____

_____

_____

_____

_____

3. WHAT WOULD YOU SUGGEST WE ADD TO MAKE THE MANUAL MORE INFORMATIVE AND USEFUL?

_____

_____

_____

_____

_____

FIGURE 1.2　*Typical feedback solicitation form*

2. The **selection of a lead person** who can motivate, coordinate, and deliberate between the members. This leader is not necessarily the senior employee, but the one who manages other people well—a "people person." He or she recognizes the skills of each team member or collaborator. For instance, one may be stronger on the actual writing, one on layout and graphics, one on revising and editing, and another on multimedia support materials, but ideally all members will participate in each activity. Generally, the leader makes the final decisions on the activity plan, the document design, the revision/editing changes, and the necessary multimedia support materials.

3. The **development of an activity plan** with specific assignments and deadlines. It must include a number of informal consultations, scheduled group meetings, and computer hookups in order to stay on track. A calendar of assignments and deadlines should be in writing. Progress reports, preferably written, should be given by all participants at each group meeting or circulated on their computers. All decisions and transactions should be written down, possibly in memo form, following each group meeting and distributed to all participants as definite reminders of the group progress.

4. A **sender/receiver awareness** of the previously discussed theories: the Rhetorical Method and the Bell Telephone Laboratories human communications model and its particulars. Keeping the audience in mind is always key to effectiveness.

5. The **development of a document design plan** that includes paper quality and size, page orientation, access clue decisions, presentation particulars, additions and/or perforations, graphics, art, print size, and so forth. In addition, the plan should indicate whether there will be a title page, a table of contents, a bibliography, an index, and other presentation inclusions. These particulars are discussed at length in Chapter 4.

6. A **revision/editing plan** including indications of who will read each draft, who will suggest changes, who will be in charge of final editing, who will check all visuals, and so on.

7. A **production plan** determining whether the document will be an in-house publication or one that is jobbed out. An assigned production manager will oversee the final production.

These considerations put to work should result in an effective document that all members endorse.

ETHICAL CONSIDERATIONS

Ethics are the rules of human behavior and values of a society with respect to the rightness and wrongness of certain actions, and to the goodness and badness of the motives and ends of such actions. You are or will be competing in a political, bureaucratic, and market-driven environment. Personal ambition, competition, and the need for sales often strain personal and company ethics. Your writing can easily affect the values and judgments of others. Therefore, you should always adhere to the strictest code of conduct in actions and writing. While you may aim for persuasion, you must avoid exaggerated statements, distorted facts, misrepresentations, and omission or suppression of important information. You have a strong ethical obligation to present the unadorned truth in its full content.

A 1996 study by the Ethics Officer Association and the American Society of Chartered Life Underwriters and Charted Financial Consultants suggests that ethical lapses on the job are frequent. The study, based on survey responses from 1,324 workers nationwide, found that 48 percent of the respondents said pressure had caused them to engage in at least one unethical or illegal action during the past year.

Personal Ethics

Personal ethics become an issue the moment you seek a job. It is unethical to exaggerate or misstate your credentials, experience, and expertise. It is also illegal to claim you have earned a degree at a college or university when in fact you have not. On the other hand, it is unethical to write an unfactual or glowing recommendation for a friend when, in fact, you know of that person's shortcomings or the superiority of another applicant for the same position. Eventually, your distortions will catch up with you and reflect negatively on your good name. Equally it is also unethical to suppress knowledge or give false information that would sabotage a colleague, such as failure to own up to the fact that another person is more responsible for work for which you are being credited. Understatement may be as unethical as overstatement. For example, if you only faintly acknowledge the contribution of another instead of being explicit, you may undermine that person's expertise.

You must guard against championing a position just because the majority of others in your group or organization do. Ask yourself if you are being fair and completely honest. If you verbally attack or maliciously ridicule someone, you may be subject to libel laws, which can hold you to verbal or monetary restitution. You must respect everyone's legitimate rights to privacy and confidentiality.

It is also unethical to give away company plans, formulas, and so forth. As an employee, you must be loyal to your company as long as it acts within ethical boundaries. Some employees have been induced by rival companies to give up development directions, specific technologies, and other confidential material. Although this may not involve unethical written communication, it is an ethical issue which you may face in the future.

Finally, writings should not be plagiarized; this is the presentation of the words, organization, ideas, and conclusions of others as if they were your own. You may draw from many sources as long as you give credit for joint authorship, obtain legal permission for the use of others' materials, and/or include formal documentation. (See Chapter 16 for documentation procedures.)

Organizational Ethics

The profit motive, the competition motive, the need for speed in delivering new technology, or the maintenance of a company's reputation can press individuals or entire companies to unethical behavior. For instance, if your company has been dropped from a rating organization, is it ethical to advertise, "For 11 years XYZ Standards has rated us among the top computer software companies in the country"? Doesn't that imply that your company is still on the current list? Or if sales in your company picked up only moderately, is it fair to write, "Sales in the first quarter showed tremendous gain"? If a prestigious person is no longer with your organization, should that person's name be retained on the company masthead?

Putting very important warnings or cautions in extra fine print that may be overlooked may be unethical. Failure to raise questions about document inclusions or exclusions that you believe are dishonest or misleading may be deemed as unethical conduct on your part.

There are numerous examples of unethical conduct in manufacturing, business, and government. Some of these are not directly unethical written communications, but all involved the written word in memos, letters, and documentation of the incidents. Some examples follow:

Incident 1. A travel agency owner issued a memo encouraging his employees to utilize the familiarization trips offered to agents by tour companies, cruise ship lines, resorts, and so on. These trips are offered at reduced rates for both agents and their travel companions and are considered perks of the industry. The owner of one agency, however, demanded that the travel companions pay the agency the commission that would have been earned at the book price had the traveler not been eligible for a price reduction. While it may be legal for the owner to demand these charges, it is surely unethical not only for him to suppress his intention in the memo but also for him to violate industry practice.

Incident 2.   Serious consequences have resulted from scientific and pharmaceutical concerns that have suppressed information and documentation on the toxic results of products. In the late 1950s a German manufacturer withheld information on research regarding birth defects in rabbits given K-27, Thalidomide. The drug, already on the market, resulted in more than 8,000 deformed human births.

Incident 3.   The makers of silicone implants for breast augmentation issued written documents on the value of their product but withheld information indicating a tie-in to cancer. After silicone implants became the number one surgery in the United States, problems became too numerous to ignore, and huge lawsuits against the manufacturers resulted.

Incident 4.   Manville Corporation, the manufacturer of asbestos, withheld information from its employees and the public on the adverse effects of asbestos, resulting in many cases of lung disease and resulting deaths. Schools, dormitories, and public work places incurred extraordinary expenses to rip out and replace all asbestos. Communiques on issues of lead in paint, toxic waste disposal, and untested automobile safety features are so frequently handled unethically that the results are death and punitive damage suits that stagger the company budget.

Incident 5.   Another devastating result of unethical behavior arose when despite written engineering evidence that the O-rings in the *Challenger* space shuttle were not safe for a launch in temperatures under 75 degrees and that exhaust leakage had occurred in testing under 53 degrees, Thikol Company top managers neither released the engineering information to NASA nor canceled the launch even though the temperature was barely above 30 degrees. The result was the explosion of the shuttle and the death of all seven crew members 43 seconds after launch.

Incident 6.   In 1995 Valujet Airline officials stated to the press that the oxygen canisters that caused the fire, explosion, and death of all 147 passengers and crew were empty and had been stored correctly. Even when top officials were informed that the canisters that read "Empty" were not, several days passed before this information was released. Both the packers who mislabeled the canisters and the airline officials behaved unethically.

Incident 7.   In 1996 Newt Gingrich, the Speaker of the U.S. House of Representatives, was investigated by the House Ethics Committee for using three tax-exempt foundations to finance his partisan, televised college course. Initially, he denied this to the committee, but months later admitted he had given it wrong information—inadvertently, he said. He agreed not to refute the charge. However, an equally unethical and illegal tape, by a Florida couple, of Gingrich's cell phone calls revealed that he intended to bend that promise. It can also be argued that the Ethics Committee was unethical by postponing the full hearing until after Gingrich was reelected Speaker. The hearing resulted in a $600,000 fine but only a verbal slap on the wrist. This incident underscores the questionable ethics of many participants in government.

Think like a lawyer (or even better, a judge or a jury) to protect yourself and your organization. Channel your memos and reports through levels of hierarchy to keep others fully informed, claim authorship of your own ideas, provide a clear written record of your activities in ambiguous situations, and avoid confrontations by always acting in the highest ethical manner.

Ethical Dilemmas

Not all disputable issues indicate clear action. Ethical dilemmas arise, according to Vincent R. Ruggiero in his book *The Art of Thinking: A Guide To Critical and Creative Thought,* whenever the conflicting obligations, ideals, and consequences are so very nearly equal in their importance that we believe we cannot choose among them, even though we must. In particular, the medical community offers many ethical dilemmas. Because we can detect birth defects in the womb, is it ethical to abort the baby even though it is legal to do so? Is it ethical to terminate the life of a baby whose brain did not develop in order to donate its other organs even though the baby will die rather quickly anyway? Is it ethical to use the medical findings of Nazi doctors who experimented on concentration camp prisoners during World War II? Because the human body's genes are now being mapped, will it be ethical to alter these genes in the uterus or after birth? These and other issues demand exacting examination and judgment. These issues will continue to arise in the twenty-first century from increasing scientific investigation. Should you write about any of these issues as a technical writer for a laboratory or as an employee of the legal or medical professions, you must present the information in full context without bending or omitting any truth.

WRITING PROJECTS

1. **Recognize Technical Writing.** Locate a sample of technical writing (memo, product description, instructions, analysis of a technical process, a magazine article, or a textbook page). Photocopy it and write two or three paragraphs to explain what makes this example

technical writing. What do you know about the "sender"? To whom is it written? What is its purpose? What is its major message? Besides the words, what other methods did the writer use to "channel" the message? What "noise" do you encounter? How does the writer help you to "decode" the message? Discuss the language clarity and document design. How does it elicit "feedback"?

2. **Future Technical Writing Demands on You.** Write a brief paper to describe the writing you will be expected to do in your chosen field. For sources, interview both an instructor and a professional in the field, look up the profession in the *Dictionary of Occupational Titles* in your library, and examine some writing pertinent to your field. How often will you be expected to write each day? What kinds of writing will it be? What specific writing skills will be required? How important will it be for your job? How important will it be for a promotion?

3. **Group Writing.**   Form a group of three or four classmates in your field of study. Select one of the technical writing examples required in Writing Project 1. As a group rewrite this sample for a junior high school student. Consider that student's knowledge and vocabulary. What technical terms must be simplified? How must the message be reworded? Do you need more or fewer graphics, numbers and/or bullets, headings, font and type size changes, and so forth? Should the document design change, and, if so, why? Decide on a group leader and divide this assignment.

4. **Multinational/Multicultural Audiences.** Form a small group whose members have similar international or cultural characteristics. Obtain a technical document (memo, letter, user's guide) written in English by a writer in another country or from another ethnic group. Call the foreign embassies and companies that you know import technical products to our country. Photocopy or attach the original document and write a one-page paper that points out stylistic differences, language differences, spelling differences, and any other linguistic or anthropological problems.

5. **Encoding/Decoding.**   Reorganize the following report to make the message more readable. Divide the material into logical paragraphs; add section headings, numbers, bullets, boldface print, or any other appropriate visual devices to make the proposal more readable.

> Here is a progress report on the in-service presentation regarding the modification of diets. The doctor's dining room has been reserved and confirmed for 30 October at 1:30 P.M. A memo to the dietary employees has been posted on the dietary information board. Refreshment arrangements have been confirmed by the catering department. The diet instruction packets are currently at the printer and should be ready by next Tuesday, 15 October. Notifications of attendance are being tallied by Paul Austin. A final tally of the

number and names of those who will attend will be confirmed, and name tags will be made for each attendee prior to the presentation. Upon completion of the in-service training, a review of evaluations will be conducted, and the results will be analyzed and circulated to all members and presenters. Expected attendance is higher than anticipated, and the committee is looking forward to much success.

6. **Ethics.** Write one paragraph about an ethical (not legal) problem you have encountered. This might be a situation that required your action or a ethical question that faced a friend or colleague of yours. Explain the situation. What choices did you (or the person involved) have? What did you (or the other person) do? What could have been done differently? How can this situation be avoided in the future?

7. **Ethics.** Interview a professor who has published scholarly books or articles. Write a brief report on how the author copyrighted his or her own materials, obtained permission to reprint any other person's or company's materials, and/or documented the words, ideas, and organization of other writers quoted or paraphrased in the project. Organize your notes into a brief report. Your audience is this class. Use short paragraphs, headings, boldface, or other print styles, numbers and bullets, appropriate and logical spacing, and any other devices that will enhance, not just decorate, your report.

NOTES

CHAPTER 2

The Writing Process

SHOE   by Jeff MacNelly

S K I L L S

After studying this chapter, you should be able to:

1. Name the four major steps in the process of writing.
2. Execute the steps of prewriting: analysis of audience/sender, etc., topic list compilation, and logical outlining.
3. Write a rough draft from an outline you have prepared.
4. Write both an inductive and deductive paragraph.
5. Avoid an emotional, flowery, judgmental, or pompous tone.
6. Write in a factual, objective, impersonal style.
7. Recognize and repair sexist language.
8. Recognize and eliminate jargon, shoptalk, and gobbledygook.
9. Recognize and eliminate wordiness.
10. Recognize and eliminate redundancies.
11. Recognize and repair nonparallel constructions.
12. Add signal words where appropriate.
13. Correct word usage mistakes.
14. Recognize and repair fragmentary, run-on, and comma-spliced sentences.
15. Use variety in sentence construction and sentence length.
16. Avoid unnecessary "There are" constructions.
17. Recognize and repair faulty grammar including subject/verb disagreements, pronoun/antecedent disagreements, faulty pronoun usage, misplaced modifiers, and split infinitives.
18. Repair errors in punctuation, abbreviations, capitalization, hyphenation, italics usage, number and symbol usage, and spelling.

INTRODUCTION

Writing is a *process* of prewriting, writing, and postwriting. All the word processing, graphic inclusions, and layout skills in the world have little or no value if your writing is not effective in itself. If your writing is to have clout, it must reflect careful planning, logical organization, exacting attention to detail, and thorough attention to revising and editing. Basically, writers perform three types of activities:

- Prewriting: analyzing the sender and audience, planning the content, and designing the document
- Writing: composing the message
- Postwriting: revising and editing

Figure 2.1 is a flowchart for the process of writing steps and substeps:

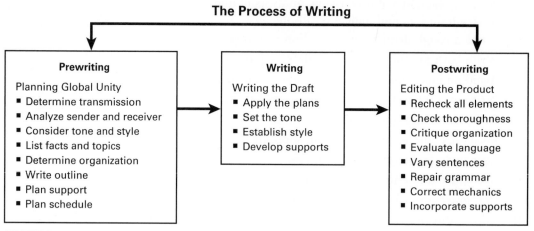

The Process of Writing

Prewriting

Planning Global Unity
- Determine transmission
- Analyze sender and receiver
- Consider tone and style
- List facts and topics
- Determine organization
- Write outline
- Plan support
- Plan schedule

Writing

Writing the Draft
- Apply the plans
- Set the tone
- Establish style
- Develop supports

Postwriting

Editing the Product
- Recheck all elements
- Check thoroughness
- Critique organization
- Evaluate language
- Vary sentences
- Repair grammar
- Correct mechanics
- Incorporate supports

FIGURE 2.1   *The writing process—a flow chart*

The experienced writer will include document design elements, graphics and other visuals, and presentation elements throughout all of the steps, but these considerations will be dealt with later. (See Chapter 4.)

PREWRITING

An effective writer seldom composes a finished product by merely typing a message off the top of the head. Global issues must be addressed concerning yourself and the audience, the purpose of your message, the topic list, the outline organization, the document design features, and the appropriate tone and style.

Global Unity

All of the considerations discussed in Chapter 1 and in this chapter should coalesce into a *global* whole, an overall unity. These issues include the analysis of the sender and the needs of the audience; the means of transmitting the message; the mechanical, semantic, and multicultural problems you may encounter; the encoding symbols and additions of support material; the document design; and feedback solicitation. Your message must use an appropriate organization, tone, and style. The facts must be correct and complete, and all possible questions must be anticipated. Your language must be concrete and free of jargon and other poor usage.

SENDER/RECEIVER

Review all the material on sender and receiver in Chapter 1. You are now ready to put this analysis to work. By careful analysis of yourself (or your writing team) and correction of any limitations you are ready to write. By targeting your exact audience, you will develop a document that appeals to the reader's needs and interests. Picture your reader as a busy person, and you will soon get the habit of coming quickly to the point. Put yourself on the receiving end. Will your reader understand your message? Are you planning the best way to transmit the message? Are you up to correcting any of its deficiencies?

TOPIC LIST

With your audience and overall purpose in mind, think over the material and consider the topics and facts you need to present. Research any missing data. List all of the facts and ideas. The list does not need to be elaborate. Brief topic entries will do to get you started. This chapter was conceived with the following loose topic list:

introduction	induction/deduction	voice
tone	wordiness	parallelism
prewriting	vivid verbs/nouns	shoptalk
global aspects	process steps	gobbledygook
emphasis	grammar	concreteness
style	spelling	roundabout phrases
organization	punctuation	redundancy
flow	usage	signal words
repairs	research	usage
jargon	sentence construction	agreements
pronoun problems	split infinitives	mechanics
spelling	sexist language	copyediting
document design	columns	visuals
draft writing	post writing	outlines

After further consideration, more topics would be added to your initial list.

OUTLINES

The next step is to organize the material into a tight presentation. Arrange your topic list entries into an outline that conveys logical order and completeness. As you organize, other important topics as well as logical groupings should emerge. Many word processors have an outline tool that helps you to organize your ideas. As you put topics together, you may easily edit, move, and subordinate information. You can move outline levels by using the icons and dragging selected text to a new location, and the outline will automatically be renumbered. An organized chapter outline with necessary subordination follows:

   I. Introduction
      A. Process
      B. Activities
  II. Prewriting
      A. Global approach
      B. Receiver/Audience
      C. Topic list
      D. Outlines
      E. Document design
         1. Space
         2. Margins, justification, and indentation
         3. Tabs and columns
         4. Headers and footers
         5. Headings, type sizes, and fonts
         6. Lists
         7. Emphatic features
 III. Writing the Draft
      A. Induction/Deduction
      B. Tone
      C. Style
         1. Sexist language
         2. Concreteness
            etc.

WRITING THE DRAFT

Writing the rough draft is simply a matter of expanding your outlined topics into sentences and paragraphs. Beyond global considerations of format, tone, and style, do not worry too much at this time about the refinements of language or the nitty gritty mechanical aspects in your draft. Refinements are the task of revision and repair. Try to write with a speedy flow, using simple conversational words and sentences.

Induction/Deduction Formats

Do decide whether to present the information in a particular paragraph inductively or deductively. *Inductive* presentation cites particular facts or individual cases followed by a general conclusion. *Deductive* presentation cites a general conclusion followed by the supporting facts or individual cases. Modern professional writing leans more and more to the deductive method without, however, entirely abandoning the other.

Draft 1 (Inductive)

Gross margins declined from 13.8% in the first quarter of 199X to 13.1% in the corresponding 199X quarter. There was a drop of 26% in sales volume between these two periods. At the branch level new unit margins dropped from 15.8% to 12.2%, but 50% of all units sold in the first quarter of 199X were from inventory made prior to 199X. Profits have been adversely affected by a decline in sales volume and a drop in gross margin.

Draft 2 (Deductive)

Company profits have been adversely affected by a decline in sales volume accompanied by a drop in gross margin. Sales in the first quarter of 199X were 26% below the volume in first quarter 199X. In the same period gross margin decreased from 13.8% to 13.1%. The principal reason for the lower gross margin in 199X was the fact that 50% of the 199X sales were from 199X inventory rather than new production.

The first draft loses the reader in a mass of figures while the second draft presents the key thought in the first sentence. It presents a global purpose first so that the reader is able to understand the statistics quickly.

The general conclusion placed at the beginning or ending of concrete supporting facts is also called the **topic sentence.** It organizes all of the other sentences. The topic sentence employs abstract generalized words expressing an opinion or conclusion. The supporting sentences are concrete, provable facts that support the opinion or conclusion stated in the topic sentence.

Tone

Almost all professional writing demands a factual and objective tone. Tone is the word choice and phrasing which expresses your attitude toward the subject. Professional writing is marked by its lack of emotionalism, editorializing, sarcasm, or even overt enthusiasm. Good writers avoid humor, satire, anger, irony, and bitterness. Consider this partial text of an incident report illustrating inappropriate tone:

On Wednesday, 26 May 199X, at 10:42 A.M., I had the misfortune of witnessing Keypunch Operator Polly Black take a nasty fall in the 5A West office area.

While not looking where she was going, Polly clumsily caught her heel on a CRT tri-stand and crashed to the floor. The paramedic on call at security rendered first aid and transported her to Mercy Hospital. The accident was due to sheer carelessness.

The tone in this report is sarcastic *(had the misfortune)*, judgmental *(while not looking where she was going, clumsily)*, emotional *(a nasty fall, crashed)*, and pompous *(CRT tri-stand, rendered, transported)*. The following, a revised version of the same incident, illustrates an objective tone:

On Wednesday, 26 May 199X, at 10:42 A.M., I witnessed Keypunch Operator Polly Black fall in the 5A West Office area.

While approaching the door, Ms. Black caught her heel on an equipment stand and fell on her left side. The paramedic on call at security treated her bruised knee and took her to Mercy Hospital.

The tone of professional communication should be factual and impartial.

Style

Technical writing uses an impersonal, simple, and nonsexist style. Style is the manner or mode of expression in language, your way of putting thoughts into words. Unless your audience is technically sophisticated, you will want to use simple, concrete words, uncomplicated sentence constructions, and short paragraphs. Avoid overtechnical terms, all jargon, shoptalk, gobbledygook, and overblown language. You must also avoid wordiness, redundancy, ambiguity, and unnecessary passive voice constructions. Inappropriate style presents more problems in technical and professional writing than does any other consideration. Some computer software offers grammar checks to help you examine and analyze your document, and some offers suggestions on style and tone. Too much reliance on these functions may prevent you from analyzing your own writing sufficiently. If your audience cannot understand your language, your message is lost.

SEXIST LANGUAGE

Sexist language favors one sex at the expense of the other. Our language tends to emphasize the role and importance of men over women. Today, sensitive writers consider carefully the usage of gender-linked nouns, titles, expressions, and pronouns.

Sexist Nouns	*Nonsexist Nouns*
manpower	human resources, work force
mankind	humankind, people
modern man	modern society, modern civilization

Sexist Titles	*Nonsexist Titles*
chairman	chairperson, chair, presiding officer
congressmen	members of Congress, representatives
fireman	firefighter
stewardess	flight attendant
policeman	police officer
salesman	sales agent

Sexist Expressions	*Nonsexist Expressions*
founding fathers	pioneers, founders
gentleman's agreement	informal agreement, oral contract

Sexist Pronouns	*Nonsexist Pronouns*
Each nurse treats *her* patients with care.	Each nurse treats patients with care.
A president sets *his* own agenda.	Presidents set *their* own agendas.
Every employee should sign *his* own clock card.	All employees should sign *their* own clock cards.

Sometimes plurals (*presidents, employees,* etc.) can be unclear.

Unclear   Drivers should study *their* manuals. (Does each driver have more than one manual?)

Clear   Each driver should study *his or her* manual.

Smoother   Each driver should study the appropriate manual.

Sometimes the use of *his or her, he or she, him and her* can be awkward, but they are, at least, clear and nonsexist. Usually a little thought can produce a nonsexist, smooth expression of the idea.

Drop the female suffix *-ess* in *authoress* and similar words. In professional letters avoid *Miss* and *Mrs.* unless you are sure; use *Ms* to indicate a female just as you use *Mr.* to indicate both single and married men. Notice that there is no period after *Ms* because it isn't an abbreviation, but a new title of respect. You will, however, often see the period used optionally as in *Ms. D. C. Grady.*

Sexism in language is not a trivial matter. Responsible writers are careful not to be offensive.

CONCRETENESS

Avoid general, abstract words which represent broad categories, unlimited number, and immeasurable concepts. Eliminate vagueness by using concrete and definite words to prevent misinterpretation and to increase readers' interest. Here are some abstract words and vague phrases along with suggested concrete terms and restructured sentences for clarification:

Vague	buildings
Concrete	bank, supermarket
Vague	devices
Concrete	laptop computer, IBM laser printer
Vague	expensive
Concrete	$70.00 $8 million
Vague	many
Concrete	5,000 to 20,000
Vague	Some of our competitors have very good businesses.
Concrete	Both Sunbelt Instruments, Inc. and Ohio Testing Laboratories grossed over $6.2 million during the fourth quarter of last year.
Vague	As we discussed recently, I have the figures on the project.
Concrete	I have the comparative costs of three word processing computers which you requested in our telephone conversation last Friday.
Vague	The policy change will affect us adversely.
Concrete	New Policy 1204.05 (Leaves) will decrease our allowable sick days from 10 to 8 per year.
Vague	We will fill your order within the next few weeks.
Concrete	We will ship C.O.D. your order for three, 4-drawer, 36 in. high by 45 in. deep by 14 in. wide, beige filing cabinets 2 September 199X. You should receive them no later than 30 September.

JARGON, SHOPTALK, AND GOBBLEDYGOOK

Jargon is the specialized vocabulary and idiom of those in the same work. The jargon of one field often spreads to the professional world at large. Other words and phrases are intelligible only to those in the same line of work and may be classified as *shoptalk.* Another type of jargon, which may be called *gobbledygook,* is characterized by unintelligible, pompous, or stiff language. General jargon and shoptalk may be acceptable in very personal oral communication, but a competent writer eliminates all jargon from written communication.

GENERAL JARGON

The following words and phrases are used rather widely in informal professional communications:

ballpark figure	mother of . . .
bottom line	optimization
finalized	output
game plan	parameters
handlers	politically correct
impacted	played
input	time frame
interface	viable

Moderate use of such terms is common in verbal transactions but should be revised for written messages. The following sentences demonstrate typical jargon exchanges and suggested written revisions:

Jargon Give me a ballpark figure on the new office furniture.

Revision Give me a price estimate on the new office furniture.

Jargon We'll use the input of each department to finalize our game plan.

Revision We will consider the suggestions of each department to complete our programming.

Jargon The bottom line is that the recession has impacted on our hiring time frame.

Revision The key point is that the recession has affected our hiring schedule.

Jargon The parameters for departmental interfacing must be viable.

Revision The guidelines for departmental boundaries must be realistic.

Jargon His running mate is a heavyweight debater, and if he doesn't dribble around in circles or suffer a late-inning letdown, he is sure to deliver the knockout punch.

Revision His vice-presidential candidate is a fine debater, and if he doesn't include too many details or get discouraged, he will win the debate.

Jargon The last vote was the mother of all elections.

Revision Over 100 million people voted last Tuesday in the largest turnout in history.

Jargon It just isn't politically correct to suggest a purchase from a company that is played.

Revision It just is not smart to suggest a purchase from a company whose sales are falling.

Jargon   Seth Deutsch is the handler of the Independent Party's candidate.

Revision Seth Deutsch is the policy adviser for the Independent Party's candidate.

SHOPTALK

The use of shoptalk, the more technical slang of those in the same profession, becomes second nature to the users but should never be used in writing. Every occupation has its own shoptalk.

Television shoptalk He shot the bridge with a minicam and then bumped up the tape.

Translation He filmed the connecting segment between the news items with a small camera and then machine-processed the videotape to a larger size.

Aviation shoptalk He checked the pax list, activated the SATCOM, and prepared the PIREP.

Translation He checked the passenger list, turned on the satellite communication system, and prepared the pilot's report on meteorological conditions.

Academic shoptalk The increase in FTE's is probably due to so many students' having clepped math.

Translation The increase in the number of full-time equivalency students is probably due to many students who waived the mathematics requirement by passing the College Level Examination Program test.

Internet shoptalk He went online to chastize the newbie who flamed a chat group last night.

Translation He connected to the Internet to chastize a new user who used abusive and rude language in a group conversation last night.

GOBBLEDYGOOK

Besides avoiding jargon and shoptalk, the skillful writer should avoid gobbledygook—unintelligible, pompous, and stiff language. Gobbledygook may sound more official or important but rarely states the message clearly.

Gobbledygook At this juncture, the aforementioned procedure should be utilized.

Plain English The plan which we discussed should be used now.

Gobbledygook	We should commence operational capabilities in systematic increments.
Plain English	We should begin the project step-by-step.
Gobbledygook	It would be prudent to consider expeditiously the provision of instrumentation that would provide an unambiguous indication of the level of fluid in the reactor vessel.
Plain English	We need a more accurate device to measure radioactivity.
Gobbledygook	The pilot deployed a vertical antipersonnel device and then activated his aerodynamic personal decelerator before the B29 underwent an involuntary conversion.
Plain English	The pilot dropped a bomb and ejected with a parachute before the bomber crashed.

WORDINESS

Effective professional writing is characterized by its brevity. The concise writer avoids roundabout phrases, redundancies, and sluggish passive-voice constructions.

Following is a checklist of shorter words and phrases to replace wordy, roundabout phrases:

Roundabout Phrases	*Concise Expressions*
a downward adjustment	cut, decrease
a great deal of	much
a majority of	most
accounted for the fact that	because
affix a signature to	sign
after the conclusion of	after
as a result	so, therefore
as a result of	because
as per your request	as you requested
as soon as	when
at which time	when
at all times	always
at an early date	soon
at a much greater rate than	faster
at the present time at this time	now
at the time of	during

avail yourself of	use
based on the fact that	because
be acquainted with	know
be of assistance to	assist, help
brief in duration	short, quick
by way of	by, to
came to an end	ended
consensus of opinion	opinion, everyone thinks
despite the fact that	although, though
due to the fact that in view of the fact that	because
enclosed please find	here is
for the purpose of	for, to
for this reason	so
for the reason that	since, because
give encouragement to	encourage
he was instrumental in	he helped
higher degree of	higher, more
in a manner similar to	like
in a position to	can
in accordance with	by, under
in favor of	for, to
in lieu of	instead
in reference to in relation to	on, about
in the amount of	of
in the nature of	like
in the vicinity	near, around
is dependent upon	depends on
is situated in	is in
it is necessary that	you must
it is recommended	we recommend
miss out on	miss
not infrequently	often
on account of	because, due to
on the part of preparatory to	from, of
prior to	before
provided that	if
pursuant to	under, with, following

referred to as	called
so as to	to
through the use of	by, with
to the extent that	as far as
until such time as	until
with reference to ⎫ with regard to ⎬	on, about
with the exception of	except
with the result that	so that

Avoid "Due to the fact that your reader has a great amount of other work to account for, it is necessary that you write so as to eliminate wordiness in your writing, through the use of concise words." Write "Because your reader is busy, write concisely."

REDUNDANCY

A *redundancy* is a phrase which says the same thing twice *(repeat again)*, contains obvious expansion *(square in shape)*, or doubles the idea *(each and every)*. Such repetition is pointless, wordy, and distracting. Consider the redundancies in the following sentences:

Redundant It is *absolutely essential* in *this day and age* to *completely eliminate bigotry and prejudice.*

Revision Bigotry must be eliminated now.

Redundant The sheriff's department and the city police *cooperated together* in *the month of May* to *devise and develop* a *totally unique drug and narcotics* control program.

Revision The sheriff's department and the city police cooperated in May to develop a unique narcotics control program.

Redundant There are *many in number* who consider the *total understanding of basic fundamentals* a *good asset.*

Revision Many consider the understanding of basics an asset.

VOICE

Verbs have two voices: active and passive. In an active-voice expression the subject of the sentence performs the action stated by the verb.

Mr. Jones *conducts* the plant tours.

The president *presented* the budget.

The firm is *spending* $50 million this year to promote the new beer.

The new price *will increase* our profit.

In passive-voice expressions the performer appears in a postponed sentence element or not at all.

The plant tours *are conducted* by Mr. Jones.

The budget *was presented* by the president.

Fifty million dollars *will be spent* this year to promote the new beer.

Our profit *will be increased* by the new prices.

Generally, the active voice suggests immediacy and emphasizes the subject. Because the passive-voice verb is always at least two words (the verb plus a form of *to be*), passive-voice expressions tend to be wordy. The passive voice may bury your main ideas in sluggish sentences.

Edit your writing to determine which voice permits your desired emphasis in the fewest number of words.

The secretary typed the report. (Emphasis on secretary)

The report was typed by the secretary. (Emphasis on report)

PARALLEL CONSTRUCTIONS

Parallel constructions are those that place related ideas in a sentence into the same grammatical forms. Adjectives should be parallel with other adjectives, verbs with verbs, phrases with phrases, and clauses with clauses.

Nonparallel Tungsten steel alloys are tough, ductile and have strength.

Parallel Tungsten steel alloys are tough, ductile, and strong.

Nonparallel He must use parallel constructions and look to avoid ambiguities.

Parallel He must use parallel constructions and avoid ambiguities.

Nonparallel The process of writing demands four activities: prewriting, message composition, the use of additions, and the ability to postwrite.

Parallel The process of writing demands four activities: prewriting, writing the message, adding graphics and other art, and postwriting.

All of these style considerations and more are discussed in Appendix A. This would be a good time to review the appendix.

POSTWRITING

Efficient revision and editing will repair your rough draft and produce a polished product. The first tip is to allow time between your draft writing and your revising and editing. Do something else for a few minutes, hours, or even days for a cooling-off period. You will find yourself more objective and critical when you return to the writing task. You are not alone in your urge to just get the paper off your computer, off your desk, and off your mind. But take the time to rethink, rewrite, and revise. The goals of postwriting are

- Rechecking all elements for audience appropriateness
- Establishing that all of the facts and topics are covered thoroughly
- Critiquing the organization and paragraphing
- Evaluating language and word choice
- Scrutinizing sentence construction and variation
- Repairing grammatical errors
- Correcting punctuation, spelling, and other mechanics

The idea here is to consider the relationship of the whole to the parts and the parts to each other. Add what is needed and purge unmercifully what is not. Review the copyediting symbols in Table 2.1 (see page 42).

As you read your entire draft, use the appropriate copyediting symbols in your text. Alternatively, put a check in the margin when you find a questionable word, an awkward or unclear sentence, an uncertain grammatical construction, a mechanical complexity, or a design discrepancy for later revision. Then begin a systematic revision of your work.

Overall Revisions

You need the basics before the shine. Study your document to assure that the first three goals (audience appropriateness, thoroughness of content, and logical organization) are met. Once again question if everything in the document is appropriate to the audience. Determine that the tone and style are consistent. Examine the facts for completeness and accuracy. Make absolutely certain that the overall purpose and emphasis are immediately clear. Establish that each paragraph or section has a clear focus and the necessary supporting information. Eliminate any digressions, needless repetitions, or padding. If you are team writing, have other members examine these first three areas and make all of the necessary revisions.

TABLE 2.1 *Copyediting Symbols*

Symbol	Meaning
㉟	Spell out
(two hundred)	Use numerals
Scientific	Use lowercase
french	Capitalize
Techncal	Insert letter
misspelled mispellede	Mark out grossly misspelled words and print the correct spelling above
saample	Delete and close up
clean clear the ribbon	Delete and join to remaining text
He typed the report ⊙	Insert period
to carefully read	Transpose words
thier	Transpose letters
. . . noise. When taping	Start new paragraph
the company's car	Insert apostrophe
run-on words	Insert space between words
to provide aide (stet)	Retain the original
know the symbols⊙	Close up or join
They will help . . .	
What did he mean?	Insert question mark
feet, inches and pounds	Insert comma
yourss	Delete letter

Language and Word Choice Revisions

The analysis of language and word choice is the next task of revising and editing. Review your language to determine if it is too technical for the audience. Consider multinational and multicultural differences that may distort your message. Review the material on sexist language, concrete wording, jargon, wordiness, redundancy, and voice in this chapter in order to apply these considerations to your revision. Revise your document accordingly. Next, examine the need for signal words, precise word choice, and usage corrections.

SIGNAL WORD REVISIONS

Add signal words—those that indicate time, additions, results, summary, comparisons and contrasts, and other indications of what is to follow—for paragraph to paragraph and sentence to sentence coherency. Consider the following paragraph:

> The writer should follow a number of steps to revise and edit a paper. The writer must recheck for audience appropriateness. The writer must establish that the facts and topics are complete. The writer should assiduously critique the organization and paragraphing. The writer must evaluate the language and word choice. The writer must scrutinize sentence construction and variation. He must make grammatical corrections. He must diagnose and correct punctuation, spelling, and other mechanics. He must reexamine and enhance the document design and additions. He must prepare the final document.

In the revision, signal words have been added for unity and coherence and some other language changes have been made.

> You should follow a number of steps to revise the paper. **First,** you must recheck the document for audience appropriateness. **Next,** you must establish that the facts and the topics are complete, **and** you should carefully recheck the organization and paragraphing. **Additionally,** you must evaluate the language and word choice. **Equally important,** you should make grammatical corrections, **and then** you must diagnose and correct punctuation, spelling, and other mechanics. **Then,** you should reexamine and enhance the document design and additions. **Finally,** you should prepare the final document.

Besides the addition of the signal words, *you* replaces *the writer* and *he* to personalize the message and to eliminate the sexist *he* or *he and she.* Furthermore, the word *carefully* replaces *assiduously* for a more consistent style.

A list of signal words to make connections and transitions follows:

Time Signals	*Addition Signals*	*Results Signals*
soon	again	hence
then	and	therefore
finally	besides	consequently
previously	furthermore	accordingly
first, second, etc.	also	thus
next	additionally	as a result
last	moreover	finally

Summary Signals	*Contrast Signals*	*Comparison Signals*
in brief	however	similarly
finally	nevertheless	likewise
in conclusion	yet, and yet	correspondingly

to conclude	but	equally
summing up	still	equally important
on the whole	on the other hand	in the same manner
lastly	though, although	in the same way

Other Signals

simultaneously	conversely
meanwhile	unfortunately
for example	to begin with
for instance	so
to demonstrate	in the past
indeed	in the future
in other words	eventually

WORD CHOICE CHECK

If you are composing and editing your document using word processing software, you probably have a **thesaurus** tool to seek out more precise meanings to words and to find synonyms for weak or overused words. You may even have a **dictionary** providing exact meanings, synonyms, and antonyms. If you do not have these tools on your computer, a desk dictionary and word thesaurus are absolutely necessary for the effective writer. Pay particular attention to your verbs; select the word with just the right nuance. For example, here are two lists of verbs requesting action. Notice how each column increases in intensity from top to bottom:

(would) appreciate	hinted at
(would) suggest, propose	implied, intimated
prefer, would like	suggested
request, ask	indicated
recommend, urge	signified
remind	demonstrated
advise	confirmed
require	substantiated
insist	proved
demand	established

Likewise, noun synonyms can suggest minute shades of meaning.

aid	sketch
assistance	plan
help	program

full support   design

championship   project

USAGE CHECK

English is a complicated language containing homonyms (words that sound alike: *there, their, they're*) and other confusing similar words *(insure/assure/ensure)*. The astute reader is disturbed over common usage errors. Check carefully such words as *its/it's, accept/except, advice/advise,* and the like. The adverb *hopefully* is frequently misused; the writer usually means "I hope." The best advice is to avoid *hopefully* altogether. Your audience may miss the impact of your message if you fail to edit these problems carefully. Familiarize yourself with the **Usage Glossary** in Appendix A, which covers 41 sets of confusing words with explanations and an exercise to test yourself. Keep this glossary on hand every time you write or revise a document in order to repair your mistakes.

Sentence Repair

Consider the parts carefully for necessary sentence repairs and variation. It is all too easy to write sentence fragments, faulty parallel constructions, and run-on or comma-spliced sentences in your draft. Read through slowly for sentence sense. Find the main sentence elements (subject, verb, and object) to ensure that your sentences convey the ideas you intend. Again, Appendix A discusses basic sentence constructions and errors and provides some exercises to test your recognition of each error. Review that material attentively. Some reminders follow:

Fragment	Lithium has many uses in bioengineering. *Especially* in the development of pacemakers.
Repair	Lithium has many uses in *bioengineering. It is especially* useful in the development of pacemakers.
Parallel problem	We need adjuncts *to handle* peak-hour activity, *to free* full-time employees from routine duties, *to relieve* assembly workers for lunch breaks, and *for the replacement of* vacationing employees.
Repair	We need adjuncts *to handle* peak-hour activity, *to free* full-time employees from routine duties, *to relieve* assembly workers for lunch breaks, and *to replace* vacationing employees.
Run-on	If the Roman government at the height of its power, and at a time when means of communication had been greatly improved, showed anxiety for the food

supply of that Italy which was dominant in the Mediterranean world, it may be imagined that in the period preceding the great economic organization introduced by the Roman Principate the peoples of the Mediterranean region peoples no one of which at the height of its power had controlled the visible food supply of the world so widely or so absolutely, had far graver cause for anxiety on the same subject, and this was an anxiety such as would be, under ordinary circumstances, the main factor, or, even under the most favorable circumstances possible in those ages, a main factor, in molding the life of the individual and the policy of the state.

Repair   If we understand that the Roman government at the height of its power showed anxiety for the food supply, we can imagine that earlier Mediterranean peoples had far graver cause for anxiety. The Roman Principate had greatly improved communications and economic organization to handle the food supply of Italy, the dominant country in the Mediterranean world. No earlier people in the region had controlled the visible food supply so widely or so absolutely as had the Romans. Under ordinary circumstances the food supply was a main factor of concern. Under even the most favorable circumstances, anxiety over the food supply would be a main factor in molding the life of the individual and the policy of the state of those earlier peoples.

Comma-splice   Cellular telephones are available in a wide variety of styles, each manufacturer offers other special features.

Repair   Cellular telephones are available in a wide variety of styles; each manufacturer offers other special features.

Comma-splice   You may have both a separate fax machine and a computer fax capability, therefore, you should consider which will be most efficient for your demands.

Repair   You may have both a separate fax machine and a computer fax capability; therefore, you should consider which will be most efficient for your demands.

Sentence variation is another area for examination and revision. Too many simple or compound sentences may not only be dull but also suggest that you do not have a good command of the alternative compound and compound/complex constructions. Review the sections on basic sentence elements and secondary elements in Appendix A. Then revise your material to lend a variety of constructions with logical participial phrases and dependent clauses. Consider the following paragraph, which employs five simple sentences:

> The writer must carefully consider the concerns of prewriting. He must consider the audience. He must then devise a topic list of the areas to be covered in the document. Next, he must organize the list into a logical outline. He may then move to the next step, writing the draft.

Consider how the use of participial phrases and dependent clauses adds variety of sentence construction and sophistication to the paragraph:

> To write well you must carefully consider the concerns of prewriting (Simple). Considering the needs of the audience, you will have direction for planning the document (Complex). Next, you must devise a topic list of the areas to be covered in the document, and then you can devise a logical outline which will guide the preparation of the draft (Compound/Complex). Finally, you are ready to write the draft (Simple).

The word *you* replaces *the writer* to offer a more personal tone and to avoid the sexist *he.*

Too many very long sentences or too many very short sentences will bore or burden the reader. An easy-to-read sentence is usually 12 to 25 words long, but variety adds smoothness and emphasis. Consider the following:

> I prepared for the Olympics by exercising my body with weights and aerobics. I repeated my routine at least 15 times a day. I asked my coach to point out even the slightest movement that could be improved. I designed and redesigned my costume for the event continually. Finally, it was my turn for my ice skating routine, and I won.

In this paragraph each sentence contains 10 to 13 words, which is acceptable, but they are boring in their construction and rhythm. Look at the following revision:

> I prepared for the Olympics by exercising my body with weights and aerobics and repeated my routine at least 15 times a day. I asked my coach to point out even the slightest movement that could be improved. Furthermore, I designed and redesigned my costume continually until finally it was my turn for my ice skating routine. I won.

In this example sentences have been combined so that the variation is from 15 words to 23 words, followed by the dramatic final two words.

Overuse of the indefinite phrases *there is* and its related forms (*there are, there will be, there may be, there might have been,* and so on) weaken sentences by delaying the appearance of the subject. Eliminating most of these constructions will make your sentences more effective.

Consider the following:

Ineffective	There are specific steps to take to make your sentences more effective.
Effective	Specific steps will make your sentences more effective.
Ineffective	There will be a fire alarm scheduled for this afternoon.
Effective	A fire alarm is scheduled for this afternoon.

Sometimes, however, the construction adds emphasis or avoids the verb *exists*.

Weak	A very good reason to rest between exercise sets exists.
Stronger and eliminate *exists*	There is a very good reason to rest between exercise sets.

Grammar Repair

Once again your computer may have a **grammar check** tool that ferrets out grammatical errors, unnecessary repetitions of words, as well as some mechanical errors. This tool is valuable for a first scan of grammar problems, but it cannot replace your obligation to know correct grammatical conventions. The most common grammar errors include subject/verb disagreements, pronoun/antecedent disagreements, general pronoun misuse, modifier misplacement, and split infinitives. Appendix A addresses these problems at length and provides some exercises to test your knowledge, but here are some tips.

SUBJECT/VERB CHECK

A verb agrees with its subject in number.

Incorrect	*Each* of the insurance companies *are* reviewing the application.
Correct	*Each* of the insurance companies *is* reviewing the application.
Incorrect	The *manager* as well as the supervisor *have* been promoted.
Correct	The *manager* as well as the supervisor *has* been promoted.

PRONOUN/ANTECEDENT CHECK

Similarly, pronouns must agree with their antecedents in number.

Incorrect	*Each* of the women bought *their* own computer software.
Correct	*Each* of the women bought *her* own computer software.

Incorrect   *Ace Company* is sending 20 of *their* employees to the seminar.

Correct   *Ace Company* is sending 20 of *its* employees to the seminar.

PRONOUN USAGE CHECK

Pronouns are grouped into cases. Case 1, the nominative, is used for sentence subjects and predicate nouns, nouns which follow a linking verb and identify the subject. Appendix A covers pronoun case fully. Some common errors follow:

It is *I*.   (not *me*)

It will be *he* who goes.   (not *him*)

He and she were appointed to the committee.   (not *him and her* or *he and her*)

Ms Johnson and *I* attended the meeting.   (not Ms Johnson and *me*)

Case 2, the objective, is used for direct objects, indirect objects, and objects of prepositions.

Direct object   Catherine's secretary took her boss and *me* to lunch.   (not *I*)

Indirect object   Ken sent Charles and *me* the document.   (not *I*)

Object of preposition   The plan divides the work between Mr. Howard and *me*.   (not *I*)

Avoid using reflexive pronouns *(myself, himself, herself, themselves, ourselves)* as subjects or objects.

Incorrect   Jane and *myself* wrote the manual.

Correct   Jane and *I* wrote the manual.

Incorrect   The work was completed by the secretary and *herself.*

Correct   The work was completed by the secretary and *her.*

MODIFIER PLACEMENT CHECK

Check the placement of modifying words, phrases, and clauses.

Incorrect   The *little brick doctor's* house.

Correct   The *doctor's little, brick* house.

Incorrect   The accountant *only* comes in on Tuesdays.

Correct   The accountant comes in on Tuesdays *only.*

Incorrect	The computer was broken *in the closet.*
Correct	The computer *in the closet* was broken.

Incorrect	*Speaking before a crowd of people,* my knees shook.
Correct	*Speaking before a crowd of people,* I had shaky knees.

SPLIT INFINITIVE CHECK

An infinitive, you remember, is a verb form using *to (to work, to write, to go).* Do not split infinitives with adverbs.

Incorrect	We need *to* further *investigate* safety measures.
Correct	We need *to investigate* safety measures further.

Incorrect	The writer needs *to* thoroughly and accurately *proofread* the report.
Correct	The writer needs *to proofread* the report thoroughly and accurately.

Mechanics Repair

Mechanics are the spelling, punctuation, abbreviations, capitalization, hyphenation, italics, and number and symbols conventions. There are some differences in acceptable styles among specific organizations and professional associations. For instance, in **standard style,** commas are used as follows:

> designers, technicians, and manufacturers

In other styles, such as newspaper writing, commas are used as follows:

> designers, technicians and manufacturers.

Most technical writers adhere to standard rather than open conventions. To establish consistency within a organization or a journal, professional organizations present their preferred styles in **style manuals.** Appendix B in this textbook reviews standard mechanics that you should review carefully. In addition, many dictionaries contain valuable mechanical style information. Following are some published mechanical style sources:

- *Webster's Standard American Style Manual,* 1995. Springfield, Mass.: Merriam Webster.
- The fourth edition of *The Modern Language Association Handbook for Writers of Research,* 1994. New York: The Modern Language Association of America.

- The revised edition of *The Associated Press Stylebook and Libel Manual: The Journalist's Bible,* 1996. New York: Addison-Wesley Publishing Company, Inc.

- The fourteenth edition of *The Chicago Manual of Style,* 1993. Chicago: University of Chicago Press.

- *Government Printing Office Style Manual,* 1993. *Supplement on Word Division,* 1994. Washington, D.C.: U.S. Government Printing Office.

The Council of Biology Editors, the American Chemical Society, and the American Psychological Association, to name a few specialist organizations, also publish their own style manuals. When writing for publication, the publisher will usually advise you of the preferred style manual. All of these manuals are updated periodically.

PUNCTUATION

As previously mentioned, Appendix B reviews the standard style in detail. Be particularly alert to the use of an **apostrophe** to indicate the plurals of letters, numbers, designated words, and symbols as in *r's,* and *7's, and's, #'s;* the use of **brackets** within a quotation to add clarifying words as in, "The author stated, 'They [writers] need to use standard mechanics'"; the use of a **colon** to introduce a phrase or clause that explains or reinforces a preceding clause as in, "The writing process consists of four main steps: prewriting, writing, adding visuals, and postwriting." **Comma usage** compels careful attention. It is best to use **exclamation points** guardedly; make your emphatic message through the use of strong words.

Very few periods are used in technical abbreviations. A list of **Common Technical Abbreviations** is on page 653. Technical usage differs from conventions in other types of writing as in *fl oz* instead of *fl. oz.* And *rpm* instead of *r.p.m.* Further, in technical writing the plural abbreviations are written in the same form as the singular as in *5 hr, 10 cc, 100 lb.* Review in the appendix **capitalization, hyphenation, italics,** and **numbers.** Table B.3 on page 659 lists some common technical symbols such as ‖ for greater than or derived from and *S* for Silurian.

SPELLING

Most word processing software includes a **spelling** tool that scans through your document, highlights misspelled words, and suggests corrections for immediate replacement. This is a valuable tool, but it will not indicate words that are correctly spelled but incorrect in particular usage. Further, some software will underline (usually in red) misspelled words at the very point where you type them for your attention and correction. But only you can ensure that your document contains correct spelling throughout. In addition to your software spell check, use a dictionary to check all doubtful spellings and refer to Table B.4: "Frequently Misspelled Words," which includes those demons *amateur, benefited, calendar, contemptible, exaggerate, foreign, knowledgeable, medieval, occasionally, pastime, rhythm, separate, sophomore,* and *weird.*

CHECKLIST

Prewriting, Writing the Draft, and Postwriting Activities

PREWRITING

❑ **1.** Have I assessed my strengths for this writing assignment?

❑ Knowledge?

❑ Adequate research?

❑ Credibility?

❑ Authority?

❑ **2.** Have I determined the audience and their needs?

❑ Are they experts or laypeople?

❑ What common denominators do the members have?

❑ What level are they in the organization hierarchy?

❑ Are they within or outside my organization?

❑ Will my tone and style be appropriate?

❑ Should I consider multinational or multicultural differences?

❑ **3.** Am I clear on the exact purpose of my message?

❑ To instruct?

❑ To persuade?

❑ To inform?

❑ To describe?

❑ To question?

❑ **4.** Have I devised a topic list?

❑ **5.** Have I devised a complete and logical outline?

WRITING THE DRAFT

❑ **1.** Is my purpose clearly evident?

❑ **2.** Have I written with a flow, knowing I will edit the work later?

❑ **3.** Is the inductive or deductive format appropriate?

❑ **4.** Have I maintained a factual, objective, and unemotional tone?

❑ **5.** Is the style effective?

❑ No sexist slips?

❑ Concrete expression?

❑ Lack of jargon, shoptalk, gobbledygook?

❑ Precise words?

❑ Correct word usage?

❑ Lack of wordiness and redundancies?

❑ Signal words?

❑ **6.** Is the sentence pattern varied?

POSTWRITING

❑ **1.** Have I considered overall revisions?

❑ Have I rechecked everything for audience appropriateness?

❑ Are all of the topics and facts covered accurately and thoroughly?

❑ **2.** Are there any errors in sentence construction?

    ❑ Fragments?

    ❑ Faulty parallelism?

    ❑ Run-ons?

    ❑ Comma-splices?

    ❑ Sufficient variation?

    ❑ Appropriate length?

    ❑ Avoidance of *there is* and *there are*?

❑ **3.** Are word revisions necessary (sexist terms, jargon and the like, wordiness, redundancy, word usage, voice)?

❑ **4.** Have I added appropriate signal words?

❑ **5.** Is the grammar correct?

    ❑ Subject/verb agreements?

    ❑ Pronoun/antecedent agreements?

    ❑ Pronoun usage?

    ❑ Modifier placement?

    ❑ Split infinitive elimination?

❑ **6.** Are there any mechanical errors?

    ❑ Punctuation?

    ❑ Capitalization?

    ❑ Hyphenation?

    ❑ Necessary italics?

    ❑ Number usage?

    ❑ Spelling?

EXERCISES

1. **Inductive/Deductive Organization.** Rewrite the following inductive paragraph to deductive organization:

> Bankall reports net income of $310 million, up from $159 million a year ago. Securities gains account for $204 million of the showing. Earnings per share were $1.29, up from 70 cents in the first quarter

of last year. Noninterest income, such as bank-card income and deposit fees, rose 12 percent. Results of the merger indicate successful operation figures.

2. **Tone.** Rewrite the following paragraph in an impersonal, objective tone:

> Last Tuesday, Ivan Brown, the arrogant, young director of personnel, brazenly fired three clerks and a computer programmer. The inexperienced computer programmer had established with the clerks an insidious plan to pilfer Inkjet computer paper, Emory address labels, pens, scissors, and other supplies from the dimly lit supply room. Brown was selling these supplies to unwary students in front of the Student Union every Monday for a huge profit.

3. **Sexist Language.** Rewrite the following paragraph to eliminate the sexist language:

> Neither the draftsman nor the designer could complete his project until he had manpower to assist him. John Ackerman became the self-appointed spokesman on how to reapportion the man hour duties. Everybody involved gave his opinion on how his work could be made easier.

4. **Concreteness.** Rewrite the following paragraph to eliminate the vague, abstract words. Make up your own details:

> The Director of Technical Writing vetoed the plan to update the software in the department. He felt that the proposal asked for too many things because the department is adequately supplied with enough software. He stated that the high expense would curtail certain other expensive plans in the works.

5. **Jargon/Shoptalk/Gobbledygook.** Rewrite the following paragraph to eliminate jargon, shoptalk, and gobbledygook:

> He accessed the Emergency Room and was asked for input on his accident. "In a nutshell," he said, "I was cruising down the Interstate when this dude tailgated me so closely that I decelerated. He deployed his pick-up into the aft of my conveyance resulting in my broken schnoz. My car was totaled.

6. **Wordiness/Redundancies.** Rewrite these sentences to eliminate wordiness, roundabout phrases, and redundancies:

   a. Pursuant to our talk and discussion, we are inclined to make the suggestion that 150 auto assemblers be furloughed until such time as the recession shows a strong indication of recovery and upward climb.

   b. Due to the fact that the seat belt broke, the passenger sustained a high degree of injury.

   c. Either one or the other of the word processing programs is absolutely essential for the secretaries and typists to work quickly and efficiently.

    d. If we plan on showing an improvement of 20 percent, we will need to advance forward in our outlay and productivity.

    e. During the month of April we will begin to package our product in boxes rectangular in shape and yellow in color.

7. **Passive Voice.** Eliminate sluggish passive voice expressions by rewriting these sentences into the active voice:

    a. A report on the HIV testing was requested by the director of the laboratory.

    b. Block moves for your word processing program are discussed in Chapter 7.

    c. For the resume to the banks, a cover letter was used.

    d. In our financial department, pricing decisions are made.

    e. Plagiarism is in the chapter in which research and documentation are covered.

8. **Parallelism.** Rewrite these sentences to reflect parallel structures.

    a. He must learn the basics of punctuation and to be efficient in abbreviation styles.

    b. He will be hired if he has the required education, if he has four years of experience, and by being recommended by his former employer.

    c. A computer allows you speed, manipulation, and makes you flexible.

    d. You must write your rough draft in a speedy flow, and errors that occur can be corrected later.

    e. Name your working copy and save the draft, errors and all. Later alterations made by you and your writing team will be edited when you revise.

9. **Signal Words.** Add signal words to the following paragraph for unity and coherence:

> There are six methods for learning to word process a document. You may arrange for instructions from a teacher or other knowledgeable person. You may study manuals which accompany purchased software. Textbooks of instructions are available. There are video and audio cassettes to teach you. You may purchase film strips of instructions. There are newsletters to update your information.

10. **Exact Word Choice.** Using a thesaurus or a dictionary, write four levels of words to increase the intensity of the following:

> cue   hesitate   encourage   collect   damage

11. **Usage.** Select the proper usage in the following sentences:

    a. Medical schools *(accept, except)* less than half of the applicants.

    b. At noon the meeting was *(already, all ready)* to begin.

    c. *(Fewer, Less)* students signed up for Speech this term.

    d. The builders of the new cruise ship *(assured, ensured, insured)* everyone that it would not sink.

    e. She felt *(bad, badly)* about the late report.

    f. He was *(eager, anxious)* to see his old friends.

    g. Jones did so *(good, well)* in the interview that he was hired.

    h. We will finish these exercises, *(hopefully, I hope),* before the end of the class.

12. **Grammar.** Repair the grammatical errors in the following:

> The word processors along with the graphic designer agrees that new formats is needed. It keeps the writer from producing their best work. Designers will have to carefully consider whether to continue to use the traditional layout or produce brighter new ones. The head designer and myself will need to often work together. The new formats will be developed by Mr. Kelly and I.

13. **Copyediting Symbols.** Photocopy the following paragraph. Use the copyediting symbols to indicate corrections of the punctuation, spelling, and capitalization errors:

> After retiring from a 30 year teaching profession John turned his favorite past-time into a full time new Profession. He became a prominent technical writing consultant, and author of related materials. He haswritten three books Wring for the technologies Approaches to Tecnical and Scientific Wring and Technical Comunications. He as in additione authored twenty one writing oriented art icles. He earns 100000 dollars a year.

WRITING PROJECT

Group Project. Form groups of three or four classmates. Select one of the following subjects to develop into a paper: the advantages of a specific word processing program, the advantages of a copy machine, the advantages of one fax machine over another, the advantages and disadvantages of a cellular telephone, the best VCR, desirable features for a CD player, or reasons to purchase a specific automobile. Collaborate on the topic list and the outline. Each group member is to write four paragraphs (brief introduction plus three body paragraphs) based on the outline. Exchange papers and allow each member of the group to critique each paper. Submit the drafts along with the final revisions.

CHAPTER *3*

Graphics

S K I L L S

After studying this chapter, you should be able to:

1. Recognize the purpose and function of tables, bar charts, line graphs, pie charts, flow charts, organization charts, drawings, maps, photographs, text art, clip art, and icons.
2. Name and practice the conventions of incorporating tables and figures.
3. Prepare random and continuation informal tables.
4. Prepare formal tables.
5. Prepare a variety of bar charts.
6. Prepare several types of line graphs.
7. Prepare several types of pie charts.
8. Prepare a flow chart.
9. Prepare an organization chart.
10. Prepare a simple, exploded, or cutaway drawing.
11. Incorporate a map or photograph into a document.
12. Recognize text art and construct it by hand or by computer.
13. Recognize clip art and incorporate a piece into a report if you use a computer.
14. Recognize a variety of icons and determine where they may be used in a report.
15. Critique the construction and incorporation of graphics and visuals in technical reports.

INTRODUCTION

Graphic illustrations (*tables, bar charts, line graphs, pie charts, flow charts,* and *organization charts*) and other visuals (*drawings* and *illustrations, maps, photographs, text art, clip art,* and *icons*) are characteristic of professional and technical report writing. Large companies may employ graphic artists to assist writers in graphic illustration, while computers enable thousands of writers to create and incorporate their own graphics and other visuals into their documents. Even if you do not have access to an expert or possess your own computer capabilities, it is your task to decide, at least, on the desired data, the appropriate graphic to convey your data, and a rough idea of the layout. In fact, many simple graphics can be handled by the novice without assistance. Graph paper, a ruler, and a compass are the tools you should have available. Computers, however, will allow you to type in the data, select the desired visual form, and then with a couple of commands print sophisticated, color displays. In addition, or in place of color, you may add shadings, hatch marks, stripes, dots, shadows, three-dimensional effects, multi-angle views, and other refinements. Many graphics can now be downloaded from the Internet. Computer graphics software offers other opportunities, such as:

- Creating visual effects of a mathematical model
- Writing equations to explain the sequences of a tornado or earthquake and the like
- Constructing models and testing for outcomes using a number of variables
- Integrating computer-assisted design (CAD) with computer-assisted manufacturing (CAM) to design architectural blueprints and to direct machinery operations
- Hypothesizing about concepts without doing the calculations
- Creating animations for presentations

This chapter addresses the uses of a wide variety of graphics and the overall and specific conventions for their construction and incorporation into your text.

PURPOSES

Graphics and other visuals are never merely random or decorative. They must serve to

- Speed up a reader's comprehension
- Add credibility to the document
- Serve as a method of quick reference
- Reveal differences at a glance
- Provide more detail than is actually discussed for full examination
- Add to the attractiveness of the report

Each type of graphic and visual serves a distinct purpose. For instance, a table displays data in vertical columns that would otherwise involve lengthy prose sentences that, in turn, might be difficult to comprehend or to interpret. A bar chart illustrates comparisons of parts, while a pie chart shows not only comparisons of parts but also the relationship of each part to the whole. Drawings, maps, and photographs can show details that words cannot describe. Clip art may serve to enhance a title or chapter page or to emphasize a concept. Icons serve to indicate warnings and directions, particularly in instructions. You must consider not only when to use a graphic or other visual but also which type will best serve your purpose.

GENERAL CONVENTIONS

Conventions of design, placement, titling, numbering, referencing, and so forth govern all graphics and visuals and include the following:

1. **Design.** Graphics should be planned ahead, well thought out, and "print ready." Do *not* try to put too much information into any one graphic. Place all explanatory notes, keys, and legends within the graphic or beneath it in the left-hand position but above the number and title. There are many graphic software programs that allow you to construct graphics and visuals with a few clicks. You may also construct graphic figures on separate paper and photocopy them into a space that you have allowed in your text. For hand-constructed figures, consider making your working graphics $1\frac{1}{2}$ times as large as you intend for the final graphic or visual; photo reduce the work by 60 percent to obtain the proper proportion for your text. Use rulers, bow compasses, protractors, templates of geometric figures, and possibly transfer or stencil letters. Buy Times Roman, Helvetica, or Univers type styles because they are easiest to read.

2. **Incorporation.** Print all of your tables and figures (all other graphics and visuals) in the final draft of your document. Prepare your graphics with the top on the vertical plane of your paper, reduce it with a copy machine, and incorporate it so that no labels, legends, or other parts extend beyond your margins. If your computer and printer are not capable of incorporating graphics, or if you do them by hand, tape or glue the properly sized graphic to the page and then photocopy each page for inclusion in your document.

3. **Placement.** Graphics and visuals that are included only as a supplement to a report or those that do not appear near the textual reference tend to lose their impact, so all graphics and visuals should immediately follow their initial references. That is, once attention is to drawn to them, they should be placed right within the text at the point of reference or no later than at the end of the section in which they are mentioned. In a document printed with facing pages (manuals and books, for instance), they may appear on a facing page. They may also be referred to later in the text without the art work being repeated. Consider it unconventional to place graphics sidewards (landscaped) on a page, but exceptions could be made. Your computer will allow you to resize your graphics.

If you do not use a computer, use a copy machine to reduce wide graphics.

4. **Titles.** Usually include a precise noun phrase with a numbered designation for each graphic. A graphic may be taken out of a report for photocopying and distribution and will obviously require a title line. Some examples follow:

TABLE 1.   Cost Comparison of Transportation Modes

TABLE 2.   Smoke Detector Ratings

Figure 1.   Cross section of a typical speed bump before and after modification

Figure 2.   Proposed Transit System Routes

5. **Numbering.** In addition to the title, number your formal graphics. Always number and title a formal table *above* the data. The word *table* is capitalized. If only one table occurs in the entire document, it will not require a number nor usually even a title. Refer to all other graphics as figures and number and title them *beneath* the graphic. Clip art and icons are neither necessarily numbered nor titled. When you use a number and title for a table, *center the data* or *place it flush left* to the margin of the document. Be consistent within any one document.

TABLE 4

MEAN SALT CONCENTRATION

IN VARIOUS SOURCES OF WATER

or

TABLE 4 Mean Salt Concentration in Various Sources of Water

6. **Continuations.** In the case of tables that require more than one full page, begin the second page with the table number and the word *Continued.*

7. **Number Sequencing.** If more than one table or more than one figure is used in a document, number each in order of the appearance throughout the material, but number tables separately from graphics. Use Arabic numbers:

Table 1

Table 2

Figure 1

Table 3

Figure 2

If the report contains numbered chapters, use a decimal numbering system to indicate both the chapter and sequential number of the graphic:

Figure 7.1

Table 7.1

Figure 7.2

Table 7.2

Figure 7.3

8. **Periods and Capital Letters.** Notice that the numbers and titles in Conventions 4, 5, and 7 indicate a variety of acceptable uses of periods and capital letters. It is essential to be consistent within a document, however. If you decide to use a period after the graphic or table number, do so for all graphic designations throughout your report. You may capitalize an entire title, capitalize initial letters of each work, or capitalize only the initial letter of the first word.

9. **Spacing of Lines.** Single space titles that require more than one line. Align second and consecutive lines under the first word of the title, not under the word *table* or *figure:*

> Figure 5 Operation Manual for a Motorola Dimension 1000 Binary GSC pager (Courtesy of Motorola, Inc., Fort Lauderdale, Florida)

10. **Referencing.** Use an introductory sentence or reference notation by graphic number to explain the purpose of each graphic before you include it. If the graphic immediately follows its sentence reference, use a colon at the end of the sentence as illustrated:

> Figure 2 shows a line drawing made by tracing a photocopy:
>
> The available equipment, features of each, prices, and warranties are shown in Table 3:
>
> As you examine Figure 4, notice the high unemployment rate for even those with some college:

Sometimes you may reference a graphic in parenthetical notation:

> Set the control dial (Fig. 1) to one of the six speeds.
>
> A document may be printed in two columns (see Figure 8):
>
> Figure 10.3 (see page 510) shows a title page for a proposal.

11. **Commentary Line.** Usually, follow the graphic with a sentence or two of comment or interpretation:

> It is evident that the introduction of the scanner product put Austin Technology in the lead for 1997 sales.
>
> The dotted line between the Comptroller and the Vice-president for Finance indicates that the Comptroller may report directly to the Vice-president for Finance without reporting first to the Treasurer.

12. **Acknowledgments.** Identify the source of borrowed graphics in parentheses after the title:

Figure 1 Sample formal proposal (Courtesy of Charles E. Smith Jr., Robert Heller Associates)

Table 7 Average Yearly Salaries by Sex and Race [Source: Catherine Brown, *Discrimination in the Work Place,* New York: Silver Press, 1998 (8)]

Figure 2 Fire Ground Injuries by Cause [From "Fire Ground Injuries in the United States during 1997" (12). Reprinted by permission.]

TABLES

Tables are visual displays of numerical or nonnumerical data arranged in vertical columns so that the data may be emphasized, compared, or contrasted. Tables may be **informal** (random and continuation) or **formal.** Certain specific conventions govern each.

Informal Random Tables

Informal random tables display brief lists of figures, dates, personnel, important points, and the like in vertical columns for visual clarity and quick reference. Figure 3.1 (see facing page) illustrates two informal random tables. The first shows an informal table of dates, places, and purposes and the second a bulleted list.

Informal Random Table Conventions

1. Use random tables only for brief data.
2. Introduce each by an explanatory sentence.
3. Indent the data 5 to 10 spaces from the left- and righthand margins of the page.
4. Include column headings, numbered data, or bullets.
5. Do not include a table designation number or title.

Informal Continuation Tables

A continuation table, another informal table, contains prose data in a displayed manner. It reads as a continuation of the text and includes the same punctuation marks that would be required if the data were presented in paragraph form. Figure 3.2 (see page 66) shows a sample informal continuation table.

The Training Center announces the beginning of a mini-course, "Write It Right—Write It Well," for senior executive secretaries and administrative assistants. Dates, locations, and purposes follow:

Date	Room	Purpose
May 11	102	to review grammar/usage
May 13	102	to review brief report forms
May 18	101	to review graphics
May 20	102	to review manual components
May 25	103	to review formal reports
May 27	101	to critique individual writing

To register, fill out the attached form and forward it to Julie Wood, Training Specialist, Room 608.

Regardless of what kind of accident is being reported, certain information must be reported objectively and specifically:

- What the accident is
- When and where the accident occurred
- Who was involved
- What caused the accident
- What were the results of the accident (damage, injury, and costs)
- What has been done to correct the trouble or to treat the insured
- What recommendation or suggestions are given to prevent a recurrence

Information required for the accident report has become so standardized that many companies have designed accident report forms.

FIGURE 3.1 *Sample informal random tables*

Our insurance policy All State #17B-445-9100K will cover the cost of the fire damage. Repairs and replacements total $619.00. This price includes

$144.00	for carpet replacement (9 sq ft @ $16.00 per ft; Carl's carpets),
75.00	for labor for removing burned carpet and replacing (5 hr @ $15.00 per hour),
15.00	for cleaning solution for wall (15 sq ft @ $1.00 per sq ft),
70.00	for paint for wall (70 sq ft @ $1.00 per sq ft),
100.00	for labor for cleaning and repainting wall (10 hr @ $10.00 per hr),
70.00	for labor for reupholstering armchair (7 hr @ $10.00 per hr),
80.00	for fabric for brown leather armchair (8 yd @ $10.00 per yd),
20.00	for new magazine stand from Pier One, and
45.00	for fire extinguisher replacement.

FIGURE 3.2 *Sample informal continuation table*

Informal Continuation Table Conventions

1. Use a continuation table to present an alignment of figures, dates, or other data.

2. Introduce each by a sentence followed by a colon if the last introductory word is *not* a verb.

3. Indent the tabular data 5 to 10 spaces from the left and right margins.

4. Punctuate the data by standard commas, semicolons, and periods as if the material were presented in paragraph form. Note the word *and*.

Formal Tables

Formal tables are used to present statistical information or to categorize and tabulate other written information. Tables 3.1 and 3.2 (see facing page) show two types of formal tables and the conventions adhered to in typing each table. Either the centered or flush left title is conventional.

Formal Table Conventions

1. For formal tables use horizontal lines from margin to margin *above* the title.

TABLE 3.1 *Time/Cost for Aerial Photograph Searches*

Time Frame	Searches #	Time @ 30 min ea (hr)	Cost @ $100 per hr wage
Daily	3	1.5	150.00
Weekly	15	9	900.00
Monthly	60	10	1,000.00
Quarterly	180	90	9,000.00
Yearly	720	360	36,000.00

TABLE 3.2 *Troubleshooting Chart for Heath Kits*

Difficulty	Possible Cause
Receiver section dead	Check V1, V3, V4, V7, and V8
	Wiring error
	Faulty speaker
	Faulty receiver crystal
	Crystal oscillator coil mistuned
Receiver section weak	Check V1, V2, and V3
	Antenna, RF, or IF coils mistuned
	Faulty antenna or connecting cable
Transmitter appears dead	Check V5 and V6
	Wiring error
	Recheck oscillator, driver, and final tank coil tuning
	Dummy load shorted on open

2. Use a **box head** of vertical column headings and symbols in parentheses [i.e., ($), (rpm), (hr), (ft)]. Notice the lack of periods in the standard abbreviations; refer to Table B.2 in Appendix B to study more uses of abbreviations in technical writing that differ from standard abbreviations.

3. Do *not* close the sides of formal tables.

4. Always use vertical columns; the first body column is called the **stub.**

5. Following modern practice, do not use **leaders** (spaced periods to aid the eye in following data from column to column).

FIGURES

As previously mentioned, all graphics except tables are referred to as figures both in textual reference and in titling. If you use a computer, construct your graphic figures on your graphics software, import them into your document, and place and size them with your cursor.

Bar Charts

Use bar charts to show differences in quantity and quality visually and instantaneously. The bars show quantities of the same item at different times, quantities and qualities of different items for the same time period, or quantities of the different parts of an item that make up the whole. They are compiled from statistical data. Conventions in addition to those that govern all graphics govern their construction.

You may plot your bars vertically or horizontally. Normally, use vertical graphs for bars that represent monetary units. If you cannot do a bar chart on a computer, use graph paper to plot your chart. The scale you select is critical to the success of your chart. Do not use grids that will not accommodate some portion of your bar. Scale all grids to equal increments, such as 0, 1, 2 or 0, 5, 10, 15, but not 0, 5, 7, 12.

If the order of the bar placement is not controlled by a sequential factor, place the longer bars to the bottom of the horizontal chart or to the right of a vertical chart. This placement avoids a top-heavy or one-sided appearance. Types of bar charts include **horizontal, vertical, multiple, stacked, deviation, creative,** and **histograms.** Three-dimensional or overlapping bars may also aid the reader. You can easily construct them, but avoid overuse of these options unless they actually contribute to the import of the graphic. Figures 3.3 through 3.10 show bar charts that are horizontal, vertical, multiple bar, stacked, deviation, and creative, as well as histogram.

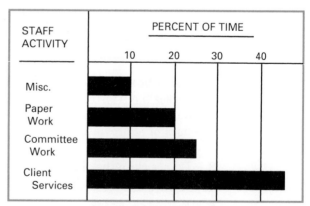

FIGURE 3.3 *Typical horizontal bar chart—hand constructed*

Bar Chart Conventions

1. If possible *box in* all of the bars, headings, legends, and so on.
2. Use bars of *equal width and design* within any one chart.
3. Use *partial cutoff* lines to separate headings from grid or tick notations. *(list continues on page 70)*

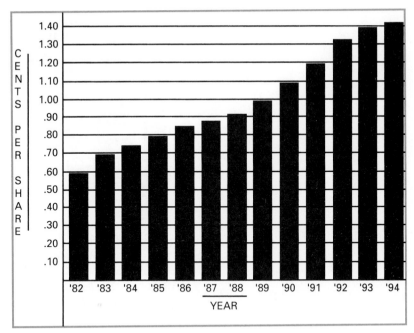

FIGURE 3.4 *Typical vertical bar chart—hand constructed*

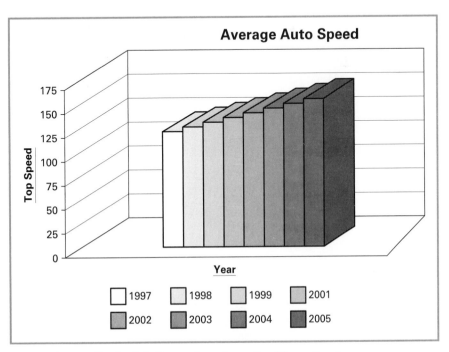

FIGURE 3.5 *Typical 3-dimensional, vertical bar chart*

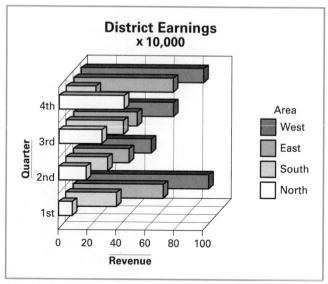

FIGURE 3.6   *Typical 3-dimensional, horizontal bar chart*

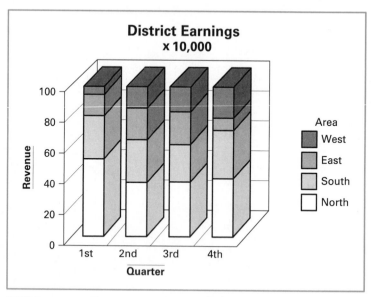

FIGURE 3.7   *Typical stacked, 3-dimensional bar chart*

4. Use vertical grid lines or tick marks for horizontal bar charts and horizontal grids or tick marks for vertical bar charts; never use both in a single chart.

5. Include a *heading* to indicate what the grids or tick marks show: hours, number of sales, amounts, activities, and so forth.

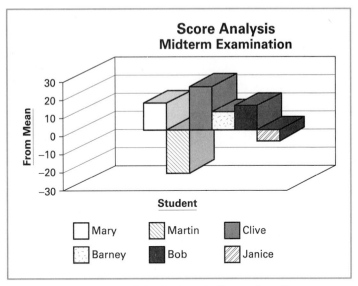

FIGURE 3.8 *Typical deviation, 3-dimensional bar chart*

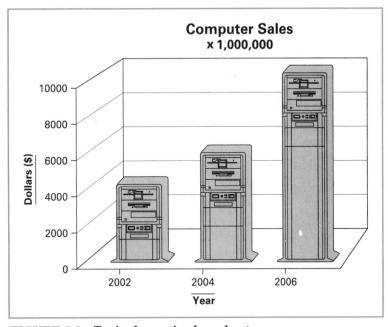

FIGURE 3.9 *Typical creative bar chart*

6. *Center grid notations* on the grid lines, not just above or just below.

7. *Center bar notations* on the bars.

8. When displaying multiple bars with various colors or texture, use *legend boxes* to distinguish the differences.

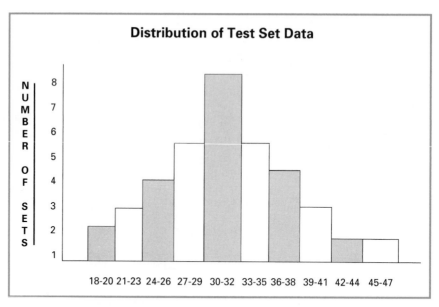

FIGURE 3.10 *Typical histogram*

Line Graphs

Line graphs, or curves, are used to show changes in two values. Most commonly, they show a change or trend over a given period or performance against a variable factor. These are standard *Cartesian* graphs, named after the French mathematician and philosopher who invented them, Rene Descartes. Points are plotted algebraically on the vertical *(y)* axis and the horizontal *(x)* axis. Figures 3.11 through 3.13 show a single line graph, a multiple line graph, and a multiple band graph:

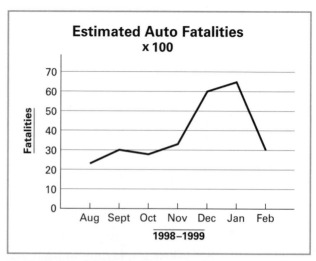

FIGURE 3.11 *Typical line graph*

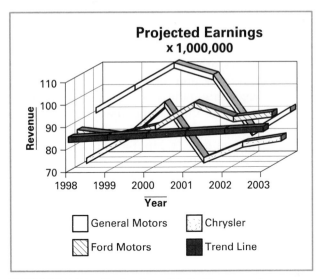

FIGURE 3.12   *Typical multiple line graph*

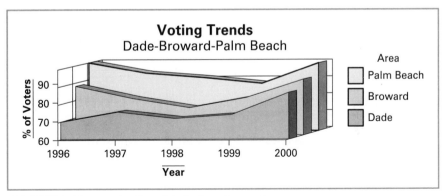

FIGURE 3.13   *Typical multiple band graph*

Line Graph Conventions

Line graphs adhere to the same conventions as bar charts with some additions.

1. Always plot your curves *from left to right.*
2. Indicate the grids with tick marks; do *not* include grid lines because they become confused with the curves themselves.
3. *Capitalize* major headings; *capitalize* only the initial letters of sub-headings and tick mark notations.
4. Use tick marks (not grid lines) on each line to indicate how many points have been used to plot the graph.

Pie Charts

Pie charts, also known as circle charts, circle graphs, pie diagrams, or sector charts, are used to compare the relative proportions of various factors to each other and to the whole. The circle represents 100 percent while the segments indicate the proportionate percentage of each factor. They are useful for illustrating financial information, survey results, and other results indicating parts as related to the whole.

If you use a computer, you can simply enter the data, and the software will construct the pie chart. If you need to figure out and construct the chart, consider the following: Since a circle contains 360 degrees, you must first convert your real data to percentages of the whole and then calculate the number of degrees needed to represent each segment or wedge. Use the following format to calculate the number of degrees for each segment:

Item	Raw Data	Frac-tion	4-place decimal	3-place decimal	Percent	Percent rounded	Degrees (% × 360)	Degrees rounded
Rent	$400	$\frac{400}{2300}$	.1739	.174	17.4%	17%	61.2°	61°
___	___	___	___	___	___	___	___	___
___	___	___	___	___	___	___	___	___
___	___	___	___	___	___	___	___	___
TOTAL: $2,300						**TOTAL:** 100%		**TOTAL:** 360°

There is a quick formula for these calculations: Multiply the 4-place decimal by 360 degrees. However this short method is not totally accurate, and you will probably have to make arbitrary adjustments to make sure your final calculations total 360 degrees.

If you are constructing your pie chart by hand, you will need a compass or circle template and protractor to draw the circle and to divide it into segments. As a rule of thumb, use a 3″-diameter circle on standard 8½″ × 11″ paper. This will make your circle large enough for emphasis yet small enough to allow space for labeling the wedges outside of the pie. Think of your circle as a clock, and plot your largest wedge in the upper right-hand quadrant from the 12 o'clock position. Then place the wedges from the next largest proportionately to the smallest clockwise.

A pie chart is effective without wedge differentiations, but color or shadings and hatches can add effectiveness. Like bar charts, pie charts may be three dimensional, or they may indicate cutaway portions. Creatively, it may be appropriate to use a round fruit, pizza, clock, or other naturally round object instead of a plain circle. Figures 3.14 through 3.16 show a typical pie chart, a chart with separated wedges, and a three-dimensional chart:

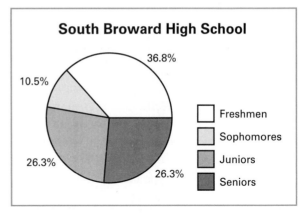

FIGURE 3.14   *Typical pie chart*

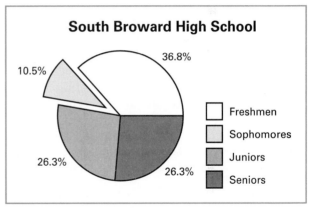

FIGURE 3.15   *Typical separated wedge pie chart*

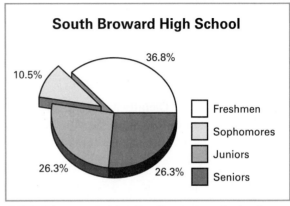

FIGURE 3.16   *Typical 3-dimensional pie chart with separated wedge*

Pie Chart Conventions

1. Normally, do not present a pie larger than 3 inches in diameter on an $8^{1}/_{2}'' \times 11''$ page.

2. Place the largest segment in the upper right-hand quadrant with the segments decreasing in size clockwise.

3. Write headings along with the percentages *outside* of each wedge to avoid crowding.

4. *Center* each label on the radius of each wedge or use a *tag line* to aid the eye.

5. Type labels on a horizontal plane.

6. Contain all labels within the left- and right-hand margins.

Flow Charts

A flow chart is used to show pictorially how a series of activities, procedures, operations, events, or other factors are related to each other. A flow chart shows the sequence, cycle, or flow of the factors and how they are connected in a series of steps from beginning to end. Such a chart condenses long and detailed procedures into a visual chart for easy comprehension and reference. The information is qualitative rather than quantitative as in bar, line, and pie charts.

The components of a flow chart may be diagramed in horizontal, vertical, or circular directions. Computer software uses templates that contain a variety of shapes symbolizing various activities, such as

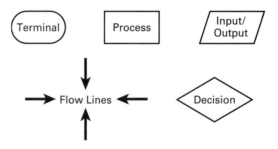

Usually boxed steps are arranged in sequence and connected by arrows to show the flow. Flow charts may be simple or pictorial. Figures 3.17 through 3.19 illustrate simple and pictorial flow charts:

Six Steps in Preparing a Work Sheet

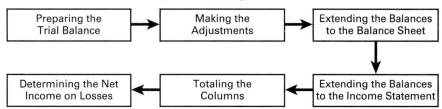

| Preparing the Trial Balance | → | Making the Adjustments | → | Extending the Balances to the Balance Sheet |

| Determining the Net Income on Losses | ← | Totaling the Columns | ← | Extending the Balances to the Income Statement |

FIGURE 3.17 *Typical flow chart*

Automated Event Handling

FIGURE 3.18 *Simple pictorial flow chart* (By permission of Smartdraw Software, Inc. http:/www.smartdraw.com)

Flow Chart Conventions

1. Employ squares, boxes, triangles, circles, diamonds, and other shapes to enclose each step.

2. Lay out your flow chart in a horizontal, vertical, circular, or combination of directions.

3. Name major activities within the shapes.

4. Use lines or arrows of various dimension to connect the shapes and to indicate the flow.

Computer future

The Broward County School Board's five-year plan for technology envisions a computer system that will allow students to talk to other students, parents to communicate with administrators and teachers, and educators to trade ideas with peers worldwide.

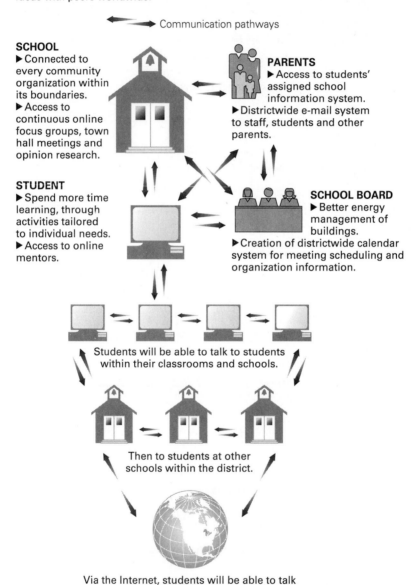

Communication pathways

SCHOOL
▶ Connected to every community organization within its boundaries.
▶ Access to continuous online focus groups, town hall meetings and opinion research.

PARENTS
▶ Access to students' assigned school information system.
▶ Districtwide e-mail system to staff, students and other parents.

STUDENT
▶ Spend more time learning, through activities tailored to individual needs.
▶ Access to online mentors.

SCHOOL BOARD
▶ Better energy management of buildings.
▶ Creation of districtwide calendar system for meeting scheduling and organization information.

Students will be able to talk to students within their classrooms and schools.

Then to students at other schools within the district.

Via the Internet, students will be able to talk to others throughout the country or the world.

FIGURE 3.19   *Complex pictorial flow chart*   (Reprinted with permission from the Sun-Sentinel, Fort Lauderdale, Florida)

Organization Charts

Like flow charts, organization charts show qualitative, rather than quantitative, material. An organization chart is used to show the relationship of the organization's staff positions, units, or functions to each other.

A **staff organization chart** shows the chain of command of the staff positions, such as president, vice-presidents, directors, comptroller, and salespersons. A **unit organization chart** depicts the relationships among such units as Public Relations Department, Research Division, Financial Office, and Personnel Section. A **function chart** shows the span of control of such functions as Planning, Engineering, Technical Writing, Marketing, and Production. These three aspects—staff, unit, functions—should not be mixed together in the same chart.

Either a horizontal or a vertical emphasis can be imparted to an organization chart by the layout as shown:

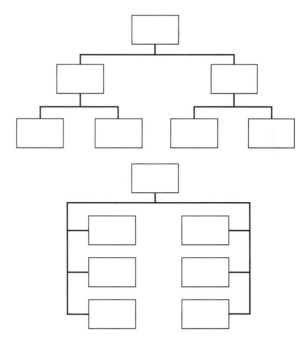

Figure 3.20 (see page 80) shows a staff organization chart for a manufacturing company. Notice within the chart that the Accounting Director is subordinate to both the Treasurer and the Comptroller. The solid line indicates that the normal chain of command is from the Comptroller to

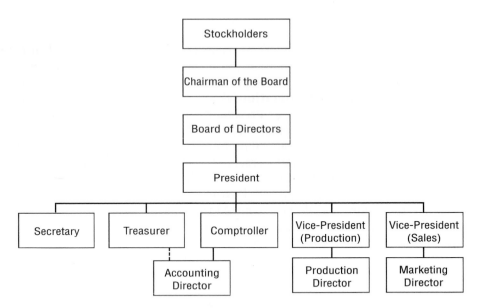

FIGURE 3.20 *Typical organization chart*

the Accounting Director; the dotted line indicates that the Treasurer may direct inquiries or report important information directly to the Accounting Director. The lines also show the normal and possible chains for the Accounting Director to direct questions or to issue information upward.

OTHER VISUALS

Besides tables and graphics, technical writing is characterized by drawings and illustrations, maps, photographs, text art, clip art, and icons. Each of these is used to emphasize and clarify points within your text. Computer software provides drawing capabilities, a variety of map types, text art, clip art, and icons. In addition, a scanner can copy photographs or other materials directly into your documents. Computer programs allow you to draw any shape imaginable and add shading, hatches, color, and labels to your work. Computer Assisted Drafting (CAD) requires special training; CAD is used primarily by drafters, engineers, and architects, but you may well find a use for such software in your writing.

Drawings/Illustrations

A variety of simple drawings may be executed by the novice. A simple line drawing is often clearer than a photograph because it can give emphasis to important features. Diagrams of procedures can clarify instructions. Exploded-view illustrations show the proper sequence in which parts fit together, and cutaway drawings show the internal parts of a mechanism or a piece of equipment. Electricians and electronic technicians use schematics and wiring diagrams to illustrate concepts.

Complicated or specialized drawings should be attempted by hand only if you are skilled in commercial art or drafting, but anyone can draw the types of art shown in this section. Poorly constructed drawings have an effect that is exactly the opposite of well-produced drawings in that they confuse the reader.

Figures 3.21 through 3.25 show typical procedural, labeled line, exploded view, cutaway, and schematic drawings that can clarify descriptions, definitions, and process analyses:

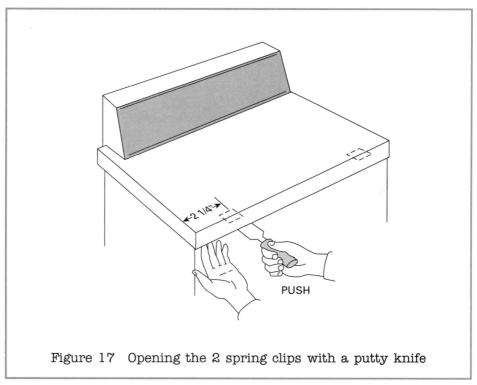

Figure 17   Opening the 2 spring clips with a putty knife

FIGURE 3.21   *Typical procedural drawing*

Drawing Conventions

1. If you do not use a computer drawing program, use grid paper and a ruler for careful drawings.

2. Keep your drawings uncluttered, properly ruled, and carefully labeled.

3. *Type,* do not hand letter, all labels and symbols.

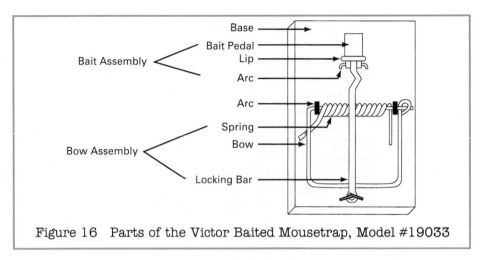

Figure 16   Parts of the Victor Baited Mousetrap, Model #19033

FIGURE 3.22   *Typical line drawing with labels*

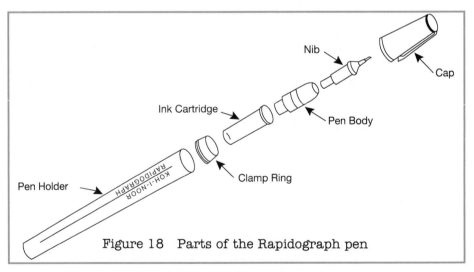

Figure 18   Parts of the Rapidograph pen

FIGURE 3.23   *Typical exploded-view illustration*
(Courtesy of student Jennifer Woper)

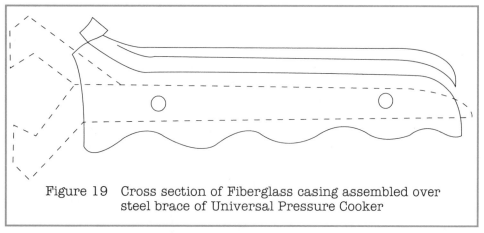

Figure 19   Cross section of Fiberglass casing assembled over steel brace of Universal Pressure Cooker

FIGURE 3.24 *Typical cutaway drawing*

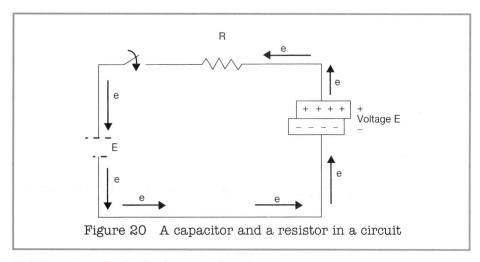

Figure 20   A capacitor and a resistor in a circuit

FIGURE 3.25 *Typical schematic drawing*

Maps

Maps can show sites, routes, and comparisons by geographical location. Maps may be **large scale** (highlighting an area) or **small scale** (showing a large area with position notations or comparisons). Computers offer all kinds of maps, but you may hand draw, photocopy, or scan others into your documents. As with other visuals you may add color, shadings, hatches, or patterns to distinguish regions from each other. Figures 3.26 through 3.28 (see pages 84 and 85) show two small-scale maps and one large-scale map:

CENSUS REGIONS AND DIVISIONS OF THE UNITED STATES

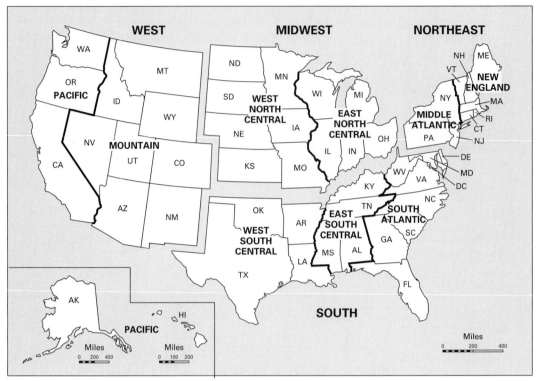

FIGURE 3.26 *Typical small-scale map* (Downloaded from maps—United States Department of Commerce—Economics and Statistics Administration. Bureau of Census)

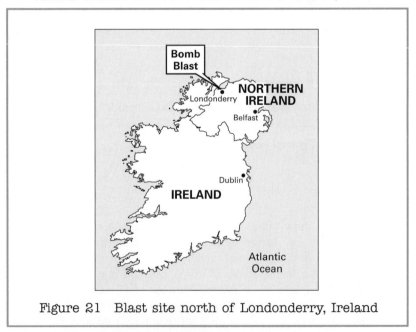

Figure 21   Blast site north of Londonderry, Ireland

FIGURE 3.27 *Typical small-scale map* (Fort Lauderdale *News/Sun Sentinel* map by Keith Robinson)

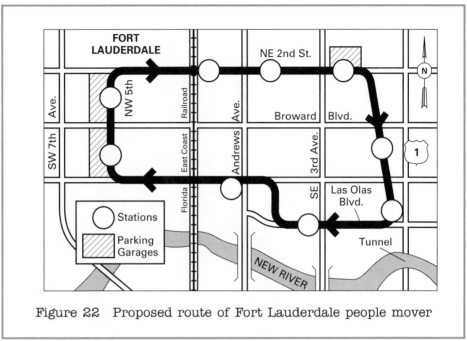

Figure 22   Proposed route of Fort Lauderdale people mover

FIGURE 3.28   ***Typical large-scale map***   (Fort Lauderdale *News/Sun Sentinel* map by Tom Alston)

Photographs/Line Art

Photographs allow you to provide overall views, and they may be cropped to focus on detail. You must be careful that the photo is not cluttered with unnecessary and distracting elements. You must choose well-focused photos for reproduction. Large companies with sophisticated technical writers and computer capabilities use scanners to produce from actual photographs digitized line art that can be incorporated directly into a document. Line art is as clear as a photograph.

It may help to provide a scale to your photo by including a person, a hand, or even a ruler. Consider ethics. For example, an insurance photograph can selectively make a $300 dent appear to be a $1,000 dent with selective light and shadow. Furthermore, digital imaging technology allows storage of photographs on a disk, which can then be unethically altered. The photographs you use must be clear and accurate and without unethical alterations. Figures 3.29 and 3.30 (see page 86) show a photograph and a digital line art production from a photograph:

FIGURE 3.29   *Photograph of a 900 MElz digital cordless telephone*   (Courtesy of Lucent Technologies, Parsippany, New Jersey)

FIGURE 3.30   *Line art produced on a computer from an actual photograph*   (Courtesy of Motorola, Boynton Beach, Florida)

Text Art

Text art may be used for titles or within manuals for special emphasis. Computers allow you to write text, contour it to a variety of shapes, use various fonts and type sizes, insert lines in a variety of formats, and add color. In technical writing, you must guard against the purely decorative, but if such art adds meaning to your message, it may be effective. Figure 3.31 shows a few words produced by the text art feature of a word processing program. Notice the different fonts, type sizes, and contours:

NOTICE>

DANGER!

Caution

FIGURE 3.31 *Typical text art in various fonts, type size, and contours*

Clip Art

Clip art is also known as quick art on some computers. There are thousands of clip art possibilities. This writer's computer contains a library of more than 40,000 items. Such a library allows you to add professional images into your documents. These images include arrows, professional supplies, business representations, equipment including computer items and science objects, faces and people, transportation and space-age pictures, symbols, borders and banners, buildings, maps, plants, animals, birds, and a multitude of others. You can even edit these pictures by adding text or facial features. Again, these images should be used sparingly, but they can enhance your documents in many ways. Images can be sized, dragged to any place on your document page, and enhanced with more detail. Figure 3.32 shows a brief sampling of simple clip art images:

FIGURE 3.32   *Typical clip art images*   (Corel Word Perfect 7)

Icons

An icon is a picture or image with a conventional meaning, such as a hand with a thumb up for approval, two clasped hands for agreement, or a round sign with a diagonal line through it meaning "not allowed." These may be carefully drawn or abstracted from your computer clip art or quick art programs as well as from fonts such as Wingdings, Milesto, Holiday Caps, and IC. These are not to be used decoratively but to convey essential meaning in documents such as operation manuals. Colors, textures, shadows, and other features may be added to icons. Figure 3.33 shows a number of easily recognizable icons.

FINAL WORDS ON EFFECTIVENESS

The level of detail, clarity, texture and color, size, and orientation on the page must all be considered when producing effective graphics and other

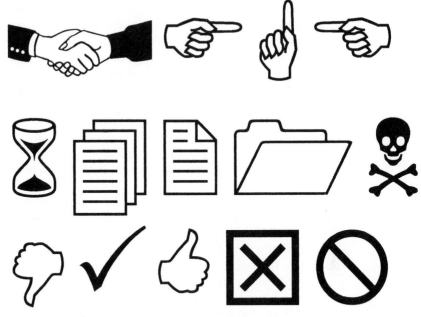

FIGURE 3.33   *Typical recognizable icons*

visuals. Be careful not to try to include too much detail in any one piece of work. Figure 3.34 shows a graphic that is both overdesigned and lacking in vital information:

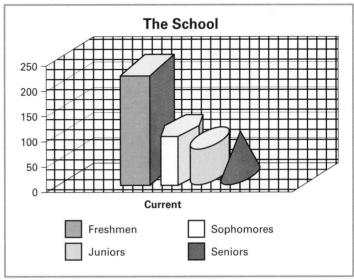

FIGURE 3.34   *Overembellished graphic with inadequate titles and headings, too many grid lines, and unnecessary multiple shapes*

Use an inkjet or laser printer for **high resolution.** A dot matrix printer is generally not effective because of its low resolution and broken lettering. Do not try to reproduce computer graphics or art on regular copier paper because the details may be fuzzy, muddy edged, and dull. Use a premium paper available from your office supply store. Figure 3.35 shows a computer-generated document on a poor quality printer with low resolution.

The **size** of any graphic should not overwhelm or underwhelm the text. It takes some practice to determine a compatible size; give size your most careful attention. If the graphic or visual is too small or too large in relationship to the type size and other elements of the document design, it will lose effectiveness.

Further, the **orientation** on the page is important. A graphic (with the possible exception of a table) should always be contained on one page. Changing the size slightly may enable you to do this. If the size or configuration will not fit well into its place of reference, consider placing it on a facing page. Decide whether your visual should be centered, flush left, or in some other position on the page. Do not exceed the margins of your document; leave some white space around the graphic or visual.

Textures and colors must also be sharp and logical. Textures and colors that "bleed" from one to another are not acceptable. Use both sparingly; remember less is more. Use texture and color, perhaps, for headings to draw attention to important or additional information, such as a checklist. Study a color wheel to determine how colors are related and to discern which colors interact. Those that lie directly opposite each other on the wheel are dynamic and exert a push/pull effect. Monochromatic color schemes (one color with variable tints and shades) tend to unify a document. Colors that lie closely together on the color wheel (yellow, orange, rust) create harmony. Use bright colors for accent and pale colors as background. Unnecessarily brash colors are disconcerting and distracting. Stick to the primary colors for most graphics, and do not overdo the number of textures in any one graphic or visual.

Textures may include background shadings (grey tone or color); patterns of horizontal, vertical, or diagonal straight lines placed close together or farther apart; a variety of wavy or curvy lines; large or small dots; checkered, brick, or beehive patterns; or other designs.

1??X Schedule of Fees and Services

Studio Fees

Service	Time Schedule	Fee	Notes
Portrait Sitting	.5	$ 59.95	pkg 35-40
Modeling Portfolio	2	99.95	sitting only
Real Estate Portfolio	2	100.00	location fee
Industrial Photography	per diem	250.00	location fee
Medical Photography	per diem	500.00	location fee
Insurance (Inventory)	per hour	50.00	2 hr min-LF (lab fee)
Insurance (Investigation)	per hour	100.00	2 hr min-LF
Collector's Catalogue	setup fee	59.95	
Commercial Catalogue	per diem & exp.	500.00	
Sports & Team Sets	sets only	no fee	
School Sets	sets only	no fee	
Clubs & Events	sets only	no fee	
Boudoir Photography	2	99.95	sets only
Post Cards / Calendars	setup fee	99.95	
Macro Photography	per hour	49.95	2 hr min-LF
Slide Shows	per hour	49.95	2 hr min-LF
Film Strips	per hour	49.95	2 hr min-LF
Public Relations (Political)	per hour	75.00	2 hr min-LF
Public Relations (Show)	per hour	75.00	2 hr min-LF
Weddings	4	300.00	$50.00 proof fee

Lab Services

Copywork	5.00
Custom Touch up	5.00 +
Restorations	.25 per inch square
Black & White Hand Coloring	1.00 per inch square
Magazine Mockups	19.95 setup fee

FIGURE 3.35   *Computer-generated document on a poor quality printer*

C H E C K L I S T

Graphics and Visuals

OVERALL CONSIDERATIONS

❏ **1.** Have I decided on a graphic that speeds the reader's comprehension, adds credibility to the document, reveals content and differences at a glance, and adds to the attractiveness of the document without being merely decorative?

❏ **2.** Have I chosen the most appropriate graphic to convey my data?

❏ **3.** Is the size proportionate to the text?

❏ **4.** Are all of the numbers correct?

❏ **5.** Are all elements (line thickness, notes, labels, legends, titles) positioned within the margins?

❏ **6.** Is my graphic/visual incorporated into the correct place in the document after the initial textual reference?

❏ **7.** Is there a sentence of reference for the graphic?

❏ **8.** Is it properly numbered and titled?

❏ **9.** Is the sequencing of graphic numbers correct?

❏ **10.** Am I consistent in my use of periods and capital letters in my numbering and titles?

❏ **11.** Is there a commentary line following the graphic?

❏ **12.** Are my sources acknowledged correctly?

DESIGN

❏ **1.** Are the elements of my graphic correct (table columns, bars, lines, pie wedges, flow chart arrows, organization boxes and arrows, drawings, photographs, maps, etc.)?

❏ **2.** Is the graphic large enough to convey the data but not overwhelming to the text?

❏ **3.** Is the material indented and spaced well within the document?

❏ **4.** Are table sides open but other graphics boxed?

❏ **5.** In a bar chart are the bars of equal width and design?

❏ **6.** In a line graph are the curves plotted from left to right?

❏ **7.** In a pie chart do the wedge percentages total 100 percent?

❏ **8.** In a pie chart are the wedges placed from the largest in the upper right-hand quadrant to the smallest clockwise?

❏ **9.** In a pie chart are the labels outside of the wedges and placed in the center of each radius?

❏ **10.** In a flow chart are the shapes appropriate and are the directions of flow clear?

❏ **11.** In an organization chart are the boxes evenly spaced and connected correctly?

❏ **12.** Are my drawings and maps accurate, sharp, and labeled in size appropriate to the document?

❏ **13.** Are my photographs and line art appropriately cropped and rid of extraneous material?

❏ **14.** Is my text art correct and integrated in design to the purpose of the document?

❏ **15.** Is my clip art well placed and helpful in conveying the message?

❏ **16.** Do my icons instantaneously convey the correct meaning?

EXERCISES

1. **Graphic Choices.** Which graphic would best illustrate the following (be specific, such as a random, continuation, or formal table; a simple, multiple, stacked, or creative bar chart; a simple, exploded, or cutaway drawing, etc.)?

   a. The changing cost of your telephone bill over a six-month period.

   b. The number of students enrolled in each department of a community college division (i.e., communications division: expository writing, technical writing, literature, speech).

   c. An indicator to continue to the next page.

   d. A comparison of three types of sneakers by cost, design, material, and durability.

   e. The installation directions for a VCR.

   f. A specific model of Sony stereo receiver.

   g. Tools and materials needed for installation of a Westinghouse clothes dryer.

   h. Assembly of a halogen lightbulb into a lamp.

   i. A list of times, room numbers, and places for class registration.

   j. The percentages of time spent in eight job-related duties.

   k. A storm-damaged roof.

   l. A breakdown of presidential vote choices by county in your state.

   m. The chain of command where you work.

   n. The steps in the process of preparing a technical instruction manual.

2. **Table.** Determine the percentages and devise a table to present the following data obtained from a corporate study on sex discrimination. Include the job titles, numbers, and percentages of the total.

   > Of 215 employees, 103 are women
   >
   > Of 36 executives, 12 are women
   >
   > Of 9 managers, 3 are women
   >
   > Of 7 assistant managers, 5 are women
   >
   > Of 18 administrative assistants, 3 are women
   >
   > Of 6 technical writers, 0 are women

3. **Bar Chart.** Determine how many hours you spend on a typical Monday on each of the following activities. Present this information in a standard bar chart. Number and title your graphic.

Travel	Grooming
Study	Class attendance
Leisure	Work
Meals	Sleep

4. **Pie Chart.** Using an imaginary monthly income of $2,180, devise a budget for the following expenses and present your data in a pie chart. Number and title your graphic.

Rent/mortgage	Insurance
Utilities	Leisure
Auto Expenses	Credit payments
Food/meals	Clothing
Education	Miscellaneous

5. **Line Graph.** Devise a line graph showing your grade point average on a term-to-term basis. Number and title your graphic.

6. **Flow Chart.** Devise a flow chart depicting the line of legislative command from the president of the United States to the members of the House of Representatives. Do not confuse this with an organizational chart. Number and title your graphic.

7. **Drawing.** Select a small mechanism, hand tool, or kitchen implement and construct a simple or exploded drawing of it. Label all parts including nuts, bolts, rivets, tabs, and handles. Number and title your graphic.

8. **Map.** Obtain a bus route map and write a brief paragraph into which to insert the map. Introduce, number, title, and comment on the graphic.

9. **Text Art.** Design a title page for a longer report you may be submitting based on the information in Exercise 3 or 4. Design some text art by hand or on your computer.

10. **Clip Art.** If you have clip art on a computer, print an image that could be used in a technical document. Write a few sentences to explain your choice and intention.

11. **Icons.** Show three icons that may be commonly used in technical writing. Draw these by hand or locate these icons on a computer. Explain what each icon indicates.

WRITING PROJECTS

1. **Graphic Evaluation/Positive.** Select a graphic that you find effective in a magazine, newspaper, textbook, manual, newsletter, or brochure. Write a few paragraphs about its effectiveness. How does it adhere to the general conventions of graphics and visuals? Is it the best type of graphic for the situation? What, if anything, could be improved?

2. **Graphic Evaluation/Negative.** Select a graphic that you find ineffective in any of the sources listed in the previous project. Write a few paragraphs about its weaknesses. How does it deviate from

the general conventions of graphics and visuals? Why doesn't it work well? What needs improvement? Would another type of graphic work better?

3. **Group Project.** Form a group of three or four students. Imagine that you are a team of marketing specialists at National Motor Corporation. Select a lead person, decide on your responsibilities, and set deadlines. Write a memorandum report to the Vice-president of Marketing concerning data on the new-model Econo automobile. Your purpose is to suggest facts that should be stressed in the new-model Econo sales promotion materials. Use an informal table, a formal table, a pie chart, and a bar chart or line chart in the text of your memorandum. Introduce your graphics and number and title them.

The Econo was the most popular car compared to four other competitive mid-sized cars on the market last year. Of total sales 30 percent of buyers chose Econo, 23 percent selected Car A, 20 percent chose car B, 17 percent chose car C, and 10 percent chose car D.

The Econo has doubled its fuel efficiency in 5 years. Four years ago the Econo was rated at 10.5 miles per gallon (mpg); 3 years ago the Econo averaged 14 mpg; 2 years ago the fuel efficiency increased to 17 mpg; a year ago it increased to 20.5 mpg; and this year it has a 21 mpg rating.

Despite base price increases, the new Econo is as economical to own and to operate as it was 4 years ago due to increased fuel economy. Four years ago the base sticker price was $5,801. A 20 percent down payment of $1,160 resulted in a $4,641 balance to finance. The monthly cost of financing for 48 months was $125 at 13.2 percent. The fuel expense per month for 1,250 miles of driving (15,000 miles annually) at $1.30 per gallon was $116 at 14 mpg. The cost per month over 48 months (finance charge plus fuel cost) was $241. The new Econo has a base sticker price of $7,301. A 20 percent down payment of $1,460 leaves $5,841 to finance. The monthly cost of financing over 48 months is $170 at 17.7 percent. But fuel expense is reduced to $77 at 21 mpg. Therefore, the monthly cost of owning a new Econo over 48 months (finance charge plus fuel cost) is only $247.

NOTES

CHAPTER 4

Document Design, Presentation, and Production

B.C.   by Johnny Hart

By permission of Johnny Hart and Creators Syndicate, Inc.

S K I L L S

After studying this chapter, you should be able to:

1. Name seven elements of document design.

2. Recognize and employ a number of layout designs with spatial variations.

3. Understand the value of and use of headings.

4. Recognize and use various type sizes.

5. Recognize and use a number of fonts, such as Eurostyle, Helvetica, Times New Roman, and Goudy.

6. Know when to use *serif* and *sans serif* fonts.

7. Recognize and employ variant margins, paragraph print justifications, and indentation choices.

8. Distinguish between a ragged-edge and justified page.

9. Recognize and employ lists.

10. Recognize the value of and employ boldface, italics, underlining, all capitals, small capitals, and drop caps.

11. Recognize and employ tabs and columns to enhance a document.

12. Recognize and employ headers and footers.

13. Be able to devise front presentation matter: a title page, a table of contents, a list of illustrations, and separate chapter title pages.

14. Be able to devise back presentation matter: a glossary, an appendix, and an index.

15. Recognize a summary, abstract, and bibliography or works cited listing.

16. Recognize factors that determine the size, quality, and orientation of paper.

17. Recognize watermarks and borders.

18. Evaluate the use of flaps, pockets, perforations, and windows in documents you own.

INTRODUCTION

Document design, final presentation, and production options are increasingly more the task of the writer or writing team than the professional printer or publisher. Awareness of the capabilities of computer enhancement of design is leading even those writers without computer literacy to pay more attention to the layout of the document, its special presentation features, and its production options. Ideally, you have considered and taken advantage of design elements as you wrote your rough draft, edited and revised it, and added graphics and other visuals. It is never too late to enhance your document.

PLANNING THE DOCUMENT DESIGN

Word processors, desktop publishing, and other page design programs are focusing more attention on text design. Heretofore, just typing a document in a Times New Roman 12-point font, using one-inch margins all around, double spacing, indenting new paragraphs five spaces, and perhaps inserting headings in regular or boldface type sufficed.

A number of text design elements may be considered to enhance the actual document content. The most common text manipulation features are

- Space
- Headings, type size, and fonts
- Margins, justification, and indentation
- Lists
- Emphatic features
- Tabs and columns
- Headers and footers

By implementing even simple design techniques, your layout can visually underscore your message.

Space

The first consideration is the use of space on the page. You already recognize the following as a memo:

XX:	ZZZZZZ ZZZZZZZZ
XXXX:	ZZZZ ZZZZZZZ
XXXX:	O ZZZZZZZ 0000
XXXX:	ZZZZZZ ZZZZ ZZZZZZZZ

The To, From, Date, and Subject lines are usually followed by a partial cutoff line to separate the control data from the text, which is usually presented in block paragraphs. The memo design is just one possibility of space design to organize materials. Consider the following possibilities of page designs and layouts.

Document Page Designs

HEADING

Text, text. Text, text.

Text, text. Text, text, text. Text, text, text, text, text, text. Text, text, text. Text, text, text, text, text.

HEADING

Text, text, text, text, text, text, text, text, text, text. Text, text, text, text, text, text, text, text, text, text, text, text, text, text, text. Text, text, text, text, text, text, text, text. Text, text, text, text, text, text, text, text, text, text, text, text, text, text, text. Text, text, text.

Text, text, text, text, text, text. Text, text, text, text, text, text, text, text, text, text, text, text, text, text, text, text. Text, text, text. Text, text, text, text, text, text, text, text, text. Text, text.

Text, text, text, text, text, text, text, text, text, text, text, text, text, text, text, text, text. Text, text, text, text, text, text, text, text, text, text, text, text, text, text, text, text, text, text. Text, text, text, text, text, text, text, text, text. Text, text, text, text, text, text, text, text.

Text, text, text, text, text, text, text, text, text, text, text,text. Text, text. Text, text, text, text, text, text, text, text. Text, text, text, text, text. Text, text, text, text, text, text, text, text. Text, text, text, text, text. Text, text, text, text, text, text, text, text, text, text, text, text, text, text, text.

HEADING

Text. Text, text, text, text, text, text, text, text. Text. Text, text, text, text, text, text, text, text. Text, text, text, text, text, text, text, text, text. Text, text.

Text, text, text, text, text, text, text, text, text, text. Text, text, text, text, text, text, text, text, text. Text, text, text, text, text, text, text.

Text, text, text, text, text, text, text, text, text. Text, text, text, text, text, text, text, text, text, text, text, text. Text, text, text, text, text, text, text. Text, text, text, text, text, text, text. Text, text, text, text, text, text, text. Text, text, text, text, text, text, text, text, text, text.

HEADING

Text, text, text, text, text, text, text, text, text, text, text. Text, text, text, text, text, text, text, text.

Text, text, text, text, text. Text, text, text, text, text, text, text.

In this example a horizontal line separates the title from the text. The text is presented in two columns. Capitalized, boldface headings separate parts of the text. The paragraphs begin flush left to the margin, and a space is skipped between paragraphs.

Document Page Layouts

Text, text, text, text, text, text, text, text, text, text, text. Text, text, text, text. Text, text, text, text, text, text, text. Text, text, text, text. Text, text, text, text, text, text, text, text. Text, text, text, text, text, text, text, text. Text, text, text, text, text, text, text, text. Text, text, text. Text, text, text, text, text, text, text, text, text, text.

Text, text, text, text, text, text, text, text, text, text, text, text, text, text. Text, text, text, text, text, text, text, text, text. Text, text, text, text, text, text, text, text, text. Text, text, text, text, text, text, text, text, text. Text, text, text, text, text, text, text, text, text, text. Text, text.

Text, text, text, text, text, text, text, text, text, text, text, text, text, text, text. Text, text, text, text, text, text, text, text, text, text, text, text. Text, text, text, text, text, text, text, text, text, text, text, text, text, text, text. Text, text. Text, text, text, text, text, text, text, text. Text, text, text, text, text, text, text, text. Text, text, text, text, text.

Text, text, text, text, text, text, text, text, text. Text, text, text, text, text, text, text, text. text, text, text, text, text, text, text, text. text.

Text, text, text, text, text, text, text, text, text, text, text, text, text, text, text, text. Text, text, text, text, text, text, text, text, text, text, text, text, text, text, text, text, text. Text, text, text, text.

Text, text, text, text, text, text, text. Text, text, text, text, text. Text, text, text, text, text, text, text. Text, text, text, text, text, text, text, text, text. Text, text, text, text, text, text, text, text, text, text. Text, text, text, text, text, text, text, text, text, text, text, text, text, text. Text, text, text, text, text, text, text, text, text. Text.

Text, text, text, text, text, text, text. text, text. Text, text, text, text, text, text, text. Text, text, text, text, text, text, text, text, text, text, text, text, text, text. Text.

Text, text. Text, text, text, text.

Text, text, text, text, text, text, text, text, text, text, text, text, text, text, text, text, text, text, text, text. Text, text, text, text, text, text, text, text, text.

Text, text.

Text, text, text, text, text, text, text, text, text, text, text, text, text, text. Text, text, text, text, text, text, text.

Text, text. Text, text.

Text, text, text, text, text, text. Text,

In this example the title is boxed and textured (shaded). The text is presented in two columns with a line separating the columns. Each paragraph is indented five spaces, and a space is skipped between each paragraph. The text is justified. One paragraph is textured to give it emphasis.

Document Page Design

Text, text, text, text, text. Text, text. Text, text, text, text, text, text, text, text, text, text, text, text, text. Text, text, text, text, text, text, text, text, text, text. Text, text, text, text.

Text, text. Text, text, text, text, text, text, text, text, text, text. Text, text, text, text, text, text, text, text.

Text, text, text, text, text, text, text, text, text, text. Text, text, text, text, text, text, text, text, text, text, text, text.

Text, text, text, text, text, text, text, text, text, text, text, text, text, text. Text, text, text, text, text, text, text, text, text, text, text.

Text, text, text, text, text, text, text, text, text, text, text, text. Text, text, text, text, text, text, text, text.

Text, text, text, text, text, text, text, text, text, text, text, text, text. Text, text, text, text, text, text, text. Text, text. Text, text, text, text, text, text, text, text, text, text, text, text, text, text, text, text, text, text.

Text, text, text, text, text, text, text, text,

text. Text, text, text, text, text, text, text, text, text, text, text, text, text, text, text, text, text, text, text, text.

Text, text, text, text, text, text, text, text, text. Text, text, text, text, text, text, text, text, text, text, text, text, text, text, text, text, text, text, text. Text, text.

Text, text, text, text, text, text, text, text, text, text, text, text. Text, text, text, text, text, text, text, text, text, text, text, text, text, text.

Text, text, text, text, text, text, text, text, text, text, text, text, text, text. Text, text, text, text, text, text, text, text, text, text, text, text. Text, text, text, text, text.

Text, text, text, text, text, text, text, text, text, text, text, text, text, text. Text, text, text, text, text, text, text, text. Text, text, text, text, text, text, text, text, text.

In this example the text is presented in two columns, and the title is bordered and placed in the first column. A drop capital T begins the first paragraph. The text is fully justified left and right. A visual is inserted into the text of the second column.

Sample Document Page Designs, Presentations, and Productions

Text, text, text, text, text, text, text, text, text, text, text, text, text, text, text, text, text, text. Text, text, text, text, text, text. Text, text. Text, text, text, text, text, text, text, text, text, text, text, text, text, text, text.

Text, text. Text, text, text, text, text, text, text, text, text, text, text, text, text, text, text, text.

Text, text, text, text, text, text, text, text, text. Text, text, text, text, text, text, text, text, text, text, text, text, text, text, text, text, text, text. Text, text, text, text, text, text, text.

Text, text, text, text, text, text, text. Text, text, text, text, text, text, text, text, text, text, text, text, text, text, text, text, text.

Text, text, text, text, text, text, text, text, text, text, text, text, text. Text, text, text, text, text, text, text, text, text, text, text, text. Text, text, text, text, text, text, text, text, text.

Text, text, text, text, text, text, text, text, text, text, text. Text, text. Text, text, text, text, text, text, text, text, text, text, text, text, text, text, text, text, text, text. Text, text, text, text, text, text, text, text, text, text, text, text, text, text, text. Text, text.

Text, text, text, text, text, text, text, text, text, text, text, text, text, text, text, text, text, text.

In this example the title is presented in two columns. It is separated from the text. The text is presented in a nearly three-quarter-page column with left and right justifications. Visuals are placed in the remaining column, which is approximately one-third of the page. The text is flush left without indented paragraphs.

Document Page Layouts

COMPUTER VIRUSES

Text, text, text, text, text, text, text, text, text, text, text, text, text, text, text, text, text, text, text. Text, text, text, text, text, text, text, text, text, text, text, text, text. Text, text, text, text. Text, text, text, text, text ,text, text, text, text, text, text.

Text, text. Text, text, text, text, text, text, text, text, text, text, text, text, text, text. Text, text, text, text, text, text, text, text, text, text, text, text, text, text, text, text, text, text. Text, text.

VIRUS REMOVAL

Text, text, text, text, text, text, text, text. Text, text, text, text, text, text, text, text, text, text, text, text, text, text, text, text, text. Text, text, text, text, text, text, text, text, text, text, text, text, text, text, text. Text, text, text, text, text. Text, text, text, text, text, text, text, text, text,text, text, text, text, text, text, text, text, text, text, text, text, text. text, text.

Text, text, text, text, text, text. Text, text, text, text, text, text, text, text, text, text, text, text, text, text. Text, text, text, text, text, text, text, text, text, text, text, text, text, text, text, text. Text, text, text, text. Text, text, text, text, text, text, text, text, text, text, text, text, text, text, text, text.

Text, text, text, text, text, text, text, text, text, text, text, text, text, text, text, text, text, text, text. Text, text, text, text, text, text, text. Text, text, text, text, text, text, text, text, text, text, text, text, text, text, text, text, text, text, text, text.

PREVENTING INFECTION

Text, text, text, text, text, text, text, text, text, text, text, text, text, text. Text, text, text. Text, text, text, text, text, text, text, text, text, text, text, text, text. Text, text, text, text. Text, text, text, text, text, text, text, text ,text, text, text. Text, text, text, text, text, text, text, text, text, text, text, text, text, text, text, text, text, text.

In this example the title is landscaped (presented sideways) and underlined. Boldfaced, capital headings separate portions of the text. A line is skipped under each heading. Paragraphs are flush left against the margin but ragged on the right. Display quotes are indicated by the inset text. Because of their indentation, they do not require quotation marks.

These are just a few of the many printing possibilities that word processors can provide with a few clicks.

Headings, Type Sizes, and Fonts

Headings are not just decorative. They can

- Break up continuous text
- Indicate to the reader coherent sections of text
- Provide a clue to the upcoming content

Look at the headings in this or other chapters. The various sizes indicate major divisions, first-level sub parts, and even third-level sub parts. Notice, too, that the headings are printed in bold type. Some are all capitals whereas others are a combination of uppercase and lowercase letters. The major section headings are all capitals. The next level of headings are the same size but are set with uppercase and lowercase.

More than three levels may tend to confuse the reader and so must be used cautiously.

You may also design your headings with white space clues. Generally, there are more blank lines before a heading than after. Judicious use of headings breaks up the "gray" text and makes the document more coherent and accessible.

The headings in this text book are noun phrases, but you may use headings that name actions, ask or answer questions, or list steps. Consider the following headings:

WHAT IS MY GOAL IN THE WRITING PROCESS?
WHAT IS MY PURPOSE?
WHAT IS MY MESSAGE?

or

PREWRITING THE DOCUMENT
Brainstorming the Topics
Writing the Outline

The top three headings ask questions whereas the next three name actions. Subordination is indicated by size and in the second grouping by changing from all caps in the major heading to combination uppercase and lowercase in the sub headings. Usually a two-point difference in size is used, in these instances from 16 point to 14 point and from 15 point to 13 point. It is difficult for the untrained eye to differentiate anything less than a two-point difference. Further, the boldface type adds weight to the headings.

A new element, typography, has been added here. Typography is a design concept that includes the selection of typefaces, called fonts on your computers. The headings in this textbook are Galliard, but the upper group of sample headings here are in Helvetica, and the bottom three are in Eurostyle. Your computer software offers perhaps 20 to 25 typeface fonts; however, you can buy software to add thousands more fonts. Familiarize yourself with your fonts. Some are irresistible for invitations, holiday cards, brochures, and the like, such as

ZAPF CHANCERY BOLD,
ROSEWOOD,

or

STENCIL

in various sizes and even color. The technical writer, however, usually uses straight-edged fonts such as Eurostyle or Helvetica for headings and titles, and the more curly fonts such as Times New Roman or Goudy for the body text. The straight-edged fonts are called *sans serif* (without a flourish at the end of the letters), and curly fonts are called *serif* (with the flourish at the end of the letters). Consider using all of these elements in your documents. The professional technical writer will study typography and use page design software.

Margins, Justification, and Indentation

If a document is to be bound, a wide, empty inner margin will make the content easier to read. If annotations are expected, a wide, empty right-hand margin is beneficial. Text can be set flush left as is this

small portion,

centered on the page like
this brief example, or

flush right as is shown in
this brief sample of text.

Further, most typewriters and word processors automatically present a **ragged right margin,** meaning the text is not flush to the right-hand margin. The writer using a conventional typewriter may try to compensate for the excessive ragged right by splitting words with a hyphen, but sometimes too many words end up hyphenated at the end lines, creating an unaesthetic appearance. Word processors automatically wrap the overlong word to the next line, although you may use a setting that allows for end-word-hyphenated divisions. Text may also be justified so that the right-hand margin is always flush.

Most of this textbook uses the **full justified block format** for regular text and ragged right in the figures. Actually, the ragged right helps readers to track the text and to keep their place. Justified text may also cause excessive white space when the extra spacing between words forms noticeable white rivers running down the page. Justification is most often used in textbooks, brochures, newspaper columns, and some flyers.

Notice also in this book that the beginning line of a paragraph under a heading is not indented. The heading indicates a new paragraph will follow, so there is no good reason to indent the first line. In many technical texts important stress and focus are signaled by structural levels in the document. Robert Kramer and Stephen A. Bernhardt demonstrate this as follows:

> Subordinated block indents from the left signal content structure and use white space effectively.
>
> Any indent pattern beyond two levels of subordination becomes merely decorative and most probably useless to the reader.

Actually, the reader is likely to lose track of the level he/she is on, so the focus is lost.

An indent cue that the reader doesn't understand only confuses that reader.

Lists

Often your material will suggest a textual list. By organizing material into lists, the content becomes emphasized and highly visual. Examine your text to extract lists and then write them in parallel structure, indent them from the rest of the text, and consider numbering or bulleting each item. In fact what I have just written suggests a list:

Organize the items

Write each item in parallel structure

Indent the list from the text margin

Consider numbering or bulleting each item

Some lists reflect chronology or time sequence:

1. Lift the handset to turn on your phone.
2. Start your conversation.
3. Place the handset in the base unit to turn off your phone.

Numbers are visual clues for your reader to indicate unquestionable order. Lists that do not necessarily indicate a chronology may be bulleted, as in

- Definitions
- Mechanism Descriptions
- Instructions
- Process Analysis

Very often a list may become headings of text sections.

Emphatic Features

Emphatic features are those that style your type, such as **boldface,** *italics,* underlining, ALL CAPITALS, SMALL CAPITALS, and drop caps (the big E) at the beginning of a paragraph such as this. The next four sentences provide a contrast for you to see the value of some of these features:

Save your text before beginning a new document.

Save your text <u>before</u> beginning a *new* document.

Save your text BEFORE beginning a new document.

<u>Save</u> your text *before* beginning a new DOCUMENT.

The first one offers no emphasis. Perhaps the second does not need the italicized *new* because it seems like overkill. The third is effective with both boldface and small caps, and the fourth might indicate that the word DOCUMENT is included in a glossary. (Glossaries are discussed later in this chapter.) Don't overdo these emphatic devices. Decide on a plan: **bold** for action verbs, *italics* or <u>underlining</u> for emphasis, and SMALL CAPS for glossary terms. Excessive use of these elements will make your document a meaningless hodgepodge.

As you become more experienced in document design, you will discover other possibilities to use sight as meaning. Quotation marks, dashes, symbols (@, #, ^, *), equation formulas, and a variety of bullets (●, ■, ★, ✓) are other possibilities.

Tabs and Columns

Subordinated text is easy to do by presetting tabs so that just a click or two moves the carriage or cursor to the correct spot. Most word processors have preset one-inch tabs all across a line saving the writer the time of establishing the indentations. Charts, graphics, tables, and other artwork can be dragged by your cursor or moved around by tabs and placed into columns easily.

Headers and Footers

Headers and footers orient information such as page numbers, author names, publication information (volume and issue, chapter number and title). Notice that this textbook layout uses headers to indicate page, part number and title, and chapter number and title. It would be just as logical to move only the page number to the bottom and to retain the other information at the top. In this text the left-hand page places the page number flush left and the part number and title set flush right. The right-hand page places the chapter number and chapter title flush left and the page number flush right.

You should consider the logical headers and footers for your documents. Page numbers are essential, but other information serves as navigational tools to your reader. Word processing programs like Microsoft Word and WordPerfect will create headers and footers in separate windows under the toolbar format function, place them in the document where you want them, and then automatically make them appear on each consecutive page. They may be set once and then forgotten. Your computer will do the work.

ADDING PRESENTATION FEATURES

Many final presentation decisions will already be made by the time you complete your text revisions and editing. The very purpose of your document (a response to a problem, a set of instructions, a manual to accompany a product, a brochure, a proposal, a professional page for publication, and the like) will dictate the final production decisions of your document. Your final presentation must consider both the front matter and back matter to be included within the document.

Front Matter

Front matter may include a separate title page, a table of contents, and possibly a list of illustrations, a summary or abstract, and separate chapter title pages.

A long document, such as a proposal, a user manual, or a professional paper for publication will require a separate **title page.** A title page usually includes at the least a centered, full title; the name of the author or corporate author; and usually a date. Other data may be included. The title is usually presented in a larger type size than the largest heading within the document. The font should usually be straight-edged *(sans serif)* and may be presented in all capital letters. If the title is the name of a product, you will include the model number, purpose, and other significant material to indicate fully the content to follow. Figures 4.1 and 4.2 (see page 108) show title pages of a proposal and a user manual. Figure 4.3 (see page 109) shows a number of sophisticated title pages using various fonts, type sizes, drop capitals, artwork, photographs, and watermarks.

A **table of contents** in a long document (eight pages or more) will not only help your reader(s) to turn rapidly to a particular section of the document, but also gives the reader an initial indication of the organization, content, and emphasis of a document. Title the list *Table of Contents*. Some software will create this table for you by allowing you to type in the chapter titles and major headings plus the page numbers in a separate tool. The computer then generates the finished table. To create your own table of contents take your headings and subheadings (or chapters and major headings if the document is divided into chapters) using indentation between the major headings and the subordinate headings. Use spaced dots to connect the heading with the beginning page number of that section.

PROPOSAL TO IMPROVE
EMPLOYEE WORK SCHEDULES
IN THE
CUSTOMER INQUIRY DEPARTMENT

Prepared for
Andrea Brooks
Assistant Vice-president
Operations Division
State Trust Charter Bank
Bigtown, New Hampshire

by
Monica Ferschke
Administrative Assistant
October 19, 199X

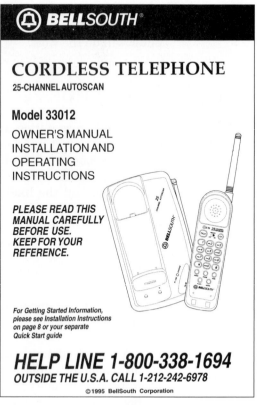

FIGURE 4.1 *Sample title page for a proposal* (Courtesy of Monica Ferschke)

FIGURE 4.2 *Sample title page for a user manual* (By permission of BellSouth Products, Inc.)

If your document contains **illustrations** (graphics, photos, maps, etc.), include a separate listing entitled *List of Illustrations* below the table of contents or on the next page. Include the illustration number and title and then connect that material with spaced dots to the exact page reference. Finally, list tables separate from figures (all other graphics, photos, maps, etc.). Figure 4.4 (see page 110) is an example of a table of contents for a proposal.

You may wish to include an **abstract** or **summary,** a short concise review of the parent document. Title this item *Abstract* or *Summary* and place it after the table of contents. Abstracts and summaries are discussed at length along with samples in Chapter 13.

If the material is divided into separate chapters, you will want to consider separate introductory **chapter title pages.** If the chapter title page is to include artwork or other significant prefatory material, use a separate chapter title page. If there is nothing other than the chapter number and title, place them on the first page of the chapter so as not to waste space. Notice the chapter title pages in this textbook include artwork.

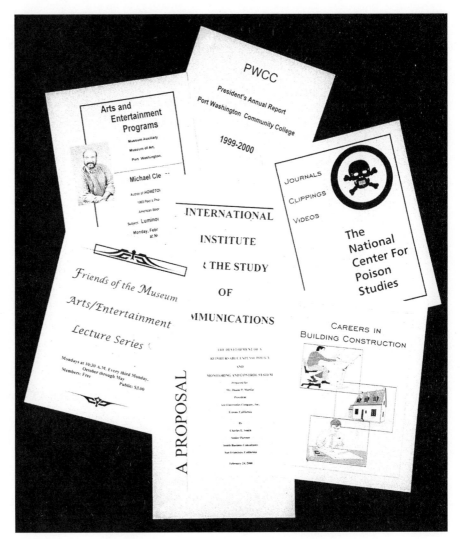

FIGURE 4.3   *Sample title pages utilizing a variety of document design options*   (By permission of Mark DeJanon)

Back Matter

Back matter, or supplements to the body of the document, may include a glossary, an appendix, a bibliography, and an index. They are placed at the back of the document.

A **glossary** is an alphabetized list of words and their definitions used within your document that may not be understood by the general reading public. Title the page(s) *Glossary* and list the unnumbered words below. If there are only a few (five or so) specialized terms used in your document, include the definitions within the text or in a footnote at the

TABLE OF CONTENTS

LIST OF ILLUSTRATIONS

FIGURE 4.4   *Sample Table of Contents for a proposal*

bottom of the page where the word occurs. If you use a separate glossary, you may want to place a star after the term within the text when it is first used.

The terms may not be all technical but may include laypersons' terms that you are using in a special sense within your document: "In this report a cell is a small group acting as a unit within a larger organization." Instructions for writing simple and expanded definitions appear in Chapter 6. Within the glossary, boldface or underline the term followed by either a colon or extra white space before the definition. Include the glossary in the listing of the table of contents. Figure 4.5 shows a partial glossary from a college textbook on business practices.

An **appendix** contains material that expands information contained in the body of your document. Parts of the appendix material may

GLOSSARY OF TERMS

Adequate protection   The protection that a debtor grants a creditor with court approval to prevent foreclosure on property. This can include a resumption of monthly payments.

Administrative creditor   All creditors holding claims for debts that were incurred after the bankruptcy was filed. This includes fees for legal and accounting professionals used by the debtor in possession.

Adversary proceeding   Matter of dispute brought before the court for determination. Many regard amount of debt, security interest, or use of cash as examples of issues to be settled by court action.

Automatic stay   Becomes effective on filling of a bankruptcy petition and prevents any creditors from taking further collection action against the debtor without court permission.

Bar date   The last date set by the court that a proof of claim can be filed to collect a debt owed by the bankrupt company.

Cash collateral   All proceeds paid or due to a company that have been pledged as collateral for a loan. Associated with the cash liquidation of inventory and accounts receivables.

Collateral   Assets that are used as security for a debt.

Confirmation   The action of the bankruptcy court approving the

FIGURE 4.5   ***Sample Glossary***   (From *Saving Your Business* by Suzanne Caplan. Copyright © 1992. Reprinted with permission of Prentice Hall.)

be discussed within the document, but the entire block of information may clutter your message. Refer to the appendix within the text if you want to alert your reader to its existence at the appropriate time.

An appendix may include tabulated interview and survey results along with the original instruments (such as questionnaires), details of an experiment or investigation, financial projections, job descriptions and resumes of new personnel, lists of references, and so on. If the document includes artwork and sample forms, you may refer to them as **exhibits** (Exhibit 1. Map of Florida Voting Results). Exhibits require numbers, whereas appendixes use letters (Appendix A. Resume of Merrill C. Tritt). An appendix (its letter and title) or a list of exhibits (their numbers and titles) is included at the end of the table of contents. This book, you will notice, includes Appendix A. Conventions of Construction, Grammar, and Usage, and Appendix B. Punctuation and Mechanical Conventions. Some of the material is included in the text with references for full discussion in the appendixes.

A **bibliography** is an alphabetized list of all works consulted in the preparation of your document. The bibliography may be titled *Bibliography, References,* or *Works Cited* depending on the documentation style appropriate to your paper. A full discussion of documentation styles along with examples of bibliographies and works cited pages are included in Chapter 16. Figure 4.6 shows a sample works cited bibliography page:

Technical Communication Quarterly   325

Works Cited

Aristotle. *The Rhetoric and the Poetics of Aristotle.* Trans. W. Rhys
    Roberts and Ingram Bywater. New York: Modern Library, 1954.
Bazerman, Charles. "How Natural Philosophers Can Cooperate: The
    Literary Technology of Coordinated Investigation in Joseph
    Priestley's History and Present State of Electricity (1767)." *Textual
    Dynamics of the Professions: Historical and Contemporary Studies of
    Writing in Professional Communities.* Ed. Charles Bazerman and
    James Paradis. Madison: U of Wisconsin P, 1991. 13-44.
—. *Shaping Written Knowledge: The Genre and Activity of the Experi-
    mental Article in Science.* Madison: U of Wisconsin P, 1988.
Brasseur, Lee E. "Contesting the Objectivist Paradigm: Gender Issues
    in the Technical and Professional Curriculum." *IEEE Transactions
    on Professional Communication* 36.3 (1993): 114-23.
Brown, Stuart C. "Rhetoric, Ethical Codes, and the Revival of *Ethos*

FIGURE 4.6   *Sample works cited bibliography listing*   (From
Gerald Savage, "Redefining the Responsibilities of Teachers and
the Social Position of the Technical Communicator," *Technical
Communication Quarterly,* Vol. 5, No. 3 [Summer 1996], p. 325.)

An **index** is an alphabetized listing of all of the key subject terms used in your document along with the page references. It differs from a table of contents in that it will contain all of the key terms and their page references occurring throughout the text, not just the chapter headings. It enables the reader(s) to locate all of the references to a particular term quickly. Some software possesses tools to help you prepare an index. You type in the terms, set the directions for a manuscript search, and the computer generates a complete index. If you generate your own index, use $3'' \times 5''$ index cards, which can be alphabetized for quick reference. Review the entire manuscript writing down each term to be indexed on a separate card along with the page number. Each time a term is used again throughout the text, include the page or pages (17, 19–24) on the appropriate card. You may even subordinate material under a particular key term. This list is then titled **Index** and included as the final back matter in your document. Figure 4.7 shows a partial index from the last edition of this textbook:

Index

FIGURE 4.7   *Sample partial index page*

PRODUCTION DECISIONS

Production decisions include the size of the paper, the quality of the paper, the orientation on the page, watermarks, borders, folds or bindings, flaps and pockets, and perforations and windows. Each of these entities must be considered before the final publication of your document.

Factors that determine the **size of the paper** include the purpose of the document, its intended distribution, and the size of packaging for the product that the document is supporting. Most in-house documents (those to be read by colleagues and management in your organization), proposals, professional articles, and book manuscripts call for the usual $8\frac{1}{2}'' \times 11''$ paper. A book or journal publisher will determine the size of a published page.

If the document is a user manual, a set of instructions, or a set of specifications, the size of the paper may vary. If the document is to accompany a product, the size is determined by the size of the packaging. You no doubt have a collection of user manuals in your possession. Note how the size varies for the manuals that accompany your cellular phone, your VCR, your fax machine, your computer and printer, and your kitchen appliances. Your cellular phone manual will probably fit into your pocket or even into the case that holds the phone.

The **quality of the paper** for in-house documents should be low-gloss, rag-bond, and white. Glossy paper produces glare and strains the eyes. Rag-bond paper has a high fiber content (25 percent minimum) and is heavier. Select a 20-pound or heavier weight paper but do not use anything as heavy as construction paper. Lower weight paper tends to be flimsy and crinkly. Out-of-house documents may require a slicker paper (like magazine pages) and a lower weight. The slicker paper allows for a sharper print, and the lower weight decreases the bulk of a document. White paper is the standard. Use color only for flyers, announcements, and the like.

Also consider the **orientation on the page.** Your document may be placed so that the narrower part of the page is at the top (called *portrait* in computer commands), or you may print out your document from left to right along the long side of your page (called *landscape* in computer commands). Landscaped lines may become so long that readers lose track as they scan back to the beginning of the next line. However, you may wish to landscape titles or other reference notations.

Watermarks and **borders** may enhance the document. The watermark, a faint picture beneath the type, is typically a logo or a picture of a product. Borders for technical papers, if used at all, are usually simple, not decorative. Figure 4.8 shows a page with both a watermark and a border:

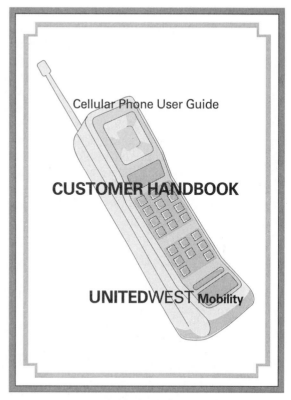

FIGURE 4.8 *Sample title page with watermark and border*

If your document is long, you may wish to consider **folds** or **bindings.** A brief, six-page manual may be folded into a three-panel brochure with printing on each of six panels. A lengthier manual could be fan folded. If some material is too extensive for a page but should be viewed all together, you may consider a foldout page. Longer documents could require a binding. A document stapled in the corner is unprofessional. You may use a binder folder, internal staples holding double pages, or a glue bond as in a book. Another type of inexpensive binding is a spiral binding put together by a professional printer. If additional pages are likely to be added over a period of time, consider a loose-leaf binder with two or three holes punched into the margins of the document so that pages may be inserted at the appropriate places. No matter how the document is to be bound, you must allow adequate margins so that the text is not crowded into the binding.

You may also consider **flaps** and **pockets.** Your separate binder may already contain front and back pockets. If you decide to include a cover harder than the content weight, you may wish to include harder internal dividers with flaps for identifying sections for easy access. You may add pockets if you foresee that additional material may be forthcoming. **Perforations** and **windows** may be decorative or functional. Consider how you might present a set of brief instructions on a cardboard format with a perforation to hang on a doorknob. Windows from one page to the next are a dramatic means of focusing attention on the framed material. You must consider all these options even if your document is being published by a professional.

CHECKLIST

Document Design, Presentation Features, and Final Production

DOCUMENT DESIGN FOR VISUAL MEANING

❑ **1.** Have I designed my document to enhance visual meaning?

❑ **2.** Does my use of space (columns, horizontal lines, borders, boxes, extra lines above a heading) make my document easier to read and access?

❑ **3.** Are my margins appropriate and do they leave enough room on the sides?

❑ **4.** Have I centered any material?

❑ **5.** Should I use a ragged right edge or full text justification?

❑ **6.** Do my margins indicate major and subordinated material?

❑ **7.** Have I added headers and footers appropriately?

❑ **8.** Have I used headings to indicate major and subordinate divisions of the text?

❑ **9.** Are my headings parallel (noun phrases, questions, naming of actions)?

❑ **10.** Have I used varied type sizes to indicate major and subordinate headings?

❑ **11.** Have I used appropriate fonts for chapter headings, major and subordinate headings, and other material?

❑ **12.** Have I organized material into appropriate lists (plain, numbered, or bulleted)?

❑ **13.** Have I used emphatic features to advantage (boldface, italics, underlining, all capitals, small capitals, and drop capitals)?

PRESENTATION FEATURES

❑ **1.** Does my document require a separate title page?

❑ **2.** Is my document long enough to require an abstract or summary before the text?

❑ **3.** Would separate chapter title pages be appropriate?

❑ **4.** Does my document require a glossary?

❑ **5.** Should I include an appendix or exhibits to expand information that is helpful but not vital to the text?

❑ **6.** Does my document require a bibliography?

❑ **7.** Does my work warrant an index?

PRODUCTION DECISIONS

❑   **1.** Is $8\frac{1}{2}'' \times 11''$ paper appropriate or should I use a different paper size?

❑   **2.** Is the quality of the paper appropriate?

❑   **3.** Should I orient any text sidewards?

❑   **4.** Would a watermark, such as a logo, or borders enhance the document?

❑   **5.** Have I considered fan folds or bindings?

❑   **6.** Does the document packaging require flaps, pockets, perforations, or windows?

EXERCISES

1.   **Headings.**   Rewrite these headings to make them parallel:

Extending Your Definition

Write the Formal Definition

Examples

You Must Include Synonyms

Providing Comparisons and Contrast

Analogy

Can You Provide the Origin of the Term?

An Etymology Should Be Provided

Are There Any Causes and Effects?

Providing an Analysis of the Process

2.   **Fonts.**   Which fonts would you consider appropriate for heading? Why?

a.   **WRITING THE DEFINITION**

b.   WRITING THE DEFINITION

c.   WRITING THE DEFINITION

d.   **WRITING THE DEFINITION**

e.   WRITING THE DEFINITION

      f. WRITING THE DEFINITION

      g. WRITING THE DEFINITION

      h. *WRITING THE DEFINITION*

3. **Justification.**   Name the major reason for using the ragged right text printing?

4. **Emphatic Features.**   Which of the following use of emphatic features do you think is the most effective? Why?

    a. *Press* the PAUSE key **to record** a cassette tape.

    b. Press the *pause* key TO RECORD a cassette tape.

    c. PRESS the **pause** key *to record* a cassette tape.

    d. Press the PAUSE key to record a cassette tape.

5. **Presentation Features.**   Answer *true* or *false* to the following statements.

    a. All technical report documents require a separate title page.

    b. A table of contents is useful for a document that is eight or more pages.

    c. A list of illustrations should be used for graphics, exhibits, and other visuals.

    d. All technical documents require a glossary.

    e. The glossary is not listed in the table of contents.

    f. If there are only a few technical terms, they may be defined within the text.

    g. A bibliography is an alphabetized list of all the works you consulted in your document preparation.

    h. An index is a list of page references chapter by chapter.

6. **Size of Paper.**   Locate a user manual that is not printed on $8^{1}/_{2}'' \times 11''$ paper. State three reasons why you believe it is the size it is. Submit it or a photocopy of a few pages with your reasons.

7. **Quality of Paper.**   Locate a document that you think has been printed on an inappropriate quality of paper. Give three reasons. Submit the original with your reasons.

8. **Orientation.**   If you are using a computer for your assignments, print out a page in portrait, and the same page in landscape. If you do not use a computer, find a partner who does. Give three advantages of one over the other. Submit the pages with your commentary.

9. **Flaps, Pockets, Perforations, Windows.**   Locate a printed publication that uses flaps, pockets, perforations, or windows. Indicate three ways in which these are either advantageous or merely decorative. Submit the publication with your critique.

WRITING PROJECTS

1. **Document Design.**   Study the following material. Add at least six document design enhancements on your computer or typewriter. Do not just be decorative. Make your design features impart meaning to the document. You may reword sentences if you have a compelling need to do so.

   > Prepare visuals to clarify and emphasize your oral report. Visual materials may include chalkboards, flip charts, posters of tables, charts and/or drawings, handout sheets, exhibits of models or equipment, filmstrips, or transparencies. The size of the room or auditorium, the people in your audience, the available monies, the available equipment (chalkboards, projectors, tables), and the nature of your speech are all factors for consideration in planning visual material. Effective visual materials are characterized by a number of factors, such as simplicity, unity, emphasis, balance, and legibility. Rehearse your speech a number of times before you actually give it. Practice before a mirror and use a tape recorder. If possible give your speech to a small group of friends. Ask them to assess your poise, eye contact, voice, gestures, and rate. Your voice should be conversational, confident, and enthusiastic. Avoid at all costs a monotonous sound. Vary your pitch and intensity. Monitor the volume and rate of your speech. Judge the quality of your voice. Are you shrill, nasal, raspy, breathy, growly?

2. **Critique a Document Design.**   Photocopy two pages from another textbook or a user manual that you believe exemplify quality document design. Write a memo explaining why you believe the format helps to convey meaning. Be specific in your references to all of the discussed design options. Use your own document design elements in the memo.

3. **Group Writing.**   Join the same students you worked with on the Writing Project in Chapter 2. Select one of the papers and reproduce it with document design elements to enhance the meaning of the document.

4. **Table of Contents and List of Illustrations.**   Write a table of contents and list of illustrations for this chapter.

5. **Index.**   Select a partner among your classmates and prepare an index for this chapter.

6. **Group Writing.**   Form a team of three or four classmates. Locate a user manual for a VCR or some other technical product. Pretend you are writing to the Director of Marketing of the firm who desires your opinions on the manual. Critique the use of document design elements, the use of presentation features, and the production deci-

sions. What would you change? Keep the same? Eliminate? Use document design elements to enhance **your** document. Do you need to include any presentation features? If one of your team is computer literate, make appropriate production decisions regarding the size and quality of the paper, the orientation on the page, and the binding.

7. **Group Writing.** Use the information in Group Writing Project 3 in Chapter 3, but instead of writing a memo design a brochure with the same information. Try to include a team member who is computer literate. Use appropriate graphics, fonts, type size, emphatic features, orientation on the paper, fan folds or bindings, and other document design options.

NOTES

PART

t w o

The Technical Strategies

Preparing Manuals

DUFFY   by Bruce Hammond

SKILLS

After studying this chapter, you should be able to

1. Recognize the purpose and uses of various types of manuals.
2. Name the four major writing strategies employed in manuals.
3. Grasp the importance of appropriate language for different manual audiences.
4. Know the procedures for preparing a user manual.
5. Identify and comprehend the appropriateness of the graphics and other visuals in a manual.
6. Recognize the publication options (paper size, quality of paper, section tabs, tiers, etc.).
7. Name at least five other supplements that may be produced in conjunction with a user manual.

INTRODUCTION

Manuals are written guides or reference materials that are used for training, organizing work procedures, assembling mechanisms, operating equipment or machinery, servicing products, or repairing products. You, no doubt, have any number of user manuals, also called user guides, owner's manuals, or operating manuals, for your computer, printer, fax machine, telephone, cellular phone, CD player, video cassette recorder, and even for your oven, refrigerator, and other appliances. Writing these manuals is the major task of technical writers. Typically a manual includes

- Precise definitions
- Descriptions of mechanisms
- Step-by-step instructions
- Analyses of processes

In a manufacturing company, professional or technical writers typically prepare manuals for the installation, operation, and repair of products or equipment. Many smaller companies develop training, procedural, troubleshooting, and job review manuals. The preparation, editing, and reviewing of component parts of a manual may be delegated to any competent writer within the organization. Those employees who use such in-house manuals are continually encouraged to review and improve the component parts.

This chapter considers the user guide because it is the most common manual. Chapters 6, 7, 8, and 9 consider the individual writing tasks of definition, description, instruction, and process analysis, which are the major component strategies of manuals.

AUDIENCE

A major consideration for manual writing is the audience (the receiver). The language and technical detail must fit the intended user. Frequently, the user of a product or mechanism is a novice or layperson; thus he or she needs a mechanism description, definitions of terms, operating instructions, and an analysis with suggested solutions of possible problems in the product's operation, as well as many graphic illustrations.

Although the product may be a highly complex mechanism, the manual information must be simple, clear, and accurate. Writers of such user guides often aim their writing at the comprehension level of a seventh- or eighth-grade student. Which of you has not found your computer manual so difficult that you have bought a book in the "Dummies" series for a simpler explanation, or relied on the imbedded "Help" tools?

The technical writer must carefully consider the formatting of the message as well as the graphics required; and the technical terms, abbreviations, symbols, and mathematical procedures employed must fit the audience's ability to decode and understand the information. The following two examples illustrate language differences in manuals intended for two different audiences. The first example shows the introductory information from an operation manual for a personal pager. The intended audience is the layperson purchaser.

Introduction

Congratulations! You are now using the world's first microprocessor-controlled 900MHz pager. Motorola's advanced technology offers unique features and benefits which provide the ultimate in performance and reliability.

The Dimension 1000 pager is a versatile unit that is designed to provide reliable communications for a variety of applications. To get the full benefit from the pager, please read these operating instructions carefully.

Coding Data Label

Dimension 1000 pagers come in several model configurations equipped with a variety of options which affect the operation of your pager. To determine how your pager operates, refer to the coding-data label located under the belt clip. The pager is capable of one, two, or three calls, depending upon how it was ordered from the factory. The coding-data label indicates the number and type of calls (Figure 2). Tone-only calls are indicated by a "T," and voice calls are indicated by a "V." If a particular call is not present, that area will be blank.[1]

---

[1] Motorola, Inc., *Motorola Dimension 1000 Binary GSC Pager* (Fort Lauderdale, Florida: Motorola, Inc., Paging Products Division, 1982), a manual.

The next example contains the introductory material from an instruction manual for servicing a walkie-talkie radio. The intended audience is skilled service technicians.

Introduction

The MX300-T "Handi-Talkie" radio described in this manual is the most advanced two-way radio available. Hybrid modular construction is used throughout, reflecting the latest achievements in microelectronic technology. The plug-in modules provide greater flexibility, greater reliability, and easier maintenance.

Each radio contains plug-in hybrid modules. These modules contain over 90% of the electronics—providing faster service and less down-time. Guide pins are provided on the modules to assist replacement and prevent incorrect insertion. Instead of complex wiring harnesses, printed flexible circuits are used in the radio. These durable, thin plastic films eliminate broken, pinched, or frayed wires—with a neat, easy-to-service interior.[2]

The first example is written with a "you" perspective. The language is general and nonspecific. The second example is less personal and contains technical terms such as *hybrid modular construction, plug-in hybrid modules, down-time,* and *flexible circuits,* terms familiar to technicians.

MANUAL PREPARATION

Let us consider the writer's procedural steps for the preparation of a user guide for a technical product.

Step 1—Determining the Audience.   The technical writer must determine if the audience, that is the potential users of the product, consists of laypersons or skilled technicians. The audience will have a bearing not only on the language but also on the complexity of the graphics, the extent and scope of the data, and the manual size and format. If the intended user is a layperson, the writer must consult the marketing department, surveys, and sales data to determine who will buy the product, what language is appropriate, and what detail is essential. Good interviewing skills are a requirement for the effective technical writer.

Step 2—Consulting the Engineering Department. In the past, technical writers were involved only after the product was finished and were given three or four months to complete a guide before the product was shipped. The modern technical writer does considerably more "front end" consultation on product development. The individual writer, or more likely a writing team, consults the engineers to determine how the product works, what instructions and warnings must be stressed, and how much detail should be included for different audiences. These dis-

---

[2] Motorola, Inc., *Motorola MX300-T Five Channel "Handi-Talkie" Portable Radio* (Fort Lauderdale, Florida: Motorola, Inc., Portable Products Division, 1982), a manual.

cussions are often tape recorded or even videotaped in order to provide exact documentation for the written materials.

Step 3—Writing the Manual.   Bearing in mind the four typical writing components of user guides (definitions, descriptions, instructions, and process analysis), the writer or team will prepare a content draft; design the document pages taking note of space considerations, headings, emphatic features, headers and footers, lists, numbering systems, and so on; and determine the need for a table of contents, the inclusion of a glossary, other appendixes, blank pages for notes, and copyright data. Figure 5.1 shows a typical table of contents for an extensive user guide:

Table of Contents

FIGURE 5.1   *Typical complex operation manual table of contents*

Step 4—Preparing the Graphics and Other Visuals.   Hand-in-hand with the manual text preparation come decisions about types of graphics and visuals and their execution. Whereas actual photographs and line art may be used to depict the overall product, drawings are used more extensively in manuals. They can emphasize parts and relationships by judicious use of exploded, cutaway, and schematic sketches and eliminate extraneous detail. Lists of items or features can be incorporated into tables for easy reference. Columnar drawings to depict steps in a process are often helpful. Print type and size decisions are important. Color, another important consideration, is usually used sparingly, but covers, borders, headings, and screens often employ color for emphasis and clarity. To protect the company and the user, cautions and warnings are strongly presented, often using graphic icons for emphasis. Figure 5.2 shows typical bold icon warnings. Be aware that the arrow for the "Note" warning is preferable to a pointing finger, which may be offensive in other cultures, for example in Asia.

FIGURE 5.2   *Typical bold warnings*

Step 5—Deciding on the Manual Production and Supplements.   The writer must make decisions on the appearance of the manual, the size and orientation of the paper, the paper quality, fonts, type size, color usage, the inclusion of blank note pages, and the number of manuals to be printed. Other considerations include pagination, section tabs or reference bands for easy accessing, a warranty inclusion, and a comment and evaluation sheet for eliciting feedback. Depending on the product and the user, several presentation formats are available: fan-folded pamphlets, which can be inserted into the product packaging; booklets with bound pages; or looseleaf pages for easy insertion and deletion in a folder.

Currently, technical writers tend to produce **supplements** in the form of, perhaps, three or more **tiers** of manuals: a fan-folded brochure, or tear-off card with perforations containing quick reference guides; a second-level instruction manual of more detailed material; and a third-level, highly technical and comprehensive manual for the sophisticated user. In addition to print material, technical writers are increasingly providing instructional audiotapes, videotapes, CD-ROMs, computer disks, large posters illustrating basic steps, and even operational information embedded in the product itself (such as the "Help" tools on your computer). Visual aids such as CAD conversions, schematics, and digitized photographs (line art) are increasingly common.

More decisions need to be made on how best to include accessory materials with the print material. Options include end pockets, accompanying plastic bags for computer disks, and boxlike casings for CD-ROMs or videotapes, to name a few.

To reinforce the written and support materials, the technical writer is responsible for a coherent library of posters, flyers, product promotional materials, and multimedia materials. It is the technical writer or team who develops the on-line materials: the company web page, product information pages, on-line manuals, and even quick-line movies showing operations for downloading. All these materials must have the same look and feel. Desktop publishing software, such as Framemaker, the standard in the industry, combines word processing, page layout, graphics, color features, and foreign language translations. It also possesses on-line distribution features, such as disseminating documents directly onto the World Wide Web, inserting hypertext links (highlighted, colored words, phrases, or pictures that can be clicked to skip directly to related documents elsewhere in a collection), and answering product questions submitted online. In fact, the technical writing departments of major manufacturers spend about 60 percent of their time writing and 40 percent developing multimedia presentations and other technical support material.

Figure 5.3 (see p. 132) shows one side of a fan-folded brochure, "Four Quick Steps to Everyday Use," produced to accompany a cellular phone. Figure 5.4 (see p. 132) shows a "Quick Reference" tear-off page for posting, which is included in the second-tier manual for the same product.

In addition to the "Everyday Use" brochure, the "Quick Reference" tear-off page from the second-tier manual, "Getting Started," there is an even more comprehensive supplement manual titled "A to Z Reference," but it is not reproduced here.

To familiarize you with a typical manual, Figure 5.5 (see p. 133) shows a complete Motorola pager user guide. It is reproduced actual size and fits the packaging and equates to the size of the pager itself. The guide is printed on both sides of the paper and bound by a staple. The print is landscaped (printed along the length of the paper) to facilitate more efficient and pleasing use of space. Graphics are combined with instructions and analysis on each page. Notice the content explanatory comments.

Four Quick Steps to Everyday Use

1. Make a Group Call:

A Group Call is an instant communication between you and others in your organization.

1. Press (Mode) until **Group Ready** is displayed. This is known as Group Call Mode.

2. Press (►) to select the group with whom you want to have a group call.

3. Press the **PTT** (Push to Talk) and speak into the Microphone. Release the PTT to listen.

2. Make a Private Call:

A Private Call is a radio call between two individuals.

1. Press (Mode) until **Prvt Ready** is displayed. This is known as Private Mode.

2. Press (►) to select the person with whom you want to have a private call.

3. Press the **PTT** (Push to Talk) and speak into the Microphone.

r370

3. Make a Phone Call:

1. Press (Mode) until **Phone Ready** is displayed. This is known as Phone Mode.

2. Dial the number. Use (◄) to correct mistakes.

   You can stop at any time by using the (··) to select the **Cancl** option.

3. Press (SEND) to place the call.

   Press (Pwr/End) to hang up.

4. Receive a Phone Call:

Press (SEND) when your phone rings to answer a phone call.

You've successfully entered the world of iDEN technology. It's time to move beyond the basics, so please refer to Getting Started to explore additional features of your r370 portable.

Volume

PTT

FIGURE 5.3 *A first "tier," fan-folded user guide for the Motorola r370 cellular telephone* (Courtesy of Motorola, Inc., Fort Lauderdale, FL)

PHONE MODE

Making a Phone Call
1. Press (Mode) until you see **Phone Ready**.
2. Enter phone number.
3. Press (SEND) to place call.
4. Press (Pwr/End) to end call.

Answering a Call
Press (SEND) to answer incoming call.

Additional Phone Features
1. Press (MENU) to view other options.

Speed Dial
2. Select **Spd#**.
3. Enter the storage number.
4. Press (SEND) to place the call.

Call Forward
2. Select **Forwd**.
3. Enter the number where you want your calls forwarded.
4. Select **On**.

Call Waiting
2. Select **Wait**.
3. Select **On**.

GROUP CALL MODE

Making a Group Call
1. Press (Mode) until you see **Group Ready**.
2. To initiate the call to the talkgroup shown, press and hold the PTT (Push to Talk) button. Wait for the chirp and then begin speaking.
3. Release the PTT to listen.

PRIVATE CALL MODE

Making a Private Call
1. Press (Mode) until you see **Prvt Ready**.
2. Enter the Private Call ID of the person you want to reach.
3. Press and hold the PTT & speak into the microphone.

Sending a Call Alert
1. Select **Alert**.
2. Scroll to the person to whom you want to send the Call Alert. You may also select the person by using Alpha Search.
3. Press the PTT to send the Call Alert.

FIGURE 5.4 *A "Quick Reference" tear-off page for posting from a user manual* (Courtesy of Motorola, Inc., Fort Lauderdale, FL)

Controls graphic with labeled parts

Side bar references throughout

Definition of each button use

Page number footers throughout

CONTROLS

Controls

Mode Button

Select Button

Read/Power On Button

1:PICK YOU UP AT 9

WORDline2.FLX

⬛	Mode	Used to scroll through available choices, or to increment a value.
▶	Select	Used to select and confirm a value.
⬭	Read/Power On	Used to read a message, to save a setting, to turn on the pager, to turn on the backlighting, and to escape from a menu.

2

Descriptive graphic showing actual pager print

Instruction with brief analysis

Inserted boldface, italic Note

Turning Your Pager On

MOTOROLA

12:35P   4/23

Press ⬭ to turn your pager on.

A start-up message is displayed momentarily and your pager activates the currently selected alert.

Note: The start-up alert can be stopped by pressing any button.

When the pager is on and no activity is taking place, the standby screen is displayed. The standby screen displays the power indicator, the time and date, the alert mode, and may display other pager indicators such as alarm status.

GETTING STARTED

3

Description and analysis

Notice menu and button icons to save words throughout

Descriptive graphics

Bulleted and numbered instructions throughout

GETTING STARTED

Menu Icons

The four menu icons on the top row (Ⓜ⬛◀🔔) correspond to the four menus: CONTROLS, DELETE ALL, ALERTS, and ALARMS.

The first time you press ⬛ these menu icons are displayed. The controls menu icon Ⓜ flashes, indicating that pressing ▶ will enter the controls menu.

Press and release ⬛ to display the CONTROLS?, DELETE ALL?, ALERTS?, and ALARMS? menu prompts. Press ▶ to enter the corresponding menus, or press ⬭ to return to the standby screen.

Turning Your Pager Off

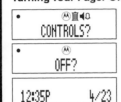

CONTROLS?

OFF?

12:35P   4/23

❶ From the standby screen, press ⬛ to display CONTROLS?. The controls menu icon Ⓜ flashes.

❷ Press ▶ to display OFF?.

❸ Press ▶ to turn the pager off. The off screen is displayed without any icons.

4

FIGURE 5.5   *Complete user guide manual incorporating definitions, descriptions, instructions, and analyses*
(Courtesy of Motorola, Inc., Boynton Beach, FL)

TIME/DATE

Setting the Time and Date

Notes, analysis graphics, and instructions continued on both pages

CONTROLS?	❶ From the standby screen, press 🔲 to display CONTROLS?. The controls menu icon Ⓜ flashes.
OFF?	❷ Press ➤ to enter the CONTROLS menu. ❸ Press and release 🔲 until TIME/DATE? is displayed.
TIME/DATE?	❹ Press ➤ to enter the TIME/DATE menu. The hour field flashes.
12:00 1/01	❺ Press and release 🔲 to adjust the hour. ❻ Press ➤ to move to the next field.
12:35P 4/23	❼ Repeat steps 5 and 6 for the minutes, AM/PM, month, and day fields. ❽ Press ⬅ from any field to save and return to the standby screen.

8

Hint: Pressing and holding 🔲 scrolls through selections quickly.

Setting the Incoming Message Alert

You can set your pager to alert with a vibrating alert (vibration with no alert tone), one of eight audio alerts, a chirp alert (short beep alert), an alert of increasing volume (Escalert), or no alert (completely silent).

ALERTS

ALERTS?	❶ From the standby screen, press and release 🔲 until ALERTS? is displayed. The alert icon ◀ flashes.
VIBRATE?	❷ Press ➤ to enter the ALERTS menu. VIBRATE? is displayed.
AUDIO?	❸ Press and release 🔲 until your choice of VIBRATE?, AUDIO?, CHIRP?, ESCALERT?, or NO ALERT is displayed.
CHIRP?	❹ Press ➤ to select the desired alert. The standby screen is displayed with the corresponding alert icon.

9

Note: If the audio alert mode is selected, your pager automatically displays the audio alert options screen.

ALERTS

Choosing an Audio Alert

If you select the audio alert mode, you can set your pager to alert with one of eight audio alerts.

If you choose no alert or vibrate, your pager emits an audio alert only if a priority message is received, or an alarm sounds.

Descriptive graphics, hints, instructions, and analysis on both pages

AUDIO?	❶ From the ALERTS menu, press and release 🔲 until AUDIO? is displayed. ❷ Press ➤ to enter the AUDIO menu.
ALERT 1	❸ Press 🔲 until your choice of audio alert is displayed. The pager emits a sample of each alert.
12:35P 4/23	❹ Press ➤ to select your choice of audio alert. The standby screen is displayed with the audio alert icon ◀.

10

FIGURE 5.5 *continued*

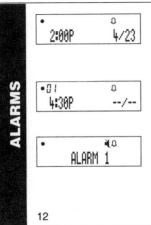

Analysis,
descriptive
graphics, notes,
and *instructions*
on each page

ALARMS

Setting the Alarm

Your pager has three alarms. Each alarm can be set for either a specific time and date, or for a specific time on a daily basis. If your pager is off when an alarm sounds, it remains off.

Note: An alarm always emits an audio alert.

❶ From the standby screen, press and release ▒ until ALARMS? is displayed. The alarm icon 🔔 flashes.

❷ Press ▸ to enter the ALARMS menu. The alarm number 01 flashes.

❸ Press ▒ to choose alarm 01, 02, or 03.

❹ Press ▸ to move to the next field. The enable 🔔 or disable 🔕 icon flashes.

❺ Press ▒ to enable 🔔 or disable 🔕 the alarm.

❻ Press ▸ to move to the hour field. The hour flashes.

11

ALARMS

❼ Press and release ▒ to adjust the hour.

Hint: Press and hold ▒ to adjust the setting quickly.

❽ Press ▸ to move to the next field.

❾ Repeat steps 7 and 8 to set the minutes, AM/PM, month, and day fields.

❿ Press ◀ in any field to save and exit.

When an alarm is enabled, the alarm icon 🔔 is displayed on the standby screen.

To alert on a daily basis, set the month and day to --/--.

When an alarm expires, the alarm icon 🔔 flashes, ALARM is displayed with the alarm number, and your pager alerts.

Press any button once to stop the alarm, and again to clear the message.

12

Read Mode

This feature allows you to choose the most comfortable way to view messages.

Setting the Scroll Speed

You can choose the speed at which your messages scroll, or read them line by line.

❶ From the standby screen, press ▒ to display CONTROLS?. The controls menu icon Ⓜ flashes.

❷ Press ▸ to enter the CONTROLS menu.

❸ Press and release ▒ until SCROLL? is displayed.

❹ Press ▸ to enter the SCROLL menu.

❺ Press and release ▒ to choose the scroll speed.

❻ Press ▸ to select the scroll speed.

There are four scroll speeds to choose from: LINE-BY-LINE, SCROLL 1, SCROLL 2, and SCROLL 3 (fastest).

READ MODE

13

FIGURE 5.5 *continued*

Instructions,
analysis,
and *descriptive*
graphics
on each page

MESSAGE FEATURES

Locking Personal Messages

By locking messages, you can save personal messages to prevent them from being replaced when the memory is full. Messages can be locked only while reading them.

❶ Press ◄▬ while reading a personal message to display LOCK?.

❷ Press ▬► to lock the message. The message indicator ▯ is displayed.

When a locked message is selected or read, the lock icon 🔒 is displayed.

Note: A maximum of eight personal messages may be locked at one time. To lock another message, you must first unlock at least one message.

Unlocking Personal Messages

UNLOCK?

❶ Press ◄▬ while reading a locked personal message to display UNLOCK?.

14

MESSAGE FEATURES

❷ Press ▬► to unlock the message. The message indicator ▲ is displayed.

Private Time

This feature allows you to turn off all pager alerts during a preselected time period. Messages received during this time period are stored. When enabled, private time works on a daily basis. When disabled, no pager alerts are turned off during the private time setting.

CONTROLS?

❶ From the standby screen, press ◄▬ to display CONTROLS?. The controls menu icon Ⓜ flashes.

OFF?

❷ Press ▬► to enter the CONTROLS menu.

PRIVATE TIME?

❸ Press and release ◄▬ until PRIVATE TIME? is displayed.

15

MESSAGE FEATURES

1:52P/ 1:52P

❹ Press ▬► to enter the PRIVATE TIME menu. The private time enable ☾ or disable • icon flashes.

1:52P/ 1:52P

❺ Press ◄▬ to enable ☾ or disable • private time.

9:00A/11:30A

❻ Press ▬► to move to the next field. The hour field of the start time flashes.

❼ Press ◄▬ to adjust the start time hour.

❽ Repeat steps 6 and 7 to adjust the start time minutes, AM/PM, stop time hour, minutes, and AM/PM.

❾ Press ◄▬▸ from any field to save and exit. While private time is active, ◄ is not displayed, and the pager does not alert.

16

FIGURE 5.5 *continued*

Instructions

Deleting Messages
Messages may be deleted one at a time or all at once.

Deleting a Single Message

❶ While reading a message press and release ◁▣ until DELETE? is displayed.

❷ Press ▶ to delete the message.

Deleting All Messages
The DELETE ALL command deletes all read and unlocked personal messages and information services. Locked or unread messages are not deleted.

❶ From the standby screen, press and release ◁▣ until DELETE ALL? is displayed. The delete icon 🏛 flashes.

❷ Press ▶ to enter the DELETE menu. The delete confirmation DELETE? is displayed.

❸ Press ▶ to delete all personal messages.

MESSAGE FEATURES

17

Analysis on
next two pages

MESSAGE FEATURES

Storing Messages
Your pager can store up to 16 personal messages. Each stored message is assigned a number, which is displayed when the message is stored. The first message received is 1, the second is 2, and so on.

Automatic Message Deletion
If all message slots are full and a new message is received, the oldest unlocked read message is automatically deleted.

When the message memory is full, MEMORY FULL is displayed. Press any button to return to the standby screen.

If all messages are unread, the oldest, unlocked message is deleted and OVERFLOW is displayed.

18

Information Services
Information services are typically news or financial reports which provide information that is relevant for a short time (a few hours). Your pager has two information service message slots. You can set each information service message slot to alert you when you receive an information service message. Contact your service provider if you are interested in receiving information services.

INFORMATION SERVICES

19

FIGURE 5.5 *continued*

Analysis,
Descriptive
graphics,
and *instructions*
on all three
pages

INFORMATION SERVICES

Reading Information Services

When an information service message is received, ▣ flashes for 12 seconds and the number of unread information services is displayed. After 12 seconds, the standby screen is displayed.

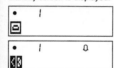

❶ Press ◠ to display the message indicator menu.

❷ Press ◠ to enter the information service menu.
The information service message icons display ▧ when selected and ▨ when unselected.

❸ Press ▶ to move to the message you want to read. The corresponding message slot number is displayed.

20

❹ Press ◠ to read the message. The time the information service was received is displayed with the first screen of the message.

▶ indicates the message is continued on an additional screen.

Press ◀ and then ◠ to display the previous screen.

Turning the Information Service Alert On and Off

You can set an information service alert for each message slot.

❶ While reading an information service message, press ◀ until CHIRP ON? or CHIRP OFF? is displayed.

❷ Press ▶ to turn chirp on or off for that information service message slot.

INFORMATION SERVICES

21

INFORMATION SERVICES

Turning the Information Service Bookmark On

You can set a bookmark to hold your place in a lengthy information service while reading it.

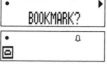

❶ While reading an information service, press ◀ to display BOOKMARK?.

❷ Press ▶ to activate the bookmark. The message menu screen is displayed.

√ is displayed in the information service menu. The next time you read that information service, the last screen displayed while the message was marked is the starting point of the display.

▧ indicates a selected bookmarked message.

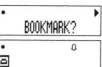

22

FIGURE 5.5   *continued*

Analysis of problems

Out of Range

If your pager is equipped with this feature, and if you are outside your paging coverage area, Ψ is displayed. As long as Ψ is displayed, your pager cannot receive messages.

Message Error Icons

If there is an error in the message received, the error icon ■ is displayed at the end of the message.

If this option is enabled, and if ⱶ is displayed at the end of the message, either the message was too long, or there was not enough memory to store the message.

Patent information

Patent Information

This Motorola product is manufactured under one or more Motorola U.S. patents. These patent numbers are listed inside the housing of this product. Other U.S. patents for this product are pending.

Cleaning instructions

Cleaning Your Pager

To clean smudges and grime from the exterior of your pager, use a soft, non-abrasive cloth moistened in a mild soap and water solution. Use a second cloth moistened in clean water to wipe the surface clean. Do not immerse in water. Do not use alcohol or other cleaning solutions.

23

OTHER FEATURES

Functionality and Use of Your Pager

For questions pertaining to the functions and use of your Motorola pager please visit our web site at www.mot.com/pagers or call 1-800-548-9954. For questions pertaining to your paging service, contact your paging service provider.

Care and Maintenance

The WORDline and WORDline FLX pagers are durable, reliable, and can provide years of dependable service; however, they are precision electronic products. Water and moisture, excessive heat, and extreme shock may damage the pager. Do not expose your pager to these conditions. If repair is required, the Motorola Service Organization, staffed with specially trained technicians, offers repair and maintenance facilities throughout the world.

Troubleshooting analysis and instructions

You can protect your pager purchase with an optional extended warranty covering parts and labor. For more information about warranties or repair, please contact your paging service provider or retailer, or call Motorola, Inc. at 1-800-548-9954.

24

USE AND CARE

Battery Information

Your WORDline or WORDline FLX pager operates with one AAA-size alkaline battery. When the battery is low, the low-battery icon ▭ is displayed between the time and date on the standby screen. Change your battery within five days of receiving a low-battery indication.

Messages are retained when replacing the battery. To retain pager alert settings, turn the pager off before removing the old battery.

Battery analysis, instructions, and descriptive graphic

Replacing the Battery

❶ Turn your pager off.

❷ To remove the old battery, slide the battery door lock towards the top of the pager to unlock the battery door.

❸ While pressing on the battery door, slide the door until the ribs on the battery door align with the ribs on the back cover.

❹ Lift the battery door to free it from the housing.

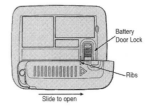

Battery Door Lock

Ribs

Slide to open

25

BATTERY

FIGURE 5.5 *continued*

Battery
instruction
continued

Double-sided
pages for notes
not reproduced
here

Next two pages
are the front
and back of
a folded quick
reference
card with
perforations to
tear it from the
manual cover

BATTERY

❺ Remove the battery.
❻ Align the new battery so the positive (+) and negative (-) markings match the polarity diagram in the battery compartment. Insert the battery.
❼ To replace the battery door, align the grooves on the battery door with the grooves on the back cover and slide the battery door closed.
❽ Slide the battery door lock toward the bottom of the pager to lock the battery door.

26

WORDline™ and WORDline™ FLX Quick Reference Card

Indicators and Icons

•	On	▶	Message continuation
◖	On, Private Time enabled	▣	Low-battery
▯	Message, selected	▣	Information service menu, selected
▲	Message, unselected	▣	Information service menu, unselected
Ⓜ	Controls menu	▮	Information service message, selected
▥	Delete All menu	▮	Information service message, unselected
◀•	Audio alert	▮	Information service, Chirp on, selected
◀	Alerts menu/Chirp alert	♪	Information service, Chirp on, unselected
◀	Vibrate or no alert (silent) icon	▮	Information service, Bookmark, selected
◺	Alarm menu/Alarm enabled	√	Information service, Bookmark, unselected
◺	Alarm disabled	▤	Locked message, selected
🔒	Locked message	🔒	Locked message, unselected
�	Out of range		

Control Buttons: Mode   Select   Read/Power On

Setting the Alarm
❶ Press 🔲 until ALARMS? is displayed.
❷ Press ▣.
❸ Press 🔲 to choose alarm 01, 02, or 03.
❹ Press ▣ to move to the next field.
❺ Press 🔲 to enable ◺ or disable ◺ the alarm.
❻ Press ▣ to move to the hour field.
❼ Press and release 🔲 to adjust the hour.
❽ Press ▣ to move to the next field.

❾ Repeat steps 7 and 8 to set the minutes, AM/PM, month, and day fields.
❿ Press ◢ from any field to save and exit.
To alert on a daily basis, set the month and day to --/--.

Locking and Unlocking Messages
While reading a personal message, press 🔲 to display LOCK? or UNLOCK?. Press ▣.

Deleting a Single Message
Press 🔲 while reading a message until DELETE? is displayed. Press ▣.

WORDline and WORDline FLX Menu Map

Controls	Delete All	Alerts	Alarms
⌐OFF	⌐DELETE	⌐VIBRATE	⌐ALARM 01
⌐TIME-DATE		⌐AUDIO	⌐ALARM 02
⌐SCROLL		⌐CHIRP	⌐ALARM 03
⌐PRIVATE TIME		⌐ESCALERT	
		⌐NO ALERT	

FIGURE 5.5   *continued*

®, Motorola, WORDline, WORDline FLX, Escalert, and ♨
are trademarks or registered trademarks of Motorola, Inc.

© 1997 by Motorola, Inc. All Rights Reserved.
Paging Products Group
1500 Gateway Blvd., Boynton Beach, FL 33426-8292
Printed in U.S.A. 2/97

6881028B95-O

Standard
copyright
information

FIGURE 5.5 *continued*

CHECKLIST

Preparing Manuals

PRELIMINARY

❑ **1.** Have I considered the audience (laypersons or skilled technicians) needs?

   ❑ Appropriate language?

   ❑ Amount of detail?

CONTENT

❑ **1.** Have I included definitions, descriptions, instructions, and process analysis as needed?

❑ **2.** Have I designed appropriate graphics?

   ❑ Actual photographs?

❏ Overall drawings?

❏ Exploded drawings?

❏ Cutaway drawings?

❏ Schematics?

❏ Tables?

❏ Warning and directional icons?

❏ Other?

❏ **3.** What other inclusions are necessary?

❏ Title page?

❏ Table of contents?

❏ Glossary?

❏ Other appendixes?

❏ Blank pages for notes?

❏ Copyright data?

❏ Warranty inclusion?

❏ Comment sheet for feedback?

DOCUMENT DESIGN

❏ **1.** Have I considered other document design elements?

❏ Will it be printed in portrait?

❏ Will it be printed in landscape?

❏ Spacing?

❏ Headings?

❏ Fonts?

❏ Type size?

❏ Boldface type?

❏ Headers and footers?

❏ Bullets or numbers?

❏ Column layout?

❏ Color?

❏ Borders?

❏ Screens?

PRODUCTION MATTERS

❏ **1.** Have I considered production matters?

   ❏ Size of package and manual?

   ❏ Paper quality?

   ❏ Bindings?

   ❏ Pockets for supplements?

   ❏ Number to be printed?

   ❏ Section tabs or reference bands?

SUPPLEMENTS

❏ **1.** What other promotional materials should be included?

   ❏ Audiotapes?

   ❏ Videotapes?

   ❏ CD-ROMs?

   ❏ Computer disks

   ❏ Posters and/or flyers?

   ❏ Imbedded "Help" tools?

EXERCISE

1. **User Guide Feature Recognition.**   Collect a variety (five or six) of fairly short user guides for such items as a pager, a telephone answering machine, a portable phone, a calculator, an alarm clock with multiple features, and so on. Include some that you deem weak. Save these for use in this and the next four chapters. Select one

for this exercise, photocopy the pages, paste each on to a separate blank page, and identify in the margins its parts: title page, table of contents, user identification, definitions, descriptions, instructions, and analysis sections. What else is included?

WRITING PROJECTS

1.   **Evaluation of a Manual.**   Select the best manual (see the previous exercise) and write a brief report on its effectiveness. Comment on its overall format (size, type of paper, page designs, headings, color, bold-face print, numbering system, etc.), its language, the clarity of its mechanism description, awareness of the need for definitions, clear instructions, and the extent of analysis of operational procedures and problems. Use headings in your report and attach the manual, or a photocopy of it, to your report.

2.   **Evaluation of Graphics and Other Visuals.**   Select another of the manuals you collected for the exercise and write a brief report on the graphics and other visuals. Concentrate on the effectiveness of the graphics and visuals. Are photographs appropriate and helpful or would drawings be more effective? Are the drawings helpful? Are there sufficient numbers of graphics and visuals? Do warnings, notes, and cautions stand out clearly? Are tables used to good effect? Are the fonts and type sizes logical and helpful? Are the graphics and visuals well labeled and titled? Could other content be presented graphically for beneficial effect? Is color used? Effectively? Other comments?

3.   **Evaluation of an Ineffective Manual.**   Select a poorly executed manual and write a brief report on its shortcomings. Comment on its overall format. Is the size appropriate for the mechanism? Is the type of paper adequate, too glossy, too thin? Is the cover design weak? Is the table of contents helpful? Are the page designs logical? Are the headings effective? Are fonts, type sizes, and color used effectively? Are the directions numbered clearly? Are boldface type, cautions, and warnings (including icons) used effectively? Are the graphics, draw-ings, and other visuals used well? Are there too many graphics? Too few? Is the placement logical and helpful? Is there an adequate pho-tograph or drawing of the mechanism and its parts. Are the labeling and explanation of parts effective? Are all necessary terms defined? Are the instructions numbered, sequential, and easy to follow? Is there an analysis section (a troubleshooting guide and/or an expla-nation of the function of the mechanism)? What other weaknesses does the guide have?

CHAPTER *6*

Defining Terms

DUFFY   by Bruce Hammond

S K I L L S

After studying this chapter, you should be able to

1. Understand the need for definition.
2. Explain the methods by which definitions may be incorporated into a document.
3. Name and explain the 17 methods of definition.
4. Recognize and avoid definition fallacies.
5. Write formal sentence definitions for a list of related terms.
6. Write an extended definition employing at least ten definition strategies.

INTRODUCTION

Because the English language contains the largest, most complex vocabulary of any known language (more than 500,000 official words, excluding the technical, scientific, and most colloquial), it is not difficult to comprehend the need for clarifying terms in every kind of document. This is especially true in professional and technical writing in which jargon, shoptalk, and formality often meet head to head. Computers and related technology have introduced an entire new "language." Words such as *cyberspace, digital cellular networks, gigabyte, hypertext,* and *encryption* bombard us. Even related jargon and argot words, such as *hacker, newbie, booting up, flame,* and *surfing,* baffle the novice.

All medical professionals must be able to recognize words *(melanoma, proteasers, cyclospora)* or a misunderstanding can result in the death of a patient. Engineers must be able to communicate with technicians *(solenoid, circuitry, torque),* and realtors must understand the terminology in their field *(escrow account, liens, easements).* Usually, students find that the most common problem they encounter in their course work is the definition of new terms pertinent to their fields of study. A student who does not fully understand the meaning of the terms *square root* and *equation* will have great difficulty in a math class, and one who cannot define *modifier* or *gerund* will encounter difficulty in English composition.

Moreover, there are two official forms of the English language: Standard English, which spans most of the English-speaking world, and Standard American English, which is used only in the United States. The British and most countries once ruled by Great Britain often spell words differently *(colour* for *color, centre* for *center)* and assign different meanings to American English words *(torch* for *flashlight, jumper* for *sweater)*. Thus, a well-developed vocabulary relative to both location and profession is essential to good writing. To define a term requires a solid understanding of what the term actually means in a given context—a difficult task in a language that is filled with words that sound alike, look alike, or are vague in meaning.

AUDIENCE

The audience determines the need for and the extent of any one definition. Writers of informal or formal reports, manuals, professional articles, and the like must define all terms that may be unfamiliar to the audience, words that may have more than one meaning, and/or those that are used in a special or stipulatory manner.

Consider the simple word *tongue,* which has at least eight distinct meanings to different audiences:

To a biologist	A tongue is a fleshy movable portion of the floor of the mouth of most vertebrates that bears sensory end organs and small glands, and that functions in taking and swallowing food, and in humans as a speech organ.
To a geographer	A tongue is a long, narrow strip of land projecting into a body of water.
To a cobbler	A tongue is a flap under the lacing or buckle of a shoe at the throat of the vamp (the part of the shoe that covers the instep and toe).
To a linguist	A tongue is a spoken language; the manner, or quality of utterance; or the intention of a speaker.
To a belt maker	A tongue is a movable pin in a buckle.
To a carpenter	A tongue is the rib on one edge of a board that fits into a corresponding groove in an edge of another board to make a joint flush (an even and unbroken line).
To a bellmaker	A tongue is a metal ball suspended inside a bell so as to strike against the side as the bell is swung.
To some religious groups	Tongue is the charismatic (divinely inspired) gift of ecstatic (extremely emotional) speech.

The parenthetical definitions of *vamp, flush, charismatic,* and *ecstatic* underscore the need for definition as an essential to meaningful communication.

Definitions may be integrated into a paper, may constitute a major portion of a paper, may be the primary purpose of a paper, or may be included in a glossary, usually at the end of a paper. An instruction manual may include an introductory list of terms to be used within. A policy handbook may begin each chapter with relevant definitions. Legal contracts may define words used within the document.

SOURCES OF DEFINITIONS

The most obvious source of a definition for a term is a dictionary. The standard dictionary for the English language is the *Oxford English Dictionary* (OED), a rather voluminous, but comprehensive, source of definitions providing complete pronunciations, spelling variations, etymologies (the origins and development of words), denotations (concrete, literal meanings), connotations (abstract, implied meanings), and historical perspectives on each different use of the term. The OED is good for an in-depth analysis of a word, but its use may represent an uneconomical use of time for someone in business or technology.

For quicker reference, there are numerous desk dictionaries (e.g., *Webster's New World Dictionary, The American Heritage Dictionary, The American Dictionary of the English Language, The Random House Dictionary*) that are more convenient to use. Though not as comprehensive as the OED, they usually provide the denotation of the term and the most common connotative meanings plus some etymology. Encyclopedic dictionaries, such as *Webster's Third International Dictionary,* also provide graphics and other data relative to words. Textbooks, technology handbooks, user guides, and documents within a given field may also be valuable sources.

METHODS

The extent to which a term should be defined depends not only on the audience but also on the complexity of the term itself. Terms may be defined by

- Parenthetical expression
- Brief phrase
- Formal sentences
- Extended sentences or paragraphs

Parenthetical Definition

The simplest way to define a term is to include a synonym in parentheses directly after the term:

> The top half of a drainage map drawing is the plan (aerial view); the bottom half is the profile (horizontal view).
>
> The ring top (round, spoked, carrying handle) of a fire extinguisher corroded.

Definition by Brief Phrase

Sometimes a defining phrase will clarify your term:

> If the body temperature is abnormal, which is above or below a range of 97.6°F to 99°F, further diagnostic procedures should follow.
>
> The cause of Reye's Syndrome, a relatively rare but serious disease that appears to be related to a variety of viral infections, particularly chickenpox and influenza, is unknown, although some studies have found an increased risk after the use of aspirin during a viral illness.

Formal Sentence Definition

A specific pattern exists for precise definition of terms. The pattern consists of three parts: the name of the term, the class of the term, and the characteristics of the term which distinguish it from all other members of its class. Some examples follow:

Term	Class	Distinguishing Characteristics
Arbitration	is a process	by which both parties to a labor dispute agree to submit the dispute to a third party for binding decision.
Sediment	is matter	which settles to the bottom of a liquid.
Assets	are items owned	such as cash, receivables, inventories, equipment, land, and buildings.
Paranoia	is a personality disorder	in which a person feels persecuted or has ambitions of grandeur.
Arson	is a criminal act	of purposely setting fire to a building or property.

| A patent | is an inventor's exclusive right | which is granted by the federal government, to own, use, make, sell, or dispose of an invention for a certain number of years. |
| Down-time | is a period | during which a computer system is inoperable due to power failure or hardware breakdown. |

Sometimes a formal definition requires more than one sentence in order to read smoothly:

A rifle is a firearm that has spiral grooves inside its barrel to impart a rotary motion to its projectile. It is designed to be fired from the shoulder and requires two hands for accurate operation.

This definition concentrates on just one meaning of the term. The term *rifle* is also a verb, *to rifle,* which is "to cut grooves," or "to ransack," "rob," or "pillage," or "to search and rob."

DEFINITION FALLACIES

A fallacy is an illogical or misleading error in a formal sentence definition. A definition fallacy is, therefore, an illogical or misleading definition including terms that are too technical, too broad, too narrow, circular, or that misuse *where* and *when.*

Too Technical.   To be useful a definition should not contain terms that are more technical or confusing than the term itself. Consider which of these two definitions is more useful to a layperson:

| Too technical | Dysgraphia is a transduction disorder that results from visual motor integration disturbance. |
| Improved | Dysgraphia is a writing disorder that results from a difficulty in writing what one sees. |

The first definition might well convey meaning to a group of learning disability specialists, but the second is more helpful to an undergraduate student.

Too Broad.   The definition writer must avoid using words that are too abstract or broad. Consider the following: "Cerumen (term) is a substance (class) in the internal ear." Is the substance waxy, hard, liquid? How does it get to the inner ear? Is it secreted by a gland or picked up externally? Is it in the canal? Is it in the earlobe? Other examples and their improvements are:

Too broad Marl (term) is earth (class) that is used in several products (distinguishing characteristics).

Improved Marl (term) is a crumbly soil (class) that consists mainly of clay, sand, and calcium carbonate and is used as a fertilizer and in the making of cement or bricks (distinguishing characteristics).

Too Narrow.

Too Narrow.   Conversely, it is also fallacious to define a term too restrictively. Consider: "A chair is a four-legged seat." Doesn't this restrict the possibility that a chair may have a single pedestal for a base, or have five or more legs arranged in a circle? Other examples and their improvements are:

Too narrow A catkin (term) is a yellow, tassel-like spike (class) on a birch tree (distinguishing characteristic).

Improved A catkin (term) is a tassel-like spike (class) that consists of closely clustered, small, unisexual flowers without petals as on a willow, birch, or poplar (distinguishing characteristics).

Circular.

Circular.   It is also important to avoid using any form of the term in the second and third parts of the definition; to do so takes your reader in a circle. Consider this statement: "A radical is a person having radical views concerning social order and systems." Does it define radical? Other examples are:

Comatose is the state of being in a coma.

Fertilization is the process of fertilizing.

A surveyor is one who surveys.

When and Where.

When and Where.   Finally, one must avoid the terms *when* and *where* in formal definition. To write "A crypt is where one is buried" fails to classify the term as "a subterranean chamber" or to distinguish it from a mausoleum or a cemetery, other places where people are buried. Other examples are:

***When* fallacy** Osmosis is when fluid passes through a membrane into a solution of higher concentration.

Improved Osmosis is a digestive process in which a fluid passes through a membrane into a solution of higher concentration.

***Where* fallacy** The Internet is where people can connect to a global network.

Improved The Internet is a global network of interconnected sites that can be accessed by people with a computer and a service provider.

In summary, when devising formal definitions do not (1) use needlessly technical language, (2) employ classes or distinguishing characteristics that are too broad, (3) employ classes or distinguishing characteristics that are too narrow, (4) use any form of the term itself in the class or characteristics citation, or (5) use the terms *where* or *when* in place of a classification.

Extended Definition

If your intent is to define a term so that your audience has a thorough understanding of it, you may need to extend your formal definition and write what is known as an *extended* or *expanded* definition. Such is the case of engineers or scientists attempting to explain their developments to the less technically oriented, or of inventors seeking a patent for a newly developed invention. The desired definition can be achieved by combining a number of different methods of definition and carefully analyzing the significance of each to the term in question.

The extended definition may be a few sentences, a paragraph, several pages, or even, for the linguist, a lengthy report or thesis. The following methods of definition can be combined to create the extended definition.

DENOTATION

Denotation is the literal, concrete definition of the word. It is usually represented in dictionaries by the Arabic numeral 1 and often seems far removed from the common usage of the word, as we tend to use language abstractly rather than literally. For example, the *Oxford English Dictionary* tells us that the word *nurse,* when used as a noun, actually means "one who gives suck to a baby." Of course, the term more commonly implies "one trained to care for the sick or infirm." Thus, the denotative meaning of the term is less commonly used than the implied meaning.

CONNOTATION

The connotative, or implied, definition of a word is often the most recognizable to a reader. This is an abstraction from the denotative meaning, sometimes to the point of losing all its original meaning. The connotative usage of a word or phrase is most often used in everyday speech, especially in informal situations. The denotative meaning of the word *grind* is "an action which crushes substances into bits or fine particles between two hard surfaces." A connotative meaning of the word *grind* is "a student who is very industrious in his studies." It is often helpful to understand and explain the denotative meaning of a word in order to comprehend and explain the connotative.

EXAMPLES

A second approach to extended definitions is to provide examples. These may be brief or extensive. Two samples follow:

> A parasite is a plant or animal which lives on or within another organism, from which it derives sustenance or protection without making compensation. Some parasites are tapeworms, sheep ticks, lice, and scabies.

> A market is a state of trade which is determined by prices, supply, and demand. In salesmanship the term *market* may refer to a trade or commerce in a specific service or commodity, such as the housing market, the stock market, or the designer jean market. Investors in a market study it carefully before investing. In a stock market a potential investor studies the stock market exchange and current prices of stocks to determine if the "market" is going up or down and to decide whether an investment would be profitable at a particular time.

DESCRIPTION

Following your formal definition, which distinguishes the item from other members of its class, it may be wise to describe the physical parts of the item, if, indeed, the term names an object. The following example begins with a formal definition and includes a brief physical description:

> An otoscope is a hand-held, diagnostic instrument which is used for examining the external canal of the ear and the ear drum. Composed of plastic, aluminum, and glass, it consists of three main parts: a dry-cell battery barrel assembly, a lens assembly, and a speculum (reflector).

To further describe its parts, subparts, dimensions, weight, and method of use would entail other specific writing strategies, which are covered in Chapter 7. A graphic illustration incorporated into the text would help to clarify the term. Figure 6.1 shows the main parts of an otoscope:

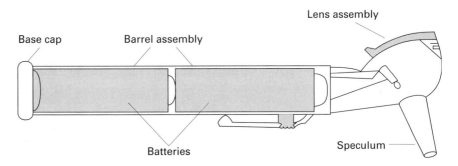

FIGURE 6.1 *The main parts of an otoscope*

SYNONYM

A third expansion of the basic definition is to list synonyms for the term. When readers are familiar with the synonym, they can better understand the definition. Here are some examples of synonym extension:

> A law is a rule of conduct which is established and enforced by the authority, legislation, or custom of a given community or other group. *Rule, regulation, precept, statute,* and *ordinance* are all synonyms for the word *law.*
>
> A motive is an inner impulse or reason that causes a person to do something or to act in a certain way. The terms *intent, incentive,* and *inducement* are sometimes used synonymously with *motive.*

One must be careful to remember that no two words are ever exactly synonymous. Meanings my overlap, but a careful examination of the differences is what conveys precise meaning.

CONTRAST/NEGATION

A fourth writing strategy for extended definitions is to contrast or to negate the term from those terms or items in the same class with which it may be confused. For example:

> In criminal investigation *motive* and *intent* are not truly synonymous. A man who provides a lethal drug to a terminally ill patient has the "motive" to alleviate suffering. Still his "intent" is to kill, making him criminally liable for the death.
>
> A stalactite is an elongate deposit of carbonate of lime which hangs from the roof or sides of a cave. It is not a stalagmite, which is a cone-shaped deposit of carbonate of lime which extends vertically from the floor of a cave.

We often understand better what a thing is by knowing what it is not.

COMPARISON

Conversely, we can better understand what some terms denote if we can examine similarities to things more familiar. Two comparisons for extending definitions follow:

> A stalactite resembles an icicle in that both are hanging, cone-shaped deposits formed by dripping water. In the case of the stalactite, the water evaporates leaving the lime deposit; in the case of an icicle, the dripping water freezes.

Resistance, the opposition to the passage of electrical current which converts electric energy into heat, may be compared to friction between any two objects. When a sulfur-coated matchstick is rubbed against an abrasive surface, the friction creates a spark which ignites the sulfur and produces heat.

ANALOGY

The extended comparison of two otherwise dissimilar things is called an analogy. Although a computer and a thermometer are basically different, the following analogy helps us to understand an analog computer:

> An analog computer is an electronic machine which translates measurements, such as temperature, pressure, angular position, or voltage, into related mechanical or electrical quantities. The operating principle of an analog computer may be compared to that of an ordinary thermometer. As the weather becomes cooler or warmer, the mercury in the glass tube falls or rises. The expansion and contraction of the mercury have a relationship to the condition of the weather. The thermometer provides a continuous measurement that corresponds to the climatic temperature. The analog computer makes continuous scientific computations, solves equations, and controls manufacturing processes.

ORIGIN

Examining the source of an item helps us to grasp the meaning of the word. The following passage briefly explains the source, the mining, and the metallurgy of tin:

> The earliest known tin is found in bronze (a copper-tin alloy) items found at Ur, dated about 3500 B.C. Tin is an element that occurs in cassiterite deposits. The ore is recovered by both opencut and underground mining. The smelting processes include roasting and leaching in acid to remove all of the impurities. The crude tin is resmelted and then refined by further heat treatments of two steps: liquation or sweating, and boiling or poling. Finally, the pure tin is cast in the form of 100 lb ingots in cast iron molds.

ETYMOLOGY

Etymology is the study of the origin and development of words, and it often provides unexpected insight to anyone attempting to gain a clear understanding of a term. English is a Germanic language, and the majority of our spoken words are derived from the highly Germanic Old English language of the early Anglo-Saxons, in which we find most of our single- and double-syllable words. On the other hand, the majority of the words used when writing formally in English are of Latin and Greek origin. Here we find many multisyllable words. In other words, we do not

always speak as we write, which is another complicating factor in the
English language.

An example of how etymology can help us to understand a word is
our word *curfew*. A curfew is usually used by governments attempting to
snuff out civil unrest, or by parents imposing a time for teenagers to be
home. The word comes from the Old French. Town criers would order people to put out their fires by a certain time (all the houses were of wood)
by shouting *couvre feu*, literally meaning "cover the fire." You can understand how the word came to mean what it does today when curfews are
imposed.

HISTORY

Another way to define a term, especially a noun, is to define it in terms
of its history. For example, a clearer understanding of the word *nurse* can
be attained by analyzing its history from the time of its known origin,
through its formalization by Florence Nightingale, to the present, noting
how it developed and expanded to apply to the professional it denotes
today.

CAUSE/EFFECT

To understand some terms completely, an examination of causes and
effects is useful. The following example describes the causes and effects
of a tornado to amplify the formal sentence definition:

> A tornado is a storm characterized by a violently rotating funnel cloud that
> has a narrow bottom tending to reach to the earth. The cloud may rotate
> clockwise or counterclockwise at approximately 100–150 mph. The funnel
> cloud results from the condensation of moisture through cooling by expan-
> sion and lifting of air in the vortex. The air outside of the funnel cloud is
> also part of the vortex, and near the ground this outer ring becomes visibly
> laden with dust and debris. Although a tornado takes only a minute or so
> to pass, it results in devastating destruction. Buildings may be entirely flat-
> tened, exploded to bits, or moved for hundreds of yards. Straws are known
> to be driven through posts. The roar of a tornado can be heard as far as 25
> miles away.

PROCESS ANALYSIS

A final strategy of expanded definition is to analyze the process in which
the item is involved:

> A skeleton is the bony framework of any vertebrate animal. It gives the
> body shape, protects soft tissue and organs, and provides a system of levers,
> operated by muscle, that enables the body to move. Bones of a skeleton store
> inorganic sodium, calcium, and phosphorous and release them into the

blood. The skeleton houses bone marrow, the blood-forming tissue. Bones are joined to adjacent bones by joints. The bones fit together and are held in place by bands of flexible tissue called *ligaments*.

A pressure cooker is an airtight metal container that is used for quick food preparation by means of steam under pressure. When the lid of the pot is secured by means of a rubber gasket and the container is placed on a heat source, fast-flying molecules of steam constantly bump against each other and the inside surface of the container. The combined blows from all molecules exert heat and pressure, "cooking" the food in approximately one-third of the time required by conventional cooking methods.

Process analysis may be extended into a separate writing strategy and is discussed thoroughly in Chapter 9.

GRAPHICS

Providing a picture, a photograph, or line drawing of the object represented by a word is probably the easiest way to define it. The graphic should be clear, free of all extraneous detail, and may include labeled parts, exploded views, or cutaways.

Figures 6.2 and 6.3 contain two sample extended definitions. Each begins with a formal sentence definition and utilizes several extended definition writing strategies. Brief graphics enhance the definitions. The writing strategies are noted in the left margin. Both samples are intended for a lay audience. Be prepared to critique these samples in class.

A DEFINITION OF VOLCANO

Formal definition

A volcano is a hill or mountain that is formed by lava, ash-flows, or ejected rock fragments that come from a central vent. The word <u>volcano</u> derives from the name of the little island in the Mediterranean Sea called Vulcano. Many centuries ago the people of this island believed that this lava and ash came from the forge of Vulcan, god of fire and metalworking.

Etymology

Cause

In actuality, the lava flows from a magma chamber close to the earth's surface. This chamber is formed from rock that has melted due to increased temperature or reduced pressure. This second cause, reduced pressure,

---

FIGURE 6.2   *Sample extended definition in expository format*
(Courtesy of student Debra Fuhrhop)

Analogy

might best be compared to a pressure cooker. While its lid is on, pressure is maintained, and the food is slowly being cooked. However, when the lid is removed, the pressure drops rapidly and steam rises violently into the atmosphere. If the pressure underneath the earth remains constant, rock will stay in its motionless, solid state, but if the pressure drops, rock will quickly melt and rise toward the earth's surface.

Description

There are three major types of volcanoes. First, the shield volcano has a gently sloping cone because it is formed from solidified lava flows. The slopes are usually between 2 degrees and 10 degrees. Second, the cinder cone consists of ejected rock fragments, called pyroclasts, such as dust, ash, cinders, and bombs. Since many of the fragments land near the central vent, a peak usually forms. Slopes are generally 30 degrees for this type. Third, the composite volcano is formed by alternating layers of pyroclasts and solidified lava flows. Because of its construction, it erodes at a slower rate than do the other two types of volcanoes and has a steep slope. Figures 1 through 3 show a shield, cinder cone and composite volcano:

Figure 1 Shield Volcano

Figure 2 Cinder Cone

Graphics

Figure 3 Composite Volcano

Examples

Some examples of composite volcanoes are Mount Rainer and Mount St. Helens in Washington State, Mount Hood in Oregon, Mount Cotaopaxi in Ecuador, and Mount Fuji in Japan.

FIGURE 6.2 *continued*

Extended Definition of <u>Accountant</u>

Denotation

The denotative meaning of <u>accountant</u>, according to the <u>Oxford English Dictionary</u>, is "one who professionally makes up or takes charge of accounts; an officer in a public office who has charge of the accounts."

Connotation

The connotative meaning of <u>accountant</u> is "one who keeps, audits, and inspects the financial records of individuals or business concerns and prepares financial and tax reports."

Etymology

The word <u>accountant</u> comes to contemporary English by way of the fifteenth-century French word <u>accomptant</u>, which derived from the earlier Old French word <u>acuntant</u>, meaning "one accountable for."

History

Humankind has been developing methods of keeping track of income and output since the dawn of time. However, a double-entry book-keeping system, which resembles modern-day accounting, is credited as having been developed by fourteenth-century merchants in Italy. This system was invented when merchants looked for a way to record the financial details of business deals which might last for months or even years with many investors involved.

Luca Pacioli, a fifteenth-century Franciscan monk, published a study on mathematics that contained a description of double-entry bookkeeping. Later, as manufacturing and trade grew, books on accounting began to be written. The profession of the accountant was soon developed, and by the late nineteenth century regulations controlled the chartering of accountants in both Europe and America.

The need for accurate financial reports and government regulations has resulted in the modern accounting system. The development of the corporation has had the greatest impact on accounting, as this allows for public scrutiny of financial records. The rise of the multinational corporation has further complicated the work of the accountant, who must deal with the commercial laws of several nations simultaneously. Finally, the increase in the number and kinds of financial instruments that have become available since the mid-1970's has significantly complicated the work of accountants.

FIGURE 6.3 *An extended definition in technical format*

EXAMPLE

There are four main types of accountants: financial accountants, managerial or cost accountants, tax accountants, and auditors.

The financial accountant gathers the figures that relate to profits, losses, costs, taxes, and other debts and presents these to the client or firm in the form of financial statements that are logical and easy to understand. These financial statements usually include a balance sheet, a statement of changes in financial position, an income statement, and a statement of changes in retained earnings and stock holders' equity.

Managerial or cost accountants provide their client or firm with the data required to evaluate costs, practice budgetary planning, and review employee and executive performance. These accountants may also be responsible for preparing a budget based on realistic estimates of what can be accomplished and then comparing that budget to actual performance at a later date.

Tax accountants make sure their firm or client is in compliance with federal, state, and local tax laws. They must keep abreast of the numerous and frequent changes in tax law; thus many hold a degree in law as well as in accounting.

Auditors are accountants who review financial statements to verify their accuracy.

ANALYSIS

The modern corporation has greatly influenced the accounting system and much of what accountants do. It is an important profession because without accountants businesses would not be able to keep accurate records of all the numbers involved with day-to-day operations.

GLOSSARY

officer	an owner or executive named in corporation documents
stock holder	a person owning a number of shares in a corporation
profits	monies earned above actual costs
balance sheet	a record of credits and debits
equity	an increase in value over initial cost

FIGURE 6.3 *continued*

CHECKLIST

Defining Terms

PRELIMINARY

❑ **1.** Have I considered the audience and field for which the definition is intended?

❑ **2.** Have I checked various dictionary definitions for the term(s)?

METHODS

❑ **1.** Will a parenthetical definition suffice?

❑ **2.** Will a definition by brief phrase suffice?

❑ **3.** Does the term require a formal sentence definition?

    ❑ Have I stated the term?

    ❑ Have I named a class for the term?

    ❑ Have I provided distinguishing characteristics?

❑ **4.** Have I avoided the fallacies?

    ❑ Too technical?

    ❑ Too broad?

    ❑ Too narrow?

    ❑ Circular?

    ❑ *When* or *where?*

❑ **5.** Are methods of extended definition required?

    ❑ Connotations?

    ❑ Examples?

    ❑ Description?

❑ Synonyms?

❑ Contrast/negation?

❑ Comparison

❑ Analogy?

❑ Origin?

❑ Etymology?

❑ History?

❑ Cause/effect?

❑ Process analysis?

❑ Graphics?

EXERCISES

1. **Fallacies.**   There is something wrong with each of the following formal definitions. Rewrite each by providing a precise class and distinguishing characteristics.

   a. A latent image is a prephotographic image on a film that cannot be seen.

   b. Cramming is when a student attempts to learn most of the contents of a course in a short period of time.

   c. Celluloid is a substance that is thin and inflammable and was formerly used for motion picture films.

   d. A bond is when two things, such as concrete and steel, are adhered.

   e. Anxiety is when one is paralyzed with fright and the source of the fear is unknown.

2. **Parenthetical Definition.**   Add a brief word or phrase of definition in parenthesis after each italicized word.

   a. The culture was studied *in vitro*.

   b. The doctor *sutured* the wound.

   c. The *loess* improved the fertility of the soil.

   d. An *implosion* occurred during the experiment.

   e. The President *vetoed* the bill.

3. **Etymology.** Look up the etymologies of the following words. How do their linguistic histories help to clarify their current denotations?

anecdote	zero	vandal	talent
candidate	snob	stocks	sock
rain-check	meander	graffiti	cobalt

4. **Formal Definitions.** Write a one-sentence formal definition for five of the following terms. Avoid the five fallacies.

flextime	veto	economic recession
kinetics	chiaroscuro	marinade
Kaplan turbine	lift (aviation)	graffiti
spectrum	plutocracy	steroid
apogee	corona	clone

5. **Strategies.** In the following extended definition, identify each method of definition employed by writing name of the method in the space provided after each sentence.

> A cyclone is a storm that may range from 50 to 900 miles in diameter and that is characterized by winds of 90 to 130 mph blowing in a circle—counterclockwise in the northern hemisphere and clockwise in the southern hemisphere—around a calm center of low atmospheric pressure while the storm itself moves from 20 to 30 miles per hour. _____ Cyclones may be called whirlwinds, hurricanes, and typhoons. _____ The term *hurricane,* however, is properly applied only to a cyclone of large extent and suggests the presence of rain, thunder, and lightning. The term *typhoon* refers to tropical cyclones in the region of the Phillipine Islands or the China Sea. _____ A tornado is not a cyclone. Although a tornado consists of whirling winds, it is characterized by a funnel-shaped cloud which is far smaller in diameter than a cyclone and by winds far exceeding the velocity of winds in a cyclone. _____ The term *cyclone* is derived from the Greek word *kykloma* which means "wheel" or "coil." _____

WRITING PROJECTS

1. **Definitions of Terms in a Specific Field.** Select five related terms from your professional field. For this assignment try to avoid terms that name mechanisms. Title your assignment by stipulating the field of the terms: for example, "Terms Used in Radiation Technology," "Terms Used in Geology" and "Terms Used in Architecture." Develop formal sentence definitions for each of the terms. Each definition must state the term, the class, and the distinguishing

characteristics that differentiate your term from all others in its class. Avoid the fallacies. Following are some suggested terms in specialized fields, but select your own if you wish or if your field is not included.

Word Processing	*Computers*
menu	byte
wordwrap	icons
block move	software
properties	modem
merge	virus

Internet Terms	*Internet Chat Group Terms*
surfing	flame
host	posting
userid	hot chat
hypertext	smileys
baud	domain

Fashion	*Electronics*
godet	electron
grommet	resistance
stonewashed	frequency
peplum	capacitance
double-faced linen	Ohm's law

Criminal Justice	*Allied Health*
larceny	emphysema
felony	atherosclerosis
manslaughter	angina
assault	escemia
battery	vasodilation

Fire Science	*Psychology*
arson	anxiety
pyromaniac	psychosis
flammable	neurosis
purple K	schizophrenia
cartridge	manic depressive syndrome

Marketing	*General Business*
a good	sole proprietorship
convenience good	partnership
shopping good	limited partnership
specialty good	corporation
unsought good	conglomerate

Surveying	*Architecture*
azimuth	fascia
stadia	cantilever
transverse	soffit
hub	strut
transit	beam

Political Science	*Astronomy*
democracy	nova
communism	black hole
socialism	albedo
oligarchy	transit
monarchy	solar eclipse

2. **Extended Definition.**   Select a broad term (one naming a field of study or a concept or a phenomenon). Write an expanded definition of the term that is suitable for first-year students in the field. Begin with a formal sentence definition and then expand it by employing at least ten extended definition strategies. One of the strategies may be graphics. Use parenthetical or phrase definitions for unusual terms within your extended definition. In the margin indicate the writing strategies that you have employed. Some suggested terms are:

political science	architecture	speech therapy
psychology	sociology	biology
cyberspace	electronics	plasma sphere
astronomy	accounting	computer science
sunspots	aurora borealis	thermal tide
law	tort	prosecution
semantics	linguistics	jargon
AIDS	hepatitis	cancer

3. **Denotations.**   Look up the denotation of a word in your field of study in four different dictionaries including the OED and compare the definitions in a brief paper. How much etymology is included? History? Denotative meanings? Connotative meanings? Examples? Description? Synonyms? Contrasts or negation? Origins? Causes and effects? Process analysis? Graphics? It is doubtful that the dictionaries will provide all of these extensions of definition. Which dictionary definition is best? Why?

NOTES

CHAPTER 7

Describing Mechanisms

BLONDIE   by Drake Young

Reprinted with special permission of King Features Syndicate.

S K I L L S

After studying this chapter, you should be able to

1. Define *mechanism.*
2. Understand the need for mechanism descriptions.
3. Name the two main purposes of a mechanism description.
4. Name types of publications that present mechanism descriptions.
5. Explain the difference between a *general* and *specific* mechanism description.
6. Write explicit and limiting titles to mechanism descriptions.
7. Understand spatial, functional, and chronological organization.
8. Name the three main sections of a mechanism description.
9. Name four types of graphics employed in mechanism description.
10. Analyze the effectiveness of a well-written mechanism description.
11. Write a general and specific mechanism description.

INTRODUCTION

Writing descriptions of the tools, appliances, apparatuses, and mechanisms we purchase or operate are an aid to understanding thoroughly their function. We may call any object—or, for that matter, a system or a location and a substance—that has functional parts a mechanism. In this chapter we are considering, then, not just the task of a technical writer, engineer, or other writer of scientific and professional material in describing a simple mechanism (a pocket knife, calculator, or louvered door) or a complex mechanism (computer hardware, an automobile, or an escalator). We are also considering the efforts of an architect, scientist, archeologist, or an anthropologist to describe a location (an office space for a new computer work station or the site of a new production plant); body organs and systems (the heart or digestive system); the method of construction of a burial tomb; or the kinship relationship system of a tribal people. Even substances (paint, aspirin, DNA, diesel fuel, etc.) may be described as a mechanism.

Technical writers, marketing specialists, engineers, and other professional writers describe mechanisms to spur sales, to explain assembly, to instruct on operating procedures, to explain functions or composition, and/or to analyze strengths and weaknesses. Mechanism descriptions may appear in textbooks, user guides, service manuals, merchandise catalogs, medical reference materials, specialized encyclopedias, specification catalogs, do-it-yourself trade books, and professional papers.

Except for sales promotion materials that may involve some subjective and persuasive writing, descriptions of mechanisms are characterized by objectivity, specificity, and thoroughness. A well-written description should enable a reader to understand the mechanism and the function of the parts. Further, the description should enable the reader to judge the efficiency, reliability, and practicality of the mechanism.

AUDIENCE

The purpose of the description and the audience for whom it is written will dictate the length and the amount of technical detail to be included. A *general* description, written for an encyclopedia or a general how-things-work book or article, emphasizes the overall appearance of the mechanism and its parts and explain its purpose, function, and operation. A *specific* description written for a user guide or service manual emphasizes not only an overall description of the mechanism and its parts, but also will include a detailed description of each part, subpart, or assembly of parts. In addition, a description of a mechanism usually discusses its strengths, limitations, optional equipment and/or similar models, the cost, and availability to allow the reader to judge the usefulness of the particular brand or model.

ORGANIZATION

Whether your description is general or specific, logical organization will aid your reader. An outline should be developed and followed carefully. There are three major sections to a general description of a mechanism:

1.0 General description, or the mechanism as a whole

2.0 Functional description, or the main parts

3.0 Concluding discussion, or assessment

An outline could be much more detailed. The components of a specific description of a mechanism could be outlined, for example, in the following manner:

1.0 The mechanism as a whole (introduction)
    1.1 Intended audience
    1.2 Formal definition and/or statement of purpose or function
    1.3 Overall description (with graphics)
    1.4 Theory (if applicable)
    1.5 Operation (if applicable, with procedural graphics)

This outline is only a guide. The purpose of your description and the intended audience will suggest the amount of detail needed for each report.

Prefatory Material

TITLES

A brief, clear, limiting title is the first writing strategy. The title *BellSouth Cordless Telephone* is not as specific as *A Description of BellSouth's Cordless, 25-Channel Autoscan Telephone, Model 33012*. *Snakes* will not do when you are writing a full description of a snake's skeleton and internal organs; a better title is *Description of the Internal and External Anatomy of a Snake*. Other examples are

Description of a Turbine Bypass Valve

Description of a Japanese K-D Socket Wrench

Description of a Universal Pressure Cooker, Model 4S

Description of a Hewlett Packard HP Deskjet 600C Printer

INTENDED AUDIENCE

An introductory statement of the intended audience and the purpose of the description may be included.

Examples

This description of a turbine bypass valve is intended for engineering students interested in the general construction, operation, and function of such valves.

This description of an Ace bit brace is intended for a junior high shop class instructional manual.

This description of a K & E pencil-lead holder is intended for a descriptive catalog of architectural, designer, and drafting supplies.

DEFINITION/PURPOSE

As the outline indicates, it is logical to include a formal definition and/or statement of the purpose and function of the mechanism.

Examples

The pressure cooker is an airtight, metal container which is used to cook food by steam pressure at temperatures up to 250°F.

The K-D socket wrench is a hand tool designed to hold and turn fasteners, such as bolts, nuts, headed screws, and pipe lugs.

It is often helpful to compare the mechanism to something similar which is likely to be more familiar to the reader.

Examples

The pressure cooker resembles an ordinary "dutch oven" pot or a large, covered saucepan.

The heart is like a pump in that both draw in liquid and then cause it to be forced away.

OVERALL DESCRIPTION

Next the physical characteristics of the mechanism are examined. Include a description of the mechanism's shape and/or dimensions, the weight, the materials from which it is constructed, the color, and the finish. Graphic illustration of the mechanism will help the reader to visualize the mechanism.

Example 1

The K-D socket wrench is made of variable grades of steel. The handle is etched to provide a firm grip. The wrench shaft is $6\frac{1}{2}$ in. long, and the head is 2 in. deep. It weighs 13 oz. Figure 1 shows the K-D wrench and its overall dimensions:

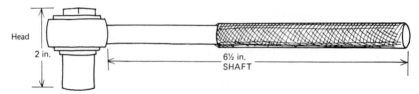

FIGURE 1 *The Japanese K-D socket wrench*

Example 2

The Hewlett Packard Deskjet 600C Printer is made of various grades of plastic with steel and other metal parts plus electric and electronic components. The overall dimensions are 450 mm (16 in) deep, 436 mm (17.2 in) wide, and 199 mm (7.9 in) high. It weighs 5.3 kg (11.6 lb). The exterior color is pale gray with darker gray components. Figure 2 shows the basic parts of the printer.

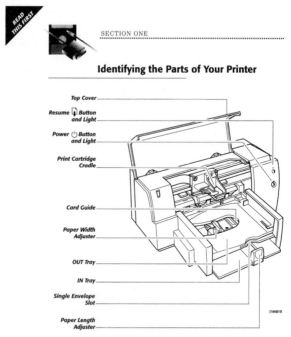

FIGURE 2 *Hewlett Packard Deskjet 600C Printer*   (By permission)

THEORY

If knowledge of theory is essential, it should be included in the introductory or prefatory material.

Examples

The functioning principle of an ordinary mercury thermometer is based on the property of thermal expansion possessed by many substances; that is, they expand when heated and contract when cooled.

The microwave oven cooks food by producing heat directly in the food. As microwaves enter the food, they cause the moisture or liquid in the food to vibrate, and the resulting friction causes the food to heat.

OPERATOR/PROCESS

If the mechanism requires an operator, the qualification or specialty of the operator should be named. If helpful, clarify when and where the process is performed. Next, the process of the mechanism in action should be explained. An explanation of process should not be confused with instructions which give commands. In explaining the mechanism's process, use third person subjects and present-tense verbs, in either the active or passive voice.

Example 1

Cooks or chefs who wish to extract fresh garlic juice without pulp and skin use the garlic press. The cook *(third person)* places *(present tense, active voice)* the bulb of garlic inside the hollow wedge section of the strainer next to the plate of the press. He *(third person)* squeezes *(present tense, active voice)* the handles together, flattening the garlic and forcing the juice through the small holes of the strainer.

Example 2

The stethoscope *(third person)* is designed *(present tense, passive voice)* to be used by doctors, nurses, and trained paraprofessionals to convey sounds in the chest and other parts of the body to the ear of the examiner. The earpieces *(third person)* are placed *(present tense, passive voice)* in the examiner's ears. The bell *(third person)* is held *(present tense, passive voice)* against the area of the body to be examined. The sounds *(third person)* are amplified *(present tense, passive voice)* through the tubing by the diaphragm assembly.

A graphic drawing may help the reader to visualize the mechanism.

LIST OF PARTS

Finally, the main parts of the mechanism are listed. The sequence should have organizational logic. You may list parts *spatially* (from outside to inside or top to bottom as you would logically "see" the mechanism), *functionally* (the order in which the parts are engaged in an operating cycle), or *chronologically* (the order in which the parts are put together).

If a part is complex—that is, it contains a number of subparts, such as nuts, bolts, springs, pins, and so forth—the part may be called an assembly. Use the following sentence pattern:

Sentence pattern	The _____ consists of _____ main parts: the _____, the _____, the _____, and the _____.
Spatial example	The K & E lead holder consists of five main parts: the casing, the push knob, the spring, the tube, and the jaws.
Functional example	The camera consists of six main parts: the housing assembly, the film feed assembly, the viewfinder, the focusing assembly, the lens, and the shutter.

The Main Parts

This section, possibly the lengthiest part of your report, should define and describe each part or assembly in detail. However, a general description of a mechanism will not require as much detail as will a specific description.

DEFINITION/PURPOSE

Each part requires a formal definition or statement of its purpose. Use one of the following sentence patterns to introduce each part or assembly:

Sentence patterns	First, the _____ is designed to _____. The _____, the first main part, supports _____. The first functional part, the _____, connects _____.
Examples	First, the etched handle is designed to provide a firm grip. The base, the first main part, supports all of the other parts. The first functional part, the pedestal, connects the base to the hole punch.

If a main part is an assembly, its subparts should be named.

Sentence pattern	The _____ assembly consists of the following subparts: the _____, the _____, and the _____.

Example The direction assembly consists of the following subparts: the tension spring, the pin, and the knob.

DESCRIPTION

Next, the shape, dimensions, and weight of the part are described. If the material and finish of a part differ from the overall description, each should be described. The strategy should be to explain how each part is related to the other parts and how each is attached to the overall mechanism. If the part is an assembly, each subpart should be described in the order listed.

It may take practice to handle the punctuation of such words as "the bulb-shaped, 7 in., clear shaft" or "a spring-loaded, S-shaped, trigger." Study the punctuation rules, particularly those for hyphens and commas, in Appendix B, "Punctuation and Mechanical Conventions." Further, to name parts accurately, refer to Figure 7.2 (p. 182), which shows terms used in mechanical descriptions.

A graphic illustration of each part or assembly may be appropriate. Such graphics may include exploded drawings, sections, or schematics.

Example 1
The container, the first main part, is designed to hold and to measure the food to be chopped. It is a round, glass bowl which is $4\frac{1}{2}$ in. high and $3\frac{1}{2}$ in. in diameter. The container is etched in 2 oz gradients, and it has a capacity of 12 oz ($1\frac{1}{2}$ cups). The top rim is threaded to receive the lid (Figure 3).

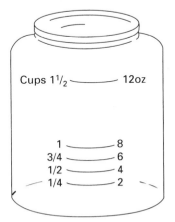

FIGURE 3 *Food chopper container*

Example 2

The plunger assembly, the second main part, consists of the following subparts: shaft, knob cap, shaft housing, shaft housing cap, spring, and blades. When the plunger assembly is depressed, the blades rotate and chop the food in the container.

The shaft is a solid piece of pot metal 8½ in. long and ³/₁₆ in. in diameter. It is slightly spatulate at the end where the blades are welded to it. Two stopper tabs protrude 2½ in. from the blade end to secure the shaft in position.

A bulb-shaped, wooden knob cap is pressed securely to the top end. The cap is ³/₈ in. long and ½ in. wide.

The shaft housing is a hollow tube 3⅛ in. long and ⅙ in. in diameter. The housing fits over the shaft and contains the spring. A threaded shaft housing cap secures the spring into position.

The steel spring coils around the shaft inside of the housing. The spring is 3 in. long.

The two blades, the final subparts of the plunger assembly, are razor-sharp steel. Each is 2½ in. long and ½ in. high. They are bent at a 45° angle and welded to the spatulate end of the shaft.

Figure 4 shows the plunger assembly parts:

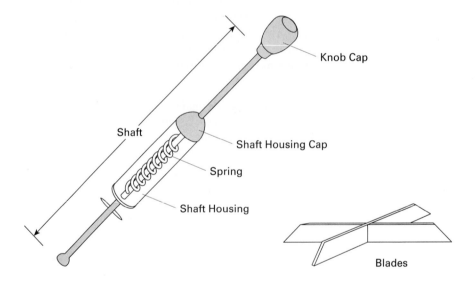

FIGURE 4   *Plunger assembly*

A complete description of a manual food chopper would include description of the other main parts: the lid and the chopping pad.

Concluding Discussion/Assessment

The concluding discussion assesses the efficiency, reliability, and practicality of the mechanism. This assessment may include an examination of the mechanism's advantages and disadvantages, its limitations, its optional uses, the comparison of one model to another, and the cost and availability.

Example 1

The Bostich B8 desk stapler is compact and lightweight, making it easy to store and to transport. The finish is scratch resistant and rust proof. It can be used as a tacker as well as a paper stapler.

No more than 20 pages of copy can be stapled at one time. The Bostich Standard stapler is recommended for larger volumes.

The recommended retail price is $8.95. A box of 5,000 staples is approximately $3.00. The Bostich Standard stapler is sold for $16.75. Bostich staplers are available in most office supply stores.

Example 2

Advantages. The HP Deskjet 600C Printer is compatible with MS Windows, an extensive range of DOS software programs, and OS/2. It will print on all plain, premium, and glossy paper plus transparency film. In addition, it will print standard U.S. and European media sizes plus index cards, postcards, and labels. The built-in feeder will handle up to 100 sheets, 20 envelopes, single envelopes, 30 index cards, and up to 25 sheets of Avery paper labels. It will hold 50 sheets in its out tray.

Depending on fonts, type sizes, color, and graphics, its color speed is from 1 to 4 minutes per page, and its black print speed is also from 1 to 4 pages per minute. It is designed to print 60,000 pages in its lifetime. It will print more than 1,000 fonts at any prescribed size and is capable of both portrait and landscape orientations. It requires 2 watts for power consumption when off and up to 12 watts maximum when printing.

Limitations. It will not operate well under 41°F or over 104°F. The recommended operating environment is 59 to 95°F within 10 to 80 percent humidity. It is programmed to print out solutions on your computer screen for problems you may have printing.

Cost/Availability. The recommended price is $250.00 but may be less if purchased in a computer/printer package. It is available at all stores that sell computers and their components, through computer mail order catalogs, and directly from the manufacturer.

Graphics

Descriptions of mechanisms should employ ample graphics. Consider an *overall drawing* with the main parts and dimensions labeled. A sketch of the *mechanism in action* also helps your reader to envision its use. As you describe each main part or assembly, picture just that portion of the mechanism with all of its detail and labeling of subparts. Sometimes this entails *exploded* or *cutaway views*. Figures 7.1 and 7.2 (see pp. 178–182) show typical drawings of mechanisms (overall, process, exploded and cutaway views), plus some gears and washers, and terms for configurations, materials, finishes, shapes, and attachments that may help you in your writing options.

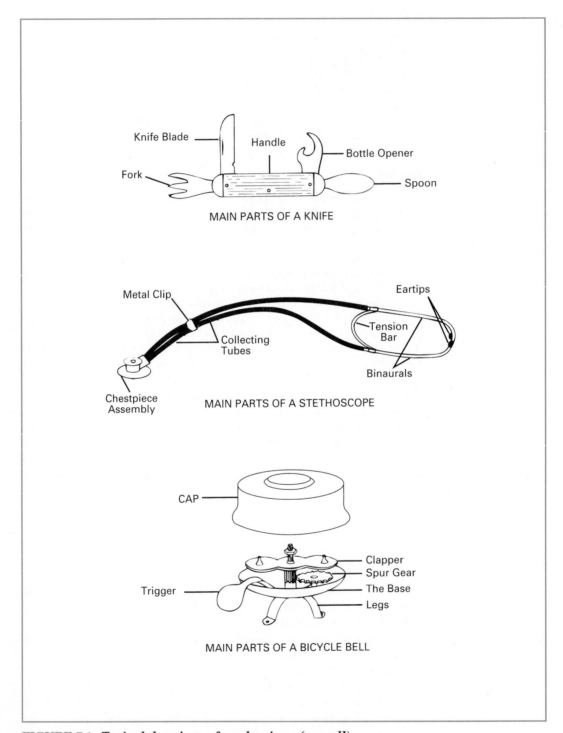

FIGURE 7.1 *Typical drawings of mechanisms (overall)*

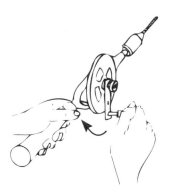

HAND DRILL IN USE

BLOOD PRESSURE CUFF IN USE

FIGURE 7.1 *continued* **(*process graphics*)** (Blood pressure cuff in use from *Coronary Bypass Surgery: A Guide for Patients.* San Ramon, CA: The Health Information Network, 1996. By permission.)

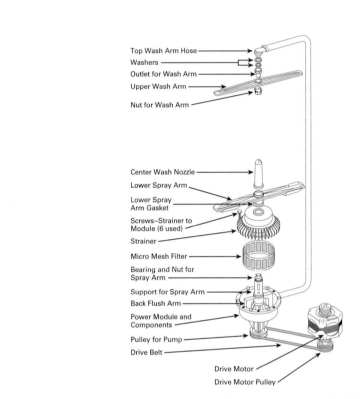

Top Wash Arm Hose
Washers
Outlet for Wash Arm
Upper Wash Arm
Nut for Wash Arm

Center Wash Nozzle
Lower Spray Arm
Lower Spray Arm Gasket
Screws–Strainer to Module (6 used)
Strainer
Micro Mesh Filter
Bearing and Nut for Spray Arm
Support for Spray Arm
Back Flush Arm
Power Module and Components
Pulley for Pump
Drive Belt
Drive Motor
Drive Motor Pulley

EXPLODED VIEW OF JETWASH SYSTEM INCLUDING MOTOR

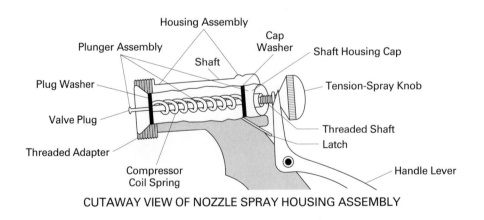

Housing Assembly
Plunger Assembly
Cap Washer
Shaft
Shaft Housing Cap
Plug Washer
Tension-Spray Knob
Valve Plug
Threaded Shaft
Latch
Threaded Adapter
Compressor Coil Spring
Handle Lever

CUTAWAY VIEW OF NOZZLE SPRAY HOUSING ASSEMBLY

FIGURE 7.1   *continued (**exploded and cutaway views**)* (Jetwash system by permission of Maytag Corporation)

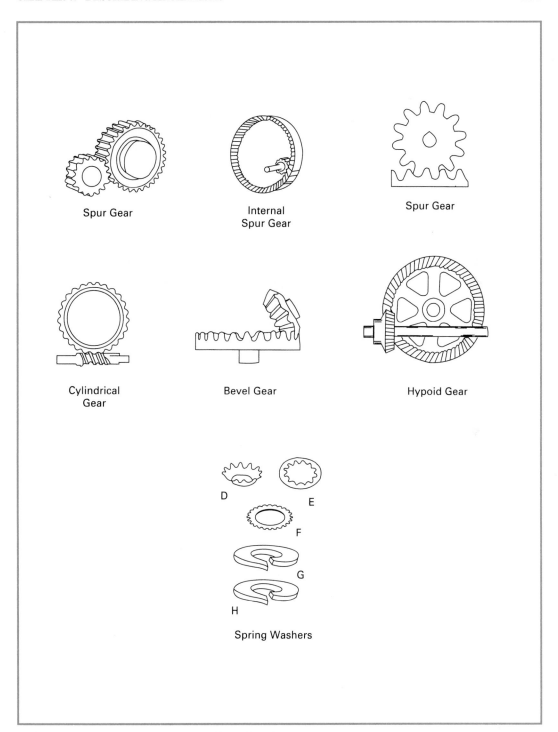

Spur Gear

Internal
Spur Gear

Spur Gear

Cylindrical
Gear

Bevel Gear

Hypoid Gear

D

E

F

G

H

Spring Washers

FIGURE 7.1 *continued (**gears and washers**)*

TERMS USED IN MECHANISM DESCRIPTIONS

Configurations
arc
arm
assembly

ball
bar
barrel
bearing
bevels
blade
bolt
bore
bow
brace
bracket
buckle
bushing

calibrations
cap
casing
channel
clamp
clip
coil
collar
cone
cotter pin

diaphragm
disk
dowel

extension arm
eye bolt

face
fin
fitting
flange
frame
funnel

gauge
gear
gradients
groove
guide

handle
hinge
hook
housing
hub

jacket

key

latch
leg
leg ring
lever
lip

marking
matting
mouth

nib
nozzle
nut
  slotted
  square
  wing

O-ring

pad
pin
plate
plug
plunger
pocket clip
point

ratchet
reservoir
ribbing

ring
rivet

screw
  metal
  recess
  wood
shell
sleeve
slot
socket
spline
spool
spring
stem
stopper
switch

teeth
threads
tip
toe plate
tray
trigger
tube

wand
washer
webbing
wedge

yoke plate

Materials
aluminum
copper
non-corroding
  metal
plastic
pot metal
steel
  anodized
  drop-forged

galvanized
stainless

Finishes
brushed
buffed
etched
glazed
lacquered
lustrous
matte (dull)
stained
semi-gloss

Shape
circular
concave
conical
convex
cylindrical
flared
grooved
hexagonal
hollow
octagonal
rectangular
solid
square
tapered
triangular
u-shaped

Attachment Methods
coiled
compressed
crimped
flange/slot
  attachment
glued
riveted
screwed
soldered
welded

FIGURE 7.2   *Some terms used in mechanism descriptions*

Figures 7.3 and 7.4 show sample specific and general mechanism descriptions. Figure 7.4 is a description of a nonmechanical "mechanism."

DESCRIPTION OF THE VENOJECT BLOOD COLLECTION SYSTEM

Intended audience

This description of the Venoject Blood Collection System, which is used for obtaining blood specimens for laboratory tests, is intended for medical laboratory students.

General Description

Definition and purpose

The Venoject Blood Collection System obtains blood specimens for laboratory tests. It is designed to obtain multiple blood sample tubes from a patient with only one puncture site required.

Overall description

Assembled, the steel, glass, and plastic parts measure approximately 6 inches long depending upon the length of the selected tube. Figure 1 shows the assembled system:

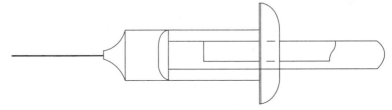

Figure 1 The Venoject Blood Collection System

Theory Process

The system operates on the principle of a vacuum in the collecting tube. First, the stopper on the top of the tube is punctured by the needle, allowing the blood sample to flow into the tube and stop when the tube is full. Second, when the full tube is removed, the needle stops the blood flow until another tube is punctured by the needle. This

FIGURE 7.3 *Specific description of a mechanism*   (Courtesy of student Joanne Fata)

procedure can be repeated for each tube needed for specific blood tests with no discomfort to the patient. Figure 2 shows the Venoject System in a venipuncture procedure:

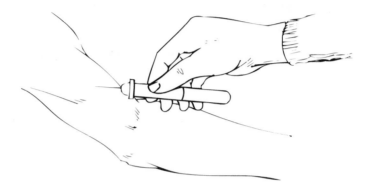

Figure 2   Venoject System in venipuncture procedure

List of parts

The Venoject System consists of three main parts: the double-pointed needle assembly, the adapter, and the blood-collecting tube.

Functional Description

Purpose of first part

The first main part, the needle assembly, functions in two ways: it pierces the skin at the site, and it closes off the blood flow when a collecting tube is not attached. The needle assembly consists of three subparts: the needle, the connector, and the cover. The 2.4-in. sterile needle is hollow steel. A 2.1-in. plastic connector fits securely over the needle at the halfway point. It has threads that screw into the holder and an extended tube to protect the end of the

FIGURE 7.3   *continued*

needle which is inserted into the collecting tube. A plastic cover protects the needle until it is to be used. Figure 3 illustrates the needle and plastic cover:

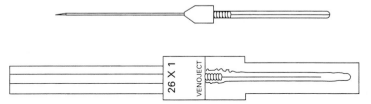

**Figure 3   Venoject double-pointed needle
with plastic cover**

Purpose of
second part

Description

The second functional part, the adapter holder, connects the needle and the collecting tube. The 2.8-in. plastic, cylindrical holder has a diameter of $^3/_4$ in. One end is threaded to receive the needle; the other end is open to receive the collecting tube. Figure 4 shows the holder:

Figure 4   Venoject System holder

Purpose of
third part

Description

The third part, the collecting tube, is designed as a vacuum to collect the blood. It consists of three subparts: the tube, a rubber stopper, and a label tape. Hollow, glass tubes are available in 3-in., $3^1/_2$-in., and 4-in. lengths; the two shorter tubes have a $^1/_4$-in. diameter while the 4-in. tube has a $^1/_2$-in. diameter. A color-coded rubber stopper is inserted into or over the open end of the tube. The color

FIGURE 7.3   *continued*

of the stopper indicates whether the tube contains an anti-coagulant which is necessary for certain blood tests. The collecting tubes in general use are not sterile; sterile tubes are available when needed for bacterial determinations. A plastic label to record the patient's name and date is taped onto every tube. Figure 5 shows three collecting tubes with rubber stoppers and labels in place:

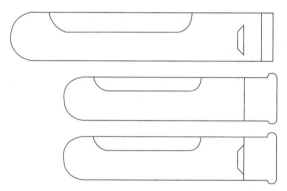

Figure 5   Venoject collecting tubes

Concluding Discussion

Assessment

The Venoject Collecting System is an efficient apparatus for collecting blood for any number of laboratory tests. It allows the technician to obtain multiple samples while pre serving the patient's vein. Because it is a disposable system, bacteria and hepatitis cannot be transmitted from one patient to another. It is available at medical supply houses.

FIGURE 7.3   *continued*

**DESCRIPTION OF THE
STRUCTURE OF THE HUMAN HEART**

*Intended
audience*

    This description of the human heart is intended for a
general audience as an introduction to a discussion of diagnostic procedures to determine heart diseases.

GENERAL DESCRIPTION

Definition and Purpose

*Formal
extended
definition and
purpose*

    The human heart is a hollow, muscular organ that is
located in the chest, slightly to the left of the body's mid-
line. By means of the heart's pumping action, blood flows
through the circulatory system to the various tissues, providing them with the oxygen and nutrients that sustain life
and removing waste for elimination through the lungs or
kidneys.

Overall Description

*Overall
description*

    The heart may be compared to a large pear about the
size of two clenched fists of an adult. It is positioned in
the middle of the chest, with its widest portion **(base)** at
the top and its smallest portion **(apex)** pointing down and
to the left. It lies between the lungs and immediately
behind the breastbone **(sternum),** resting upon the
diaphragm. It is shielded by the rib cage in front and the
spinal column in back. Figure 1 shows two views of the
heart: the anterior view of the exterior of heart and a cut
view showing the inside:

FIGURE 7.4 *Description of the human heart, a nonmechanical mechanism*

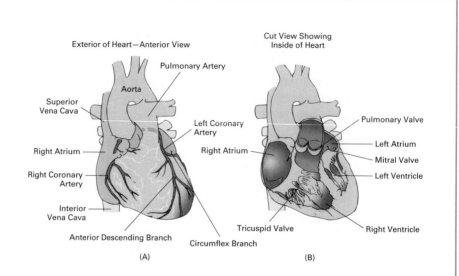

Figure 1   Two views of the normal heart (From *Columbia University Complete Home Medical Guide* 1989 *Edition* by Columbia University College of Physicians and Surgeons. Copyright © 1989 by The Trustees of Columbia University in the City of New York and The College of Physicians and Surgeons of Columbia University. Reprinted by permission of Crown Publishers, Inc.)

The heart weighs 11 to 16 ounces in the average adult.

Operation

The heart begins beating within three months of conception and may continue for 100 years or more. Normal cardiac rhythm is maintained by the heart's electrical system, centered primarily in the group of specialized **pacemaker cells.** Blood that is depleted of oxygen and loaded with carbon dioxide flows into the right ventricle, which pumps it through the pulmonary artery into the lungs. In the lungs, the carbon dioxide is removed, and a fresh supply of oxygen is added. The oxygenated blood then travels through the pulmonary vein into the left atrium and on to the left ventricle. This chamber is the heart's major pump, responsible for pumping the oxygenated blood into the

FIGURE 7.4 *continued*

aorta (the great artery that forms the main trunk of the arterial system) and eventually to all parts of the body through a vast network of arteries, arterioles, and capillaries before it returns via the venules and veins—a total of about 60,000 miles of blood vessels. A system of valves keeps the blood moving in the right direction through the heart.

The heart consists of five main assemblies: the pericardium, three layers, four chambers, five great vessels, and four valves.

THE MAIN PARTS

The first main assembly, the **pericardium,** is a sac that envelops the entire organ and consists of two layers and the pericardial cavity. The first layer, the **fibrous pericardium,** is a tough, dense, outer membrane that protects the heart and anchors it within the chest cavity. The second layer is a thin, smooth, inner membrane, the **serous pericardium,** that forms the outer surface of the heart. Between these two layers is a slight space known as the **pericardial cavity,** which contains a watery fluid that prevents friction when the heart contracts and relaxes.

The second assembly consists of three layers: the epicardium, the myocardium, and the endocardium. The **epicardium,** the outer layer, is the same as the serous pericardium. The inner layer, the **myocardium,** consists of thick bands of cardiac muscle tissue and forms the bulk of the heart, doing most of the work. By alternately contracting and relaxing, the myocardium draws blood into the heart and then propels it outward again in a steady rhythm. The inner layer, the **endocardium,** is a thin membrane that lines the heart's inner surface covering the cardiac valves and is continuous with the inner lining of the major blood vessels of the heart.

The chambers of the heart, the third assembly, consist of four cavities: two atria (upper chambers) and two ventricles (lower chambers). They are arranged so that one pair sits on top of the other pair. See Figure 7.4a to locate these chambers.

FIGURE 7.4 *continued*

Definition
and purpose

Subparts
description

Shape

Subparts
description

Definition
and purpose

Shape and
dimension

Relationship
of parts

Description
of fourth
assembly

Definition
and purpose

Subparts
description

Purpose

Subparts
description

Shape and
dimension

Subpart
description

The **left and right atria,** two cavities, serve as momentary storage reservoirs for receiving blood from the veins. They act as booster pumps for filling the ventricles. Because they do not need to exert much force, their walls are fairly thin. The right atrium receives blood from all portions of the body except the lungs and is slightly larger than the left atrium. The upper front portion of each atrium features a small conical pouch called an **auricle,** which increases the chamber's surface area.

The two ventricles propel blood out of the heart, so their walls are two or three times thicker than those of the atria. The **left ventricle** has the heaviest workload since it must pump blood at high pressure to all portions of the body except the lungs, and its wall may be as thick as 0.5 in. It is also longer, narrower, and more cone-shaped than the right ventricle. The **right ventricle** receives blood from the tissues and returns it to the lungs to eliminate carbon dioxide and to take on fresh oxygen. The left ventricle receives purified blood from the lungs and pumps it through the arteries to the body tissues.

The five great vessels, the fourth assembly, include the inferior vena cava, the superior vena cava, the aorta, the pulmonary artery, and the pulmonary veins. The pulmonary vessels transport blood to or from the lungs while the other great vessels carry blood to or from the rest of the body. See Figure 7.4a to locate these vessels.

The **inferior and superior vena cava,** the body's largest veins, empty into the right atrium. The superior vena cava returns blood from the head, neck, upper limbs, and thorax. The inferior vena cava returns blood from the lower part of the body.

The **aorta,** the body's largest artery, transports blood to all parts of the body except the lungs. Measuring about 1.2 in. in diameter, the aorta extends upward as the ascending aorta, curves to the back and left to form the aortic arch, and then passes down as the descending aorta.

The **pulmonary artery** originates from the right ventricle and divides into right and left branches, which carry blood to the right and left lungs, respectively. In turn, the

FIGURE 7.4 *continued*

Purpose

Description
of fifth
assembly

Subparts
description

Shape

Relationship
of parts

Purpose

Relationship
of parts

Concluding
discussion

Efficiency

Reliability

four **pulmonary veins** transport blood back to the left atrium.

The four valves, the fifth assembly, keep blood moving in the desired direction by opening and closing in regular sequence due to pressure changes within the cardiac chambers. The atrioventricular valves are the **mitral valve** and the **tricuspid valve.** The mitral valve lies between the left atrium and the left ventricle and has two pointed cusps, or leaflets. It is sometimes called the biscuspid valve. Its counterpart, the tricuspid valve, is located between the right atrium and the right ventricle. The **aortic valve** lies between the left ventricle and the aorta, and the **pulmonary valve** lies between the right ventricle and the pulmonary artery. See Figure 7.4a to locate these valves.

Both of these half-moon-shaped valves have three leaflets that spread apart in response to pressure from the bloodstream so that their pointed, free edges extend in the normal direction of the blood flow. To prevent the leaflets from opening backward, the two valves have fine, tendon-like cords called **chordae tendineae,** which anchor to the small muscles on the inner surface of the ventricles. When a valve closes, the free edges of its leaflets normally come together to form a seal preventing blood from leaking backward. The chordae tendineae of the atrioventricular valves stretch tightly against the leaflets, holding them in the desired positions.

CONCLUDING DISCUSSION

Efficiency

During an average lifetime of 74 years, the heart beats more than 2.5 billion times. Each minute it beats between 60 and 80 times and pumps about 5 quarts of blood, almost all of the body's blood supply. During exercise, the pumping action automatically increases three- to fourfold in response to the tissues' demand for increased oxygen.

Reliability

During fetal development, the heart originates as a tubelike structure that gradually enlarges and twists back upon itself forming ridges that partition it into the four

FIGURE 7.4 *continued*

Assessment

chambers. The heart is fully formed and beating months before birth although the exchange of nutrients, oxygen, and waste products occurs through the umbilical vessels in the placenta (an organ that connects the fetus with the mother).

The heartbeat depends on electrical impulses generated and transmitted by the heart's conduction system. Because the conduction system does not depend on outside stimuli, generation of the heartbeat is entirely automatic. The rate of the heartbeat is affected by the autonomic nervous system, body temperature, and various hormones and other chemicals, particularly potassium and sodium. During exercise or intense emotion, the heart automatically speeds up. Because athletes have exceptionally strong, efficient hearts, their heart rate is usually slower than that of nonathletes.

To sustain its activities, the heart must have a constant supply of blood. This requirement is met by a special vascular system that encircles the heart and reaches deep into its walls, providing the coronary circulation. If a coronary artery becomes obstructed by disease, the flow through that artery will be reduced or completely blocked. The result is that the lack of oxygen to the myocardium will result in chest pain or myocardial infarction (death of the affected tissue), commonly called a heart attack. If a person survives a heart attack, the dead myocardial cells will be replaced by scar tissue that is incapable of contracting. Therefore, the remaining healthy portions of the heart will be forced to work harder than before the heart attack.

Disorders of the heart are due to hereditary, environmental, and infectious processes that damage the heart muscle, the coronary arteries, the valves, or the conduction system. If cardiovascular disease is suspected, there are many diagnostic methods that can help to clarify the type and extent of the disorder and medicines, procedures, and surgeries to control heart disease and disfunction. Controllable risk factors include high blood pressure, cigarette smoking, high blood cholesterol, Type A personality, environmental stress, obesity, diabetes, and a sedentary lifestyle. Uncontrollable risk factors include age, sex, and heredity.

FIGURE 7.4 *continued*

CHECKLIST

Writing a Mechanism Description

INITIAL CONSIDERATIONS

❏ **1.** Have I selected a mechanism with main and subparts?

❏ **2.** Have I determined my audience?

❏ **3.** Does the mechanism warrant a general or a specific description?

❏ **4.** Have I decided upon a spatial, functional, or chronological order of parts?

❏ **5.** Have I provided a precise and limiting title?

PREFATORY DESCRIPTION

❏ **1.** Have I named the mechanism precisely?

❏ **2.** Have I defined and/or stated the purpose?

❏ **3.** Have I provided an overall description?

❏ **4.** Have I discussed the operational theory?

❏ **5.** Have I stated by whom, when, and where the mechanism is operated?

❏ **6.** Have I provided a list of the main parts?

FUNCTIONAL DESCRIPTION OF MAIN PARTS

❏ **1.** Have I described the parts in the order listed?

❏ **2.** Have I defined and/or stated the purpose of each part?

❏ **3.** Have I listed the subparts of assemblies?

❏ **4.** Have I described each part adequately?

  ❏ Shape?

  ❏ Dimension?

  ❏ Weight?

  ❏ Materials?

  ❏ Finish?

  ❏ Relationship to other parts?

  ❏ Method of attachment?

❏ **5.** Have I avoided wordiness?

CONCLUDING DISCUSSION/ASSESSMENT

❏ **1.** Have I assessed the strengths of the mechanism?

❏ **2.** Have I discussed the limitations?

❏ **3.** Have I explained optional uses and equipment?

❏ **4.** Have I compared the mechanism to other models?

❏ **5.** Have I addressed the cost?

❏ **6.** Have I commented on the availability?

❏ **7.** Will the reader be able to judge the reliability, practicality, and efficiency of the mechanism?

GRAPHICS

❏ **1.** Have I used adequate and appropriate graphics?

❏ **2.** Have I referred to the graphics in my text?

❏ **3.** Have I numbered and titled each graphic clearly?

❏ **4.** Have I given credit to the graphics sources if they are not my own?

EXERCISES

1. Reorganize these sentences into a **logical description.**

a. The air pump consists of three main parts: barrel assembly, the plunger assembly, and the hose.

b. It is compact, lightweight, and portable.

c. The hand-operated air pump is designed for inflating bicycle tires and sporting goods, such as basketballs, footballs, rubber rafts, and so on.

d. It has a 60 psi (pounds per square inch) rating.

e. The rustproof, steel construction will ensure many years of useful service.

f. The brass, octagonal barrel cap allows access to the pump mechanism diaphragm. It is threaded to attach to the housing and has a $1/4$-in. hole in its center to slide over the shaft. A $1/8$-in. hole in the side of the cap allows air to enter the housing.

g. The barrel assembly consists of three subparts: the housing, a barrel cap, and a toe plate.

h. The operator clamps the hose nozzle on to the filler stem of a tire to be inflated, stands on the toe plate, and pumps the plunger up and down to inflate the tire with air. If the item to be inflated is a sporting good, the supplied filler needle is inserted into the nozzle clamp.

i. The housing is a hollow, $4^{1}/_{2}$-in. by 17-in. steel barrel. The top end is threaded to receive the barrel cap.

j. The 18-in., fabric-covered, rubber hose screws into the barrel housing 1 in. above the base with an air-tight brass fitting.

k. A 6-in., wood handle threads onto the top of the rod.

l. Welded to the bottom of the barrel housing is a $4^{1}/_{2}$-in. long toe plate base on which the operator stands during the operation of the pump's plunger mechanism.

m. The plunger assembly consists of three subparts: a rod, a handle, and a diaphragm.

n. The locking clamp nozzle is inserted into the hose end and is secured with a $1/8$-in. metal band. A thumb chuck allows quick release for regular and high-pressure use.

o. The rod is a $1/4$-in. by $16^{1}/_{2}$-in. threaded steel shaft.

p. The plunger assembly fits into the housing and is secured by the barrel cap.

q. The pump is 16½-in. high and is constructed of steel with rubber hosing and brass fittings.

r. A diaphragm, a leather washer, is secured to the lower end of the rod by two ¼-in. nuts.

2. Rewrite the following portion of **a mechanism description** to eliminate the instructional commands. Use third person subjects and present-tense verbs in the active or passive voice.

> The operation of a socket wrench is simple. First, select the proper socket size for a specific fastener. Second, lock the socket into place on the driving lug. Third, fit the socket end of the wrench over the fastener. Fourth, set the direction control to the right or left by moving the fastener counterclockwise. Fifth, move the handle in a right to left or left to right motion to twist the fastener. Simultaneously, place your free hand over the wrench and fastener to secure the wrench to the fastener.

3. Correct the **mechanics** of these sentences by adding hyphens and commas.

   a. The frame is nine and one half centimeters long.

   b. The six in long handle connects to the barrel with two 12 mm long copper rivets.

   c. The top portion has a centered 25 centimeter circular cutout.

   d. Figure 9 illustrates a spring loaded L shaped latch.

   e. The base is bolted to the cylinder by 2-in. diameter bolts.

4. Examine a simple mechanism (a mechanical pencil sharpener, a pocket knife, a garlic press, a bicycle seat, or the like) and name every part including screws, nuts, bolts, clamps, and so forth. **Classify the parts** into several major categories.

5. Determine a logical order of presentation (spatial, functional, or chronological) for the following mechanisms. Explain why you chose each order.

a.	a flute	**e.**	a soccer ball
b.	a basketball team	**f.**	a college fraternity
c.	paint thinner	**g.**	a paper stapler
d.	a child's toy truck	**h.**	aspirin

WRITING PROJECTS

1. **General Description Project.** Write a *general* description of one of the following mechanisms or a mechanism used in your field of study. Do not concentrate on a particular brand. Select a nonelectric

mechanism that consists of four or five main parts of which at least three are assemblies. Use graphics wherever possible.

a.	deadbolt lock	**i.**	eggbeater	**o.**	drawing compass
b.	skateboard	**j.**	folding lawn	**p.**	pencil sharpener
c.	paper punch		chair	**q.**	manual can opener
d.	Rolodex file	**k.**	toggle switch	**r.**	kerosene lamp
e.	tape cassette	**l.**	stethoscope	**s.**	sink trap
f.	technical pen	**m.**	butane hair	**t.**	lawn sprinkler
g.	bicycle seat		curler		
h.	hand drill	**n.**	garlic press		

2. **Individual Project.** Select a mechanism from the following list or choose one of your own and write a *specific* description of it; that is, describe a specific model or brand. Refer to the same outline and samples in this chapter. Do not stint on details, and use graphics wherever possible.

  a. Specific model of a Cuisinart food processor

  b. Specific make and model of a toaster oven

  c. Specific make and model of an automobile jack

  d. Staedtler/Mars drafting compass

  e. Particular make and model of a cellular phone

  f. Other?

3. **Group Project.** Select a **nonmechanical "mechanism"** from the following list and write a description of it. Refer to the sample outline in this chapter. Again use copious detail and appropriate graphics for a complete description.

a.	bicycle rack site	**e.**	human ear	**i.**	human knee
b.	storage shed site	**f.**	human tongue	**j.**	an amoeba
c.	wood putty	**g.**	flatworm	**k.**	a goldfish
d.	housefly's eye	**h.**	aspirin	**l.**	plaster

Giving Instructions

TIGER   by Bud Blake

Reprinted with special permission of King Features Syndicate.

S K I L L S

After studying this chapter, you should be able to

1. Define *instructions*.
2. Appreciate the need for instructions in all enterprises.
3. Appreciate the legal requirements of instructions.
4. Describe the difference between *instructions* and *process analysis*.
5. Discuss the differences between instructions for a novice, an intermediate operator, and a technical operator.
6. Write precise, limiting titles for instructions.
7. Organize instructions by a variety of numbering systems: Arabic/letters, two decimal systems, and the digit-dash-digit system.
8. Name an intended user and imply or state his or her expected knowledge and skills.
9. State a behavioral or instructional objective for a given set of instructions.
10. Emphasize the importance and benefits of a given set of instructions.
11. Define key terms in instructions.
12. Provide appropriate notes, precautions, cautions, and warnings.
13. List required tools, materials, and apparatus in the order of usage.
14. Sequence the steps chronologically.
15. Use appropriate language (avoid wordiness, use articles *a, an,* and *the,* and eliminate pronouns).
16. Provide visuals and appropriate document designs.
17. Critique sets of instructions.

INTRODUCTION

Instructions are the most common form of technical writing and are seen so often and in so many forms that we often do not realize that they are present. They may be as simple as "Wash your hands before proceeding" or as complicated as those in a computer repair manual. Recipes are sets of instructions as are the contents of the thousands of *how to* books on the market today. Almost every mechanism that we purchase, from analog computers to food processors to zylophones, includes a set of instructions for assembly and procedural steps as well as tips and warning about the care, potential problems, and improper usage of the item. Which of you has not labored over the instructions for your computer applications, your printer, a programmable telephone, a VCR, and the like? Consider the instructions, advice, and notes and warnings given in the user guide for the pager illustrated in Chapter 5.

Instructions may be oral, such as request to do something from a supervisor, parent, or instructor, or even gestural, such as those used by a police officer directing traffic. They are not, however, as simple to construct as they may seem.

DEFINITION

Instructions are a chronological set of steps, usually numbered, to be performed in a procedure by a specific operator and include the desired outcomes, notes, precautions, cautions, and warnings, and sometimes analysis of outcomes. Instructions often combine with the strategy of *process analysis.* That is, they are the result of a task analysis (a study of steps in a process), but, as you will see in the next chapter, they differ from the formal process analysis in one fundamental way—instructions direct and process analysis informs. Instructions focus on *how to perform a task,* whereas the formal process analysis focuses on *how a task is performed by a third party* and emphasizes *what the results are* for the steps within the process. In other words, instructions require a personal relationship between the writer and the reader whereas the formal process analysis maintains a more impersonal rapport. Though both may be used to define and/or inform, only instructions are expected to be performed by the reader. Whereas a scientific writer, for example, might write a formal process analysis on how a nuclear bomb is developed and assembled, it is more likely the task of a group of nuclear engineers and their technical writers to write a set of instructions on how to build the bomb.

LEGAL CONSIDERATIONS

Because we live in an era of litigation with lawsuits almost a way of life, it should not surprise you that many of the instructions we read daily are provided primarily to protect manufacturers from lawsuits. Instructions are drawn up with the utmost consideration to warning and caution statements, and some companies even have attorneys working with their technical writing teams to ensure accuracy. Leaving a critical detail out of a set of instructions can lead to loss of time, damage to property, or even death, as in the case of failing to warn about the consequences of improper usage or the dire consequences of noxious and toxic fumes, explosions, electrical shocks, burns, and fires.

AUDIENCE CONSIDERATIONS

As with all professional writing, the language of instructions should be slanted toward the intended audience. The language in instructions intended for novices (beginners) should be simple and detailed and provide definitions wherever knowledge of a term is questionable. Instructions for intermediate users are generally briefer, and those for skilled technicians are more cryptic.

For the novice	1. Locate the tray (for storing paper) beneath the copier assembly as shown in Figure 1.
	2. Lift the copier assembly and pull open the paper tray.
For the intermediate	1. Lift and pull open the paper tray.
For the technician	1. Disassemble the paper tray at lock A-1.

Titles

Another tactic to aid your audience is to provide a precise, specific, and limiting title. Your title should indicate to the reader exactly what your instructions cover. Consider the differences in the following:

Weak	How to Repair a Circuit Board
Precise	Instructions for Removing a Faulty Transistor from a Printed Circuit Board
Weak	How to Perform a Digital Block
Precise	How to Inject Lidocaine into a Finger to Produce Numbness

Organization

You audience will not read instructions as a continuous narrative, but as one-by-one steps. Instructions usually require numbering of each separate step to stay on track. In simple instructions an Arabic numbering system is usually appropriate. Complex instructions may involve main steps divided into substeps. Preliminary outlining with attention to logical groupings will reveal the main steps and their parts. Several numbering systems may be considered: Arabic with alphabetical letters, a variety of decimal systems, or a digit-dash-digit system. The most reliable are as follows:

Arabic/Letters		*Decimal System A*	
1.	Section	0.0	Preliminary section
	a. Component	1.0	Section
	b. Component	1.1	Component
2.	Section	1.1.1	Subpart
3.	Section	1.1.2	Subpart
	a. Component	1.2	Component
	b. Component	1.2.1	Subpart
	c. Component	1.2.2	Subpart
4.	Section	2.0	Section

Decimal System B		*Digit-Dash-Digit System*	
0.0	Preliminary section	0–0	Preliminary section
1.0	Section	1–1	Section
1.01	Component		1–2 Component
1.02	Component		1–3 Subpart
1.02.1	Subpart		1–4 Subpart
1.02.2	Subpart		1–5 Component
2.0	Section	2–1	Section

THE INSTRUCTIONS

Well-written instructions require informational preliminary material; attention to notes, precautions, cautions and warnings; listings of tools, materials, and apparatus required for the procedure; proper sequencing; careful attention to language and visuals; and careful page layout.

Preliminary Material

Because instructions emphasize *what to do* but not *why,* it is helpful to include preliminary statements, such as

- Naming the intended operator
- Delineating the expected prior knowledge and skills of that operator
- Stating the behavioral or instructional objective
- Stressing the importance and benefits

- Defining key terms
- Providing notes, precautions, preliminary cautions and warnings
- Listing the materials required in the order of use (if the list is short)

Following are two preliminary statements incorporating consideration of the preceding list. These may be included in a numbered or unnumbered introductory statements, such as

For a novice 0–0 These instructions on how to inject lidocaine (a pain-blocking, local anesthetic) into a finger to produce numbness are intended for an allied health student learning minor surgical techniques. If followed properly, the student can anesthetize (that is, cause loss of pain and temperature sensation in) a finger. Anesthesia is necessary prior to finger surgery or before manipulation of broken finger bones. The anesthetizing of a finger or toe is termed a *digital block.*

Warning Failure to inject the lidocaine properly may result in undue pain.

Have at hand cotton swabs, alcohol, a packaged sterilized syringe, and the lidocaine.

Instructions for a professional technician would not require as many definitions, warnings, or material assembly.

For a professional These instructions are to be used by field service personnel to install a repaired DA-1203 antenna. Alignment of the DA-1203 to the aircraft's horizontal position gyro is necessary for proper operation of a stabilized weather radar.

Notes, Precautions, Cautions, Warnings

Notes, precautions, cautions, and warnings differ in intent but are all related to the safety of the reader performing the task and the protection of the manufacturer. Because such notations are often not part of the instructions proper, it is advisable to underline, boldface, capitalize, and/or box them.

Notes. Notes are usually unboxed, supplemental material providing the reader with hints, tips, or information that may be of use, but the information is not necessary to complete the task. Notes are provided wherever needed, especially where a reminder to the reader or a reference to a previous instruction is indicated. Visuals include the following:

<u>NOTE</u>, **NOTE**, or ▶

Precautions.   Precautions are light warnings that need to be understood prior to beginning a task or a set of tasks, such as telling the reader to place newspaper on the table prior to applying glue to an object. Precautions do not usually cover health- or life-threatening situations but are simple tasks the reader can perform to avoid later problems. They may be highlighted in the same manner as notes:

PRECAUTION, **PRECAUTION,** or

Cautions.   Cautions are used to emphasize actions that may result in damage to equipment or other property. They should be strategically placed either at the beginning of a set of instructions and/or just before the steps that have a potential for trouble. Caution statements are similar to precautions but are considered stronger. They may be highlighted as follows:

CAUTION! or ⚠ or

> **CAUTION:** TIGHTENING CYLINDER HEAD BOLTS TO THE IMPROPER TORQUE WILL RESULT IN DAMAGE TO THE ENGINE.

Warnings.   Warning statements indicate potential hazards to health or even life. These warnings are very strong and should be placed strategically wherever necessary. The reader must be informed about the possibility of toxic fumes, explosions, fires, and electrocutions. They may be highlighted as follows:

WARNING!! or  or

> **WARNING!!!**
> **THIS PRODUCT PRODUCES NOXIOUS FUMES!**
> **USE ONLY IN WELL-VENTILATED AREAS.**

Tools, Materials, Apparatus

Logically, the first step in any procedure is to collect the necessary tools, materials, or apparatus. If the list is short, it may be included in a preliminary statement or section (see the previous prefatory material for injecting lidocaine). If the list calls for more precision, list the items under step one or the first task section. Do not forget to include the obvious such as old newspapers, running water, and the like. Number each item. The following is a list of tools and materials needed to complete a set of instructions on how to develop a black and white photographic print:

1.0   Collect the following equipment and materials:

    1.1   Three developing trays

    1.2   Enlarger and easel

    1.3   Print tongs

    1.4   One liter of developer

    1.5   One liter of stop bath

    1.6   One liter of fixer

    1.7   Black and white photographic paper

    1.8   Negatives to be printed

Sequencing

Each of the actual steps or commands must be in chronological order. In the following excerpt from instructions on soldering a circuit board connection, Step 9 is obviously out of place; the joint would have to be secured prior to the actual soldering:

6.   Preheat the joint to melt the solder.

7.   Apply the solder to the joint.

8.   Place the soldering iron tip against the solder and the joint for 2 or 3 seconds.

9.   Secure the joint with a vise to avoid motion.

The writer of instructions is usually quite knowledgeable about the procedure, but the user is not. Therefore, not only is careful chronological order essential but also the inclusion of all steps is a must for effective operation. It is just as important to instruct users to turn on a word processing computer CRT as to instruct them on how to perform a global search of a text.

Language

Instructions demand precision, clarity, parallel construction, simplicity, and thoroughness. Each step usually begins with a command word, which is an active-voice verb stated in the imperative mood, such as *switch, disconnect, lift, depress.*

> **Incorrect**   The following tools *should be collected.*
>
> **Correct**   *Collect* the following tools.
>
> **Incorrect**   *You cut* out the premarked damaged section.
>
> **Correct**   *Cut* out the premarked damaged section.

Further, the command verbs should be precise.

> **Vague**   Remove the bolt.
>
> **Precise**   Unscrew the bolt by rotating the wrench in a counterclockwise motion.
>
> **Vague**   Turn on the computer.
>
> **Precise**   Depress the ON/OFF key to the ON position.

Occasionally you must precede the action command with explanatory words, such as:

> While depressing the RECORD button with your left forefinger, push . . .
> Using a straightedge, outline the damaged portion . . .

Similarly, avoid all other vague terms.

> **Vague**   Check the patch to ensure that a good bond has been obtained.
>
> **Precise**   Pack the edges of the patch with stiff spackling compound to eliminate wobbling.
>
> **Vague**   Allow the glue to dry adequately.
>
> **Precise**   Allow the glue to dry for six hours.
>
> **Vague**   Screw the woodscrews only partially into the anchors.
>
> **Precise**   Place a stack of three nickels (approximately $7/32$ in.) against the mounting surface next to one of the screw locations and turn the woodscrew in until the head touches the coins.

In a set of simple instructions that are not numbered, words—such as *first, second, next, following,* and so forth—mark time and sequence.

Short sentences are easier to understand and to execute than are wordy commands.

Wordy Insert the mounting post of the breaker arm into the recess in the cylinder so that the groove in the mounting post fits the notch in the recess of the cylinder.

Short Fit the mounting post groove into the notch of the cylinder's recess.

Component parts of your instructions should be expressed in parallel (identical) grammatical form.

Nonparallel 1.0 Remove the damaged wall board.

          2.0 Prepare the patch.

          3.0 The patch should be spackled and painted.

Parallel 1.0 Cut out the damaged wall board.

          2.0 Fit a wall-board patch into the hole.

          3.0 Spackle and paint the patch.

In your effort to eliminate wordiness, do not eliminate articles *(a, an,* or *the)*. Also, do not use pronouns *(it, them, that)*.

Poor Push them through circuit board holes.

Improved Push the leads A and B through the circuit board holes.

VISUALS AND DOCUMENT DESIGN

No set of rules covers where and when to place visuals into a set of instructions; common sense is the key to success. Whenever and wherever a drawing or other visual would make the procedure clearer, include one.

Visuals are usually placed below each step or to the left or right of each step. Use plenty of white space to aid the reader and eliminate crowding and confusion. Figure 8.1 presents three basic page layouts for the same instructions:

1. Lift and pull open the paper tray.

2. Push down on the shiny plate until it locks into position.

3. Adjust the paper guides to the desired paper size.
   - Squeeze the side guide.
   - Lift and insert the rear guide.
   - When adding 14" paper, remove the rear guide and store it in the pocket in front of the side guide.

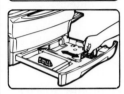

1. Lift and pull open the paper tray.

2. Push down on the shiny plate until it locks into position.

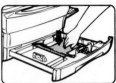

3. Adjust the paper guides to the desired paper size.
   - Squeeze the side guide.
   - Lift and insert the rear guide.
   - When adding 14" paper, remove the rear guide and store it in the pocket in front of the side guide.

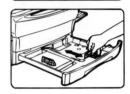

1. Lift and pull open the paper tray.

2. Push down on the shiny plate until it locks into position.

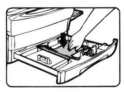

3. Adjust the paper guides to the desired paper size.
   - Squeeze the side guide.
   - Lift and insert the rear guide.
   - When adding 14" paper, remove the rear guide and store it in the pocket in front of the side guide.

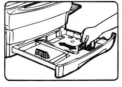

FIGURE 8.1   *Three basic page layout options for loading a copier tray*
(By permission of Xerox Corporation)

Figures 8.2 and 8.3 (see p. 211) present two sets of instructions. Figure 8.2 uses a simple Arabic numbering system and right-column graphics. Figure 8.3, because it involves several main steps and several sub steps, illustrates a complex numbering system.

Memory Dialing

Your AT&T Cordless Telephone 5455 can store nine different phone numbers that you can dial just by pressing MEM and one of the number buttons.

Programming a Number into Memory

The handset must be turned OFF.

1. Press PROG (Figure 1). The PHONE Light will blink to show that you are in the programming mode.

2. Dial the phone number you want to store. The number can be up to 16 digits long.

3. Press MEM.

4. Press any number button from 1 to 9. This assigns the phone number to the memory location you selected.

After pressing the number button, you will hear a three-part tone that means the number was stored properly. If you hear a long buzzing tone, or nothing at all, press OFF, then follow the steps above to program the number again.

Follow the steps above for each phone number you want to store, assigning each one to a different number button.

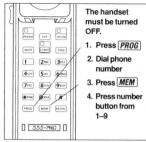

The handset must be turned OFF.

1. Press PROG

2. Dial phone number

3. Press MEM

4. Press number button from 1–9

Figure 1

NOTE: The numbers stored in memory may be lost when you change the handset batteries, or if the batteries run down completely. Follow the steps above to store the numbers again.

FIGURE 8.2 *Instructions for using the Memory Dial feature of the AT & T Cordless Telephone 5455* (By permission of Lucent Technologies Inc.)

Memory Dialing *(continued)*

Directory Cards

1. To use the directory card concealed in the back of your handset, press the arrow above the word DIRECTORY (Figure 1) and slide the door toward the top of the handset. The directory card has an erasable surface. If you write in pencil, you'll find it easy to change names when necessary.

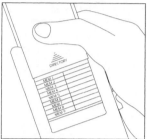

Figure 1

> **NOTE:** If the door slips off the handset, slide it onto the track and back in place.

2. To write on the directory card on the Portable Handset Cradle (Figure 2), remove the plastic cover by inserting a pointed object in the hole and gently prying the cover up until it pops out.

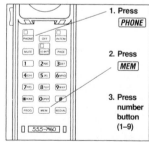

Figure 2

Dialing a Number Stored in Memory

1. Press [PHONE] to get dial tone (Figure 3).
2. Press [MEM].
3. Press the number button (1-9) you assigned to that phone number.

For example, to dial the phone number you assigned to button "6," press [PHONE] [MEM] [6].

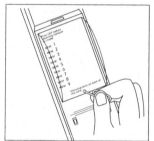

Figure 3

FIGURE 8.2   *continued*

INSTRUCTIONS FOR TAKING AN ORAL TEMPERATURE WITH AN ELECTRO:THERM

0.0 These instructions are intended for a medical assistant who is working in a doctor's office. If followed properly, the assistant will obtain an accurate temperature reading in less time than if one used a mercury type thermometer.

> **PRECAUTION:** DO NOT take an oral temperature with the electro:therm if the patient
> 1. has a mouth injury,
> 2. is an infant or young child who is not old enough to hold his lips closed when told to do so, or
> 3. has had something hot or cold in the mouth in the last five minutes.

1.0 Collect the following apparatus and place on the counter:
  1.1 an electro:therm thermometer
  1.2 a box of sterile plastic covers that are made to be used with the electro:therm
  1.3 a box of clean tissues
2.0 Familiarize yourself with the electro:therm by studying the parts as diagrammed in Figure 1:

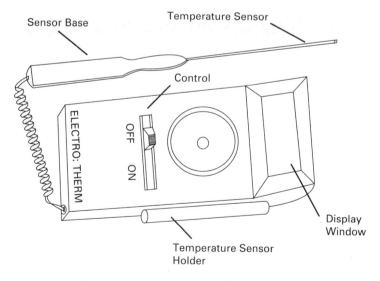

Figure 1.   Parts of an Electro:therm

FIGURE 8.3   *Sample instructions with complex numbering system*

211

3.0 Prepare the electro:therm for operation.

    3.1 Remove the temperature sensor from the control base by pulling the temperature sensor back toward the cord with your left thumb and first finger.

    3.2 With your right hand, pick up one sterile plastic cover by its paper wrapping.

> **WARNING:**   Your hands should NOT come in contact with the sterile plastic covering.

    3.3 Insert the temperature sensor into the sterile plastic cover until the plastic cover is completely engaged over the narrow sensor rod.

    3.4 Gently pull off the paper wrapping leaving the sterile plastic cover on the temperature sensor.

Figure 2 shows what the temperature sensor looks like with the plastic in place:

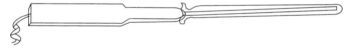

Figure 2.   Temperature sensor with plastic cover in place.

    3.5 Switch the base of the temperature sensor to your right thumb and first finger.

    3.6 Pick up the control base with your left hand.

    3.7 With your left thumb, turn the control base to the ON position.

    3.8 Observe the electro:therm's flashing numbers in the display window.

> **NOTE:**   **The numbers are flashed every second and give the temperature in degrees to the nearest tenth of a degree, as illustrated:**
>
> **99.5**   **99.8**   **99.8**   **99.8**

FIGURE 8.3   *continued*

4.0 Take temperature reading.

> **CAUTION:** **Unlike the mercury thermometer, the temperature sensor should be held continually in the patient's mouth while taking the temperature. DO NOT remove the sensor from the patient's mouth until the same degree displays at least two times in a row. The electro:therm is so sensitive that a patient taking a breath can lower the reading.**

4.1 Insert the sterile, plastic-covered temperature sensor into the patient's mouth making sure it is under the tongue.

4.2 Instruct the patient to keep lips closed tightly and not to talk.

4.3 Leave the temperature sensor in the patient's mouth for one minute.

4.4 After the minute has lapsed, observe the temperature as displayed on the control base.

4.5 When the same numbers have flashed for two times, remove the temperature sensor from the patient's mouth.

5.0 Deactivate the electro:therm.

5.1 Turn the control base switch to the OFF position with your left thumb.

5.2 Place the control base on the counter.

5.3 Switch the base of the temperature sensor to your left thumb and first finger.

5.4 With a tissue in your right hand, pull the used plastic cover off of the temperature sensor and discard the tissue and the plastic cover.

5.5 Slide the temperature sensor back into its holder on the side of the control base.

6.0 Record the patient's temperature on the chart.

FIGURE 8.3 *continued*

CHECKLIST

Writing a Set of Instructions

PRELIMINARY CONSIDERATIONS

❑ **1.** Have I determined if the audience is a novice, intermediate, or technical operator?

❑ **2.** Have I provided a precise, limiting title?

❑ **3.** Have I chosen an appropriate numbering system?

 ❑ Arabic/letters?

 ❑ Decimal system A?

 ❑ Decimal system B?

 ❑ Digit-dash-digit system?

THE INSTRUCTIONS

❑ **1.** Have I provided an appropriate preliminary statement?

 ❑ Does it name the intended user?

 ❑ Does it state or imply the desired knowledge and/or skills of the operator?

 ❑ Does it include a behavioral or instructive objective?

 ❑ Does it stress the importance and benefits of the instructions?

 ❑ Does it define key terms?

 ❑ Does it include necessary notes, precautions, cautions, and warnings?

❑ **2.** Have I provided a list of necessary tools, materials, and apparatus?

❑ **3.** Have I organized each step chronologically?

> ❑ **4.** Have I used proper sequencing?
>
> ❑ **5.** Have I used appropriate language?
>
> > ❑ Have I eliminated wordiness?
> >
> > ❑ Have I use the articles *a, an,* and *the?*
> >
> > ❑ Have I avoided pronouns?
>
> **VISUALS AND DOCUMENT DESIGN**
>
> ❑ **1.** Have I provided visuals for notes, precautions, cautions, and warnings?
>
> ❑ **2.** Have I provided visuals for steps or parts of the apparatus where necessary?
>
> ❑ **3.** Have I presented the instructions in the best possible document design?

EXERCISES

1. **Analysis of Instructions.**   Locate both a simple and a complex set of instructions for the assembly, procedures, or care of a mechanism. Make a copy of both for each member in your class. Discuss whether the directions are effective or not. Consider titles, organization format and numbering, prefatory material (intended operator, desired skills of the operator, final objective, discussion of importance and benefits, key term definitions, lists of required materials, cautions and warnings), chronology of steps, sequencing, language considerations, visuals, and page layouts.

2. **Cautions and Warnings.**   Photocopy at least five cautions and warnings from sets of instructions. Bring them to class. Be prepared to discuss their uses, importance, and effectiveness.

3. **Edit Instructions.**   Edit the following set of instructions for language problems. Look for verbs that are not expressed in the imperative mood or active voice. Also look for vague verbs and other terms, overlong sentences, nonparallel constructions, eliminated articles and questionable pronoun references.

INSTRUCTIONS FOR THE HOLGER-NEILSON (BACK PRESSURE-ARM LIFT) METHOD OF MANUAL ARTIFICIAL RESPIRATION

1–1 Positioning the victim

    1–2 Check the victim's mouth for foreign matter and wipe it out quickly.

    1–3 WARNING: <u>CHECK</u> it for obstructions <u>every 30 seconds</u>.

    1–4 Place the victim face down, bend elbows, and hands are upon the other.

    1–5 The victim's head should be turned slightly to one side with head extended and chin jutting out.

2–1 Administering the respiration

    2–2 Kneel at his head.

    2–3 You place your hands on the flat of victim's back.

    2–4 Rock forward until your arms are vertical to his back.

    2–5 The weight of the upper part of your body should now be forced down to exert a steady, even pressure downward upon your hands which are already placed on the back of the victim.

    2–6 Slide your arms to the arms of the victim just above elbows and draw the arms upward and toward you.

    2–7 NOTE: Enough lift to feel resistance and tension in his shoulders should be applied.

    2–8 The victim's arm must be lowered to the ground.

    2–9 Repeat this cycle 12 times a minute.

4. Text for Instructions. Convert the following paragraph into a set of instructions.

> To assemble an ABC vacuum cleaner connect the hose by inserting it into the opening of the machine by lining up the largest projection on the hose with the largest notch of the opening. Turn the hose to the right to tighten it. By depressing the latch and turning the hose to the left, you may disconnect the hose. To attach the extension wands, cleaning tools, and nozzles, the other end of the hose connects by turning the plastic latch ring on the right hand grip until the outer slot lines up with the inner slot. Pushing your hand down hard onto the wand or tool will push the button projection into the slot. The latch ring must be turned to lock it into place. To remove the wand, tool, or nozzle, the procedure is reversed.

WRITING PROJECTS

1. Individual. Select one of the following subjects and write a *simple* set of instructions of only 10 to 15 steps using an Arabic numbering system. Use visuals wherever you are able. Consider your page layout carefully. Refer to the chapter checklist.

  a. How to carve a turkey

  b. How to tune a guitar

  c. How to extinguish a campfire

  d. How to cuff pants

  e. How to back up data on a disk drive

  f. How to wax a car

  g. How to take an oral temperature

  h. How to wash windows

  i. How to select a golf club

2. **Group Project.** In groups of four write a *complex* set of instructions from the following list. Use the one of the decimal numbering systems or the digit-dash-digit system. Use visuals wherever you are able. Consider your page layout carefully. Refer to the chapter checklist.

  a. How to operate scuba gear

  b. How to do a stock inventory

  c. How to care for in-line skates

  d. How to check out a book from the library

  e. How to apply for financial aid

  f. How to replace a lost driver's license

  g. How to apply for a passport

  h. How to dispute a grade

  i. How to lock a bicycle to a permanent rack

NOTES

Analyzing a Process

BLONDIE by Drake Young

Reprinted with special permission of King Features Syndicate.

S K I L L S

After studying this chapter, you should be able to

1. Understand the purposes of a process analysis.
2. Understand for whom process analyses are written.
3. Distinguish between a process analysis and a set of instructions.
4. Name and identify the five types of process analysis.
5. Use indicative mood verbs in the active or passive voice to describe the steps in analysis.
6. Write a precise, descriptive, and limiting title for a process analysis.
7. Name the logical divisions of a process analysis.
8. Employ suitable graphics and visuals in a process analysis.
9. Write a process analysis embodying logical organization, significant detail, and appropriate language.

INTRODUCTION

Process analysis is a method of explaining how something occurs, how it is accomplished, or how it is organized by separating the process into its parts to examine their natures, proportions, functions, and relationships. Such an analysis concentrates not on instructions but on an ordered sequence of events such as occurs in glass blowing, wastewater management, human digestion, court decision appeals, tornados, or the enactment of laws. The processes may be **linear** (enacting laws), **cyclical** (human digestion), **independent** (glass blowing), or **interdependent** (court decisions appeals).

A process analysis may be complete in itself or part of a longer technical document, manual, or textbook. Actually, process analyses are everywhere: in our manuals explaining how our mechanisms function or what will happen under different situations in a set of instructions; in our newspapers and magazines explaining the process of HDTV (high definition television) or how broadcast television bounces off of relay satellites and into our homes; and in our textbooks explaining how diseases are transmitted or bills are passed into laws.

Scientists and engineers read analyses to understand the processes vital to their research and development. Technicians and operators read them to understand better the procedures they perform. People in business read them to promote their products or services and to perform their duties effectively. Students read them to learn the fundamental processes relevant to their fields of study.

A process analysis may contain elements similar to instructions or mechanism descriptions but should not be confused with either. Instructions emphasize *what to do,* and mechanism descriptions emphasize *how something is put together;* neither often explains *why*. An effective analysis of a process emphasizes

- *What* occurs
- *Where* it occurs
- *When* it occurs
- *How* it occurs
- *To what extent* it occurs
- *Under what conditions* it occurs
- *Why* it occurs

Perhaps the *why* is the most important element. The reader of an analysis must be able to judge the reliability, the practicality, and the efficiency of the process. The reader should be able to estimate with great accuracy the difficulty of the process, the problems likely to occur, and successful solutions to these problems.

TYPES OF ANALYSIS

An informational process analysis informs the reader about a particular process for the purpose of increasing his or her general knowledge. There are three basic types of process analysis:

1. The *historical analysis* explains how and why an idea or event occurred or an institution originated. Historical analysis explores subjects such as how the microcomputer was developed or how American teachers unionized.

2. The *scientific, mechanical, or natural analysis* explains how such processes occur or should occur. Subjects appropriate to scientific, mechanical, or natural analysis include how chemotherapy cures cancer, how a computer printer functions, or how smog occurs.

3. The *organizational analysis* explains the steps, pitfalls, and methods of efficiently performing a human process. Organizational analysis examines subjects such as how a plant is propagated or how a manager motivates a staff of workers.

PRELIMINARY CONSIDERATIONS

Preliminary considerations include audience analysis, titles, and language appropriateness.

Audience

The reader of any given process analysis is usually a novice seeking to understand the particulars of a process; therefore, the language should not be highly technical. For such an audience provide ample background material, definitions of all terms, simple language, and graphics and visuals. However, trained technicians, engineers, and scientists also read and review process analyses in order to increase their knowledge about the advances in their fields. When an analysis is written for such audiences, the language may include far more technical terms, equations, and so on.

Titles

The title should be precise, descriptive, and limiting. Avoid a *how to* title that implies that instructions, rather than an analysis, will follow. Some examples are:

> The Process of Taking Blood Pressure with a Sphygmomanometer
> Preparing an Income Statement for a Small Business
> How a Congressional Bill Becomes a Law
> How Fossil Fuels Are Created

Language

Instructions are written in the imperative mood to give a series of commands (he *puts;* the gardener *spreads;* you *apply*). A process analysis is written in the indicative mood to explain steps (the technician *applies;* the worker *spreads;* the layer of resin *is smoothed*). These indicative mood verbs may be active *(the worker spreads)* or passive *(the layer is smoothed)*. Avoid the words *you* and *your* in a process analysis. Some sample statements follow:

> **Incorrect** All banks recommend when you receive your monthly bank statement that you reconcile your records immediately.

> **Correct** All banks recommend that their depositors reconcile their records as soon as they receive their monthly bank statements.

> or

> All banks recommend checking the monthly statements immediately to rectify personal records.

ORGANIZATION

A process analysis consists of three parts: prefatory material, an analysis of the steps, and a concluding discussion. This organization is similar to that used in a description of a mechanism (see Chapter 7). The following is a typical process analysis outline:

1.0 Prefatory material (introduction)

    1.1 Intended audience

    1.2 Definition/purpose

    1.3 Background and theory

    1.4 Who, when, where

    1.5 Special considerations

    1.6 Tools, materials, supplies, apparatus (if applicable, with graphics)

    1.7 List of chronological steps (or graphic flow chart)

2.0 Analysis of the steps (body)

    2.1 First main step (with appropriate graphics)

        2.1.1 Definition and/or purpose

        2.1.2 Theory (if applicable only to the specific step)

        2.1.3 Special considerations (if applicable only to the specific step)

        2.1.4 Substeps (if applicable)

        2.1.5 Analysis of the step

    2.2 Second main step . . . (etc.)

3.0 Concluding discussion/assessment (closing)

    3.1 Results evaluation

    3.2 Time and costs (if applicable)

    3.3 Advantages

    3.4 Disadvantages

    3.5 Effectiveness

    3.6 Importance

    3.7 Relationship to larger process

A process analysis is usually written in a narrative format although it may employ outline, decimal numbering, or digit-dash-digit techniques (see Chapter 8). The writing strategies of process analysis may be interspersed with instructions and descriptions of mechanisms in an operator's or service manual (see the sample in Chapter 5).

Prefatory Material

AUDIENCE STATEMENT

You must consider the purpose and audience of your process analysis. Your audience will determine the extent of your analysis and the degree of technicality in it. State exactly what the purpose is and for whom the analysis is intended.

Examples

The process analysis of taking a patient's blood pressure is designed for a beginning nursing student with no prior experience.

This analysis of how fossil fuels are created is intended for laypersons who are interested in geology.

DEFINITION/PURPOSE

A formal definition of the process or a clear statement of the purpose of the process is necessary before an analysis of the parts. The prefatory material may include definitions of other key terms to be used within the body of the report to avoid later interruption.

Example 1

Blood pressure is the force exerted by the blood against the walls of the blood vessels. It is created by the pumping action of the heart. The greatest pressure, known as *systolic pressure,* occurs during the contraction of the heart. The lowest pressure, known as *diastolic pressure,* occurs during the relaxation or rest period of the heart. The purpose of taking a patient's blood pressure is to relate it to other health factors, to determine if the patient is healthy, or to determine the cause of illness.

Example 2

Fossil fuels are energy producers, such as coal, oil, and gas, that have been formed deep underground by heat and pressure on dead vegetation and plankton (plant and animal organisms that float or drift in fresh or salt water) over millions of years.

BACKGROUND/THEORY

It may be necessary to explain the historical or scientific background, theory, or principle of a process. Such discussion also belongs in the prefatory material.

Example 1

A brief understanding of the conditions under which blood circulates in the body is necessary. Blood passes from the heart throughout the body by way of a system of vessels that eventually return the blood to the heart. This journey is so rapid that a single drop of blood usually requires less than one minute to move from the heart through the body and back to the heart.

A single tube leading from the heart divides into smaller and smaller vessels, the arteries. The smallest arteries branch into capillaries, the most minute blood vessels. Through the thin walls of the capillaries, the blood supplies the body with oxygen from the lungs and collects the waste products of the body for subsequent removal by the kidneys and other excretory channels.

Beyond the capillaries, the branching process is reversed. The capillaries join to make slightly larger tubes, which next unite to form larger vessels, the veins. Eventually the blood is returned to the heart by two large veins. Therefore, pressure is greatest in the arteries and least in the veins.

One records blood pressure as a fraction, such as 120/80 mm Hg (the chemical symbol for mercury). The normal blood pressure for a healthy, resting adult ranges from 100 to 140 mm Hg systolic and from 60 to 90 mm Hg diastolic.

Example 2

As dead vegetation in swamps or decaying microorganisms on the floors of oceans are compressed by increasing layers of earth and sediment, such vegetation and microorganisms are transformed by heat and pressure into fossil fuels.

WHO/WHEN/WHERE

If your analysis involves an operator, the qualifications of the operator should be specified in the prefatory material. When and where the process is performed should also be explained.

Example 1

Nurses, doctors, medical assistants, and other paraprofessionals trained in the use of a sphygmomanometer determine blood pressure in the daily routine of patient care for diagnostic purposes. Because blood pressure will vary at different times of the day and because readings are usually taken for the purpose of comparison with previous readings, readings should be taken at the same time every day. Generally, only a doctor is qualified to evaluate blood pressure readings in relation to a patient's sickness or health. The readings are taken in a doctor's office, a hospital, at the scene of medical rescue operations, or in other clinical settings.

Example 2

The highest grade and most abundant coal comes from forests that grew in hot swampy areas during the Permian and Carboniferous periods 280 million to 360 million years ago. Crude oil and natural gas began as decaying microorganisms on the floors of oceans in the late Tertiary and Cretaceous periods between 30 million and 180 million years ago.

SPECIAL CONSIDERATIONS

Special conditions, requirements, preparations, and precautions that pertain to the entire process, and not just to one step, should be indicated in the prefatory material.

Example 1

It should be noted that many factors influence blood pressure. A patient should be asked if any of the following factors could be influencing his or her blood pressure at the time of the reading:

1. Increasing blood pressure factors
   a. eating
   b. stimulants
   c. exercise
   d. emotional stress
2. Decreasing blood pressure factors
   a. rest
   b. fasting
   c. depression

Other factors that should be considered are pain, climate variation, tobacco, bladder distension, hemorrhage (blood loss), blood viscosity (thickness), and the elasticity of the arteries.

Example 2

For an oil field to form, the oils must be trapped between layers of impermeable rock such as shale.

TOOLS/MATERIALS/SUPPLIES/APPARATUS

A precise, detailed list of all tools, materials, supplies, and apparatus used in the performance of the process should be provided. It may be necessary to write a brief description of an unusual mechanism used in the process (see Chapter 7).

Example

The following supplies are used to take an accurate blood pressure reading:

1. Stethoscope, an instrument used to magnify the sounds of arterial pulse.
2. Sphygmomanometer, a three-part instrument consisting of a mercury pressure gauge, an arm band with an inflatable rubber bladder, and a pressure bulb to control the flow of air going through connecting tubes in and out of the bladder.
3. Alcohol to clean the earpieces of the stethoscope.
4. Cotton balls to apply the alcohol.

LIST OF CHRONOLOGICAL STEPS

After dividing the process into its main steps, each based on completion of a stage of work or action, the steps should be listed in chronological sequence: first, second, third, and so on. Use the following form or a flow chart:

Pattern   The process consists of _____ main steps; first, _____ing the _____; second, _____ing the _____; third, _____ing the _____; and finally, _____ing the _____.

Example 1   The process of taking a blood pressure reading consists of nine main steps: first, preparing the patient; second, assembling the sphygmomanometer; third, attaching the arm cuff to the patient; fourth, placing the stethoscope over the brachial artery; fifth, closing the pressure control valve; sixth, inflating the cuff; seventh, opening the control valve; eighth, taking the reading; and ninth, removing the apparatus.

Example 2   The process of coal formation consists of four main steps: first, the formation of peat; second, the compression into lignite; third, chemical reactions creating bituminous coal; and fourth, the transformation into anthracite.

Analysis of the Steps

The body of the process analysis thoroughly examines and analyzes each step. Your emphasis here should be to explain why the process is performed in a particular manner. Analyze what would happen if the process were not performed in the correct manner.

DEFINITION/PURPOSE

Identify each step and write a formal definition and/or a statement of purpose of the step.

Example 1

The fourth step, placing the stethoscope over the brachial artery, is done so that the clinician can hear the rhythmical, thumping sounds of the blood. The brachial artery is the large artery of the arm at the inner crease of the elbow.

Example 2

The first main step, the formation of peat (partially carbonized vegetable matter used as fertilizer and fuel), is necessary before peat can be compressed into lignite (a low-grade, brownish-black coal).

THEORY

If a particular step is based on a theory, explain how the theory applies to the step.

Example 1

The brachial artery is used because it is near the heart, large enough for specific recognition, and near the surface of the skin.

Example 2

For an oilfield to form, the oils must be trapped between layers of impermeable rock such as shale.

SPECIAL CONDITIONS

Describe in detail any special considerations, requirements, apparatus, preparations, and precautions that apply to this step only.

Example 1

In most patients, the brachial artery is found quite simply. If there is any difficulty in locating it, the opposite arm may be more yielding. An injured arm or one that contains an intravenous injection should not be used.

Example 2

Bacteria and fungi are necessary to attack dead vegetation in order to break down the vegetation and form a mixture rich in hydrocarbons.

SUBSTEPS

If a main step contains substeps, list them chronologically prior to explaining each.

Example 1

The third step, attaching the arm cuff to the patient, consists of three substeps: checking the cuff bladder, positioning the cuff, and attaching the cuff to the arm.

Example 2

The first step, the formation of peat, consists of three substeps: bacteria and fungi attacking dead vegetation, the vegetation forming into a mixture rich in hydrocarbons, and the squeezing out of water.

ANALYSIS OF STEPS

Finally, explain each step and its substeps with attention to the reasons each is performed in a specific manner.

Example 1

The cuff bladder should not contain any air at the time of positioning the cuff. If it does, a secure fit will not be affected. The armband is wound around the arm above the elbow at a level with the heart, allowing room beneath it for the stethoscope bell. The band is fastened by means of the hooks, snaps, or Velcro material provided for this purpose. If no means of fastening are provided, the end of the band may be tucked securely under the top of the band. If the band is wound too tightly, it will bind the arm, create extra pressure, and cause discomfort to the patient. If the band is wound too loosely, the sounds will be deadened by the cushion of air required to tighten the band sufficiently to compress the brachial artery.

Example 2

Coal buried deeper than three miles beneath the surface, where the temperature is 400°F, is transformed into anthracite. It requires between 25 tons and 75 tons of plant matter to make one ton of anthracite.

Each main chronological step should be analyzed by the same writing strategies.

Concluding Discussion/Assessment

One or more of the following factors should be discussed in the conclusion:

- Time and cost
- Advantages
- Disadvantages
- Effectiveness
- Importance
- Relationship to a larger process

Some of these points may already have been covered in the introduction. Keep in mind that your reader is seeking to assess the efficiency of the process.

The following concluding discussion explains time and cost, advantages, importance, and relationship to the process.

Example 1

Taking a blood pressure reading requires only a few minutes. Because such readings are routine in regular medical check-ups and patient care, the cost is included in the consultation fee. Many health associations provide free blood pressure readings to the public in such places as shopping centers, libraries, and schools.

A blood pressure reading is the one sure method of detecting hypertension, the silent killer. Early detection can prevent strokes, coronaries, and kidney failures.

It is important to have an ongoing record of readings so that variations can be detected. By itself, a reading will not tell the doctor what is wrong, but along with other diagnostic procedures, it will help to determine a patient's condition.

Example 2

All of the energy produced in the United States in 1991 totaled 67.5 billion BTU. One BTU equals the energy released in burning a wooden match. The breakdown of U.S. energy sources in 1991 was coal at 32 percent, natural gas at 31 percent, crude oil at 23 percent, nuclear electric power at 10 percent, hydroelectric power at 4 percent, and other sources at 0.3 percent.

Allowing for population growth, known coal reserves should last about 300 years. Without a major switch to alternative fuels, known petroleum reserves probably will last about 40 years.

Graphics and Visuals

Graphics, such as a flow chart or pictograph of the steps in a process, drawings or schematics of individual steps, and charts and tables of costs and/or time requirements, are helpful. Figure 9.1 (see p. 230) shows a variety of sample graphics that will help to organize and analyze a process. Notice also the graphics in the sample process analyses in Figure 9.2 (see p. 232) and Figure 9.3 (see p. 234). Figures 9.2 and 9.3 show a scientific and organizational process analysis, respectively. Figure 9.2 shows a scientific analysis of a tidal wave/tsunami written for a science encyclopedia. Figure 9.3 shows an organizational analysis of a procedure (stopping a felony suspect's vehicle) designed for a police trainee manual. Either margin notes or headings within the text of each explain and identify the analyses particulars. Be prepared to critique the examples in class.

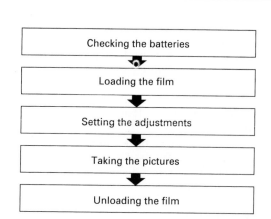

Figure 1.   Steps in the process of operating a camera.

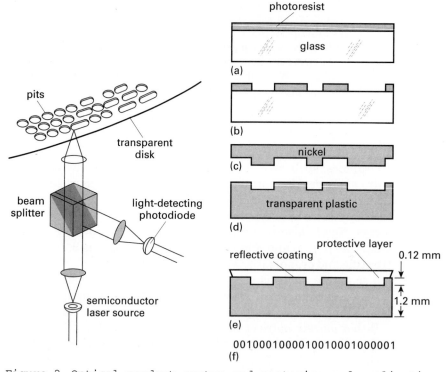

Figure 2. Optical readout system and mastering and replication process for an audio compact disk. (From McGraw-Hill Encyclopedia of Science and Technology, 7th ed., 1992. By permission of The McGraw-Hill Companies.)

FIGURE 9.1   *Sample process analysis graphics (typical flow chart of steps, pictograph, schematic, and entire process)*

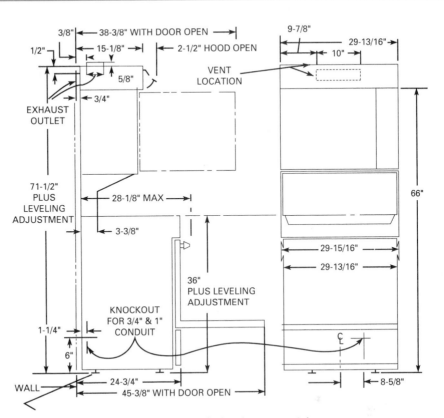

Figure 3. Schematic for positioning a cabinet
microwave oven into a wall.

Making electricity and steam together

Exxon's cogeneration systems typically use a
gas turbine to generate electricity. Exhaust from
the turbine heats water in a boiler to make steam
for use in refinery and chemical plant processing.

The hot exhaust gases from the turbine
are passed through a boiler which
generates steam.

Natural gas
combustion
turns turbine
blades

Gas

Water ➡ ⇨ **STEAM**

Turbine
turns shaft
to generator

ELECTRICITY

Shaft

Hot
exhaust
gases

⇨ Exhaust

Generator

Turbine

Boiler

Figure 4. Process of Exxon's cogeneration system
(By permission of Carol Zuber-Mallison, Brook and Company, Dallas, TX)

FIGURE 9.1 *continued*

Tsunami

A radially spreading, long-period gravity-wave system caused by any large-scale implosive sea-surface disturbance. Being only weakly dissipative in deep water, major tsunamis can produce anomalous destructive wave effects at transoceanic distances. Historically, they rank high on the scale of natural disasters, having been responsible for losses approaching 100,000 lives and uncounted damage to coastal structures and habitations. Because of their uncertainty of origin, infrequency (10 per century), and sporadicity, tsunami forecasting is impossible, but the progressive implementation, since 1946, of the effective International Tsunami Warning System has greatly reduced human casualties. Present efforts, aided by advances in the fields of geomorphology, seismicity, and hydrodynamics, are directed toward an improved understanding of tsunami source mechanisms and the quantitative aspects of transocean propagation and terminal uprush along remote coastlines. These efforts are abetted by the ever-increasing coastal utilization for nuclear power plants, oil transfer facilities, and commercial ports, not to mention public recreation. *SEE GEOMORPHOLOGY; SEISMOLOGY.*

Tsunami generation. While minor tsunamis are occasionally produced by volcanic interruptions or submarine landslides, major events are now recognized to be generated by the sudden quasi-utilized dislocations of large fault blocks associated with the crumpling of slowly moving sea-floor crustal plates, where they normally abut against the continental plates. Such sources are predominantly confined to techtonically active ocean margins, the majority of which currently ring the Pacific from Chile to Japan. As opposed to secular creep, tsunami-producing block dislocations are invariably associated with major shallow-focus (< 18 mi or 30 km) earthquakes of intensity greater than 7 on the Gutenberg-Richter scale, followed by swarms of aftershocks of lesser intensity that decay in frequency over a week or two. Independent lines of evidence indicate that the aftershock perimeter defines the dislocated area, which is usually elongate, with its major axis parallel to the major fault trend. That only about 20% of large earthquakes fitting the above description produce major tsunamis raises the additional requirement that the dislocations have net vertical displace-

ments, a view supported by seismic fault-plane analysis and confirmed by direct observations in the case of the 1964 Alaskan earthquake. *SEE EARTHQUAKE.*

Because the horizontal block dimensions are characteristically hundreds of miles, any such vertical dislocation (a few yards) immediately and similarly deforms the water surface. The resulting tsunami whose energy is concentrated at wavelengths corresponding to the block dimensions and whose initial heights are determined by the local extent of vertical dislocation. *SEE FAULT AND FAULT STRUCTURES.*

Deep-sea propagation. Having principal wavelengths much longer than the greatest ocean depths, tsunamis are hydrodynamically categorized as shallow-water waves; to good approximation, their propagation speeds are proportional to the square root of their local water depth (400–500 or 600–800 km/h in the Pacific). Thus, after determination of the source location by early seismic triangulation, real-time warnings of wave arrival times can be complied from precalculated travel-time charts **(Fig. 1).**

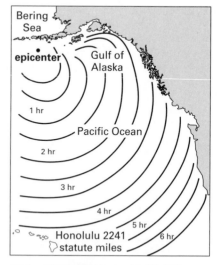

Fig. 1. Advance of tidal wave of April 1, 1946, caused by an earthquake with epicenter southeast of Unimak Island. *(After L. D. Leet and S. Judson, Physical Geology, 2nd ed., Prentice-Hall, 1958)*

FIGURE 9.2 *Scientific process analysis of tidal wave/tsunami* (William G. Van Dorn, *McGraw-Hill Encyclopedia of Science and Technology*, 6th ed., 1987. By permission of The McGraw-Hill Companies)

Because the initial wave energy imparted by the source dislocation is spread ever thinner as the wave pattern expands across the ocean, the average wave height everywhere diminishes with travel distance, amounting to only a few centimeters (virtually undetectable in deep water) halfway around the globe. Beyond this point, energy converges again toward the antipole of the source, and wave heights increase significantly. This convergence accounts, in part, for the severity of coastal effects in Japan from Chilean tsunamis, and conversely. Additionally, azimuthal variations in local wave height are caused by source orientation and eccentricity because, as with a radio antenna, the energy is radiated more efficiently normal to the longer axis. Lastly, further variations of wave height arise from refractive effects associated with regional differences in average water depth. *SEE OCEAN WAVES; WAVE MOTION IN FLUIDS.*

Coastal effects. After the waves cross the continental margins, the average wave height is greatly enhanced, partly by energy concentration in shallow water, and partly by strong refraction which, like a waveguide, tends to trap and further concentrate energy against the coastline. Ultimately, the shore arrests further progress. Here, the tsunami is characterized by swift currents in bays and harbors, by inundation of low-lying areas as recurrent breaking bores, and by uprush against steep cliffs, where watermarks as high as 60 ft (20 m) above sea level have been observed. *SEE NEARSHORE SEDIMENTARY PROCESSES.*

Prediction methods. While most of the above behavior has been predicted qualitatively in theory, the increased accuracy required for engineering design and hazard evaluation has led to the development of numerical modeling on large computers. Taking as input a representative time- and space-dependent source dislocation, as inferred from seismic or observational evidence, and the oceanwide distribution of water depths from bathymetric charts, the computer generates the initial surface disturbance **(Fig. 2),** propagates it across the sea surface, and yields the time history of water motion at any desired location. Where the source motion is known (Alaska, 1964), computer simulations agree quite accurately with observations at places where the linearized equations of motion can be expected to apply. But localized nonlinear effects, such as breaking bores, are more realistically modeled hydrodynamically, using the computer-generated wave field offshore as input.

William G. Van Dorn

Bibliography. L-S. Hwang and D. Divoky, Tsunamis, *Underwater J.,* pp. 207–219, October 1971; T. Iwasaki and K. Ilida (eds.), *Tsunamis: Their Science and Engineering,* 1983; National Academy of Sciences, The great Alaskan earthquake of 1964, *Oceanography and Coastal Engineering,* 1972.

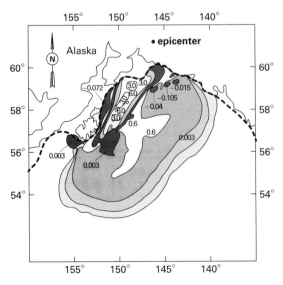

Fig. 2. Computer simulation of tsunami of March 24, 1946, in Gulf of Alaska, 18 min after earthquake. Surface contours in centimeters; wave height in meters; t = 1100 s. 1 cm = 2.5 in., 1 m = 3.3 ft. *(This figure was first published in Underwater Journal, October 1971, pp 207–219, and is reproduced here with the permission of Butterworth-Heinemann, Oxford, UK)*

FIGURE 9.2 *continued*

STOPPING THE FELONY SUSPECT'S VEHICLE

Audience

This analysis is designed for police academy trainees with no prior police experience.

Purpose/Definition

The purpose of this analysis is to familiarize police academy students with the correct, efficient, and safe way to stop a felony suspect in a vehicle. A felony is any crime, defined by law, for which the culprit could receive imprisonment in a penitentiary or the death sentence.

Background

During a tour of duty a police officer may observe thousands of vehicles. The officer may observe a license number of a wanted or stolen vehicle, or he may recognize a wanted criminal inside a vehicle. In any such encounter, the police officer must immediately distinguish between proper procedures and carelessness. At times, officers have failed to make full use of the advantages which proper procedures offer. Fear of being called "overcautious" or of "crying wolf" have led some officers to disregard the responsibility of self protection.

Precaution

Before stopping a suspected felon in a vehicle, the officer should ascertain that a supporting or back-up officer will be available for assistance.

Who/When/Where

This process could be performed by any police officer in any city, county, or state when a felony suspect is spotted in a vehicle.

Tools

The officer requires a patrol car, a two-way radio, and a service pistol.

Steps

Stopping a felony suspect in a vehicle consists of five main steps:

FIGURE 9.3   *Sample process analysis* (Courtesy of student Madalyn McGown)

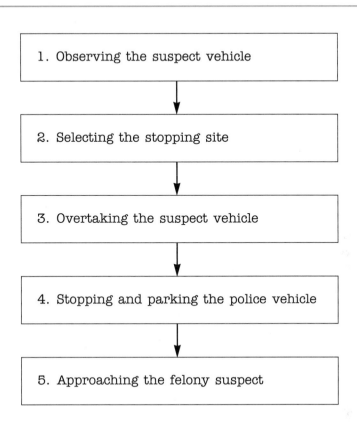

Analysis of Steps

The first step, observing the suspect's vehicle, requires the officer to notify the radio dispatcher at once. The officer gives the following information:

1. Identification of the police unit
2. Location of the contact
3. Description of the vehicle and license number
4. Description and number of occupants
5. Direction of travel and name of the last cross street passed

The last cross street is broadcast at intervals to aid the dispatcher in predicting the suspect's course of travel, thereby hastening the arrival of supporting police units. The officer writes the license number, description of the car, and the number of occupants on a memo pad inside the police car. This is necessary in case the

FIGURE 9.3   *continued*

dispatcher did not receive all of the officer's radio transmission. The suspect's vehicle is trailed until the supporting units are close. The officer should be alert for sudden stops, turns, or other evasive action on the part of the suspect's vehicle.

The second step, selecting a stopping site, requires the officer to select a location with which he is familiar. Familiar surroundings will be to the officer's advantage as he will be able to direct additional assistance to the location more quickly. The officer will also be in a better position to make an apprehension if the suspect attempts to flee. Stopping near alley entrances, openings between buildings, vacant lots, and other easy escape should be avoided. At night, a well-lighted area will enable the officer to observe whether or not the suspect is disposing of any evidence or a weapon.

In the third step, overtaking the suspect's vehicle, the officer maintains constant vigilance to guard against any evasive action on the part of the suspect. At this time, the officer operates the siren and vehicle emergency lights. The police car is driven directly to the rear of the suspect's vehicle. Overtaking the suspect's vehicle is illustrated in Figure 1:

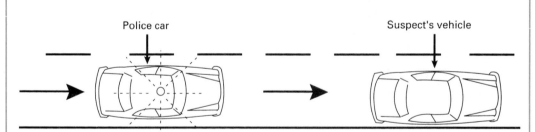

Figure 1   Overtaking the suspect's vehicle

The distance between the two vehicles will vary with the speed. Usually a distance of one car length for every ten miles per hour provides a safety zone for the officer. The officer should constantly consider the safety of other motorists and pedestrians in overtaking and stopping the suspect's vehicle. The suspect's vehicle should be stopped as far as possible to the right side of the roadway or off of the roadway altogether.

FIGURE 9.3   *continued*

In the fourth step, stopping and parking the police car, the officer notifies the dispatcher of the location of the stop. The officer should park his vehicle about 10 feet behind the suspect's vehicle with the front of the police vehicle pointing toward the center of the street. The police vehicle should be on a 45° angle to the suspect's vehicle. The police vehicle position is illustrated in Figure 2:

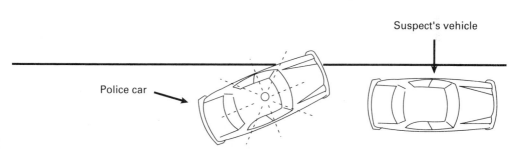

Suspect's vehicle

Police car

Figure 2   Parking position of stopped vehicles

The police vehicle emergency lights are left on so that other motorists will be warned of danger and the supporting police units will be able to find the officer quickly.

In the fifth and final step, approaching the suspect, the officer making the stop should take complete command of the situation. The officer should get out of the police vehicle and with gun drawn proceed to the left front fender of the police vehicle. Using the body and engine of the vehicle as protection, the officer identifies himself in a loud, clear voice: "Police officer! You are under arrest! Turn off your motor and drop the keys on the ground." The officer may order the suspect to place both of his hands out the driver's window or to place his palms against the inside of the windshield. The officer should stay by his police vehicle until a supporting or back-up officer arrives. The back-up officer should park his police vehicle to the rear and a little to the right of the first police car. The back-up officer should proceed to a position off the right rear of the suspect's vehicle with his gun drawn. The position of the police vehicles and officers are illustrated in Figure 3:

FIGURE 9.3   *continued*

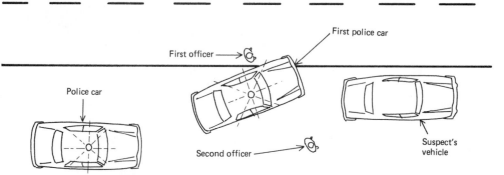

Figure 3   Position of vehicles and officer

The suspect should be made aware of the second officer's presence by the first officer to reduce the possibility of sudden attack from the suspect. The suspect is ordered out of his vehicle by the first officer. The first officer then orders the suspect to assume a search position.

Conclusion

Searching a suspect is another process and is not discussed here. No two felony stops are identical in nature. The unpredictable actions of the felony suspect make the officer rely on his training and judgment in each case. By following the steps in this process, the danger to the officer is minimized.

FIGURE 9.3   *continued*

C H E C K L I S T

Writing a Process Analysis

PRELIMINARY CONSIDERATIONS

❑ **1.** Have I determined what type of process analysis I am writing?

    ❑ Historical?

    ❑ Scientific?

    ❑ Mechanical?

    ❑ Natural?

    ❑ Organizational?

❑ **2.** Have I determined my audience?

    ❑ Laypeople?

    ❑ Trained technicians, engineers, and scientists?

❑ **3.** Have I provided a precise, descriptive, limiting title?

❑ **4.** Will the language I use be appropriate?

    ❑ Indicative mood, active verbs?

    ❑ Indicative mood, passive verbs?

    ❑ Elimination of *you* and *your?*

ORGANIZATION AND CONTENT

Preliminary Considerations

❑ **1.** Have I included an intended audience statement in my prefatory material?

❑ **2.** Have I provided a definition and/or purpose statement of the process?

❑ **3.** Have I explained the historical or scientific background, theory, or principle of the process?

❑ **4.** Have I specified the who, when, and where of the process?

    ❑ If there is an operator have I stated who that operator is and what his or her qualifications should be?

    ❑ Is there a specific time to perform the process or a time period when the process occurs?

    ❑ Have I stated where the process takes place?

❑ **5.** Have I spelled out the special conditions, requirements, preparations, and precautions that pertain to the entire process?

❑ **6.** Have I provided a precise, detailed list of all the tools, materials, supplies, and apparatus that are required for the process?

❑ **7.** Have I divided the process into main steps based on a completion of a stage or work or action and listed them in chronological order?

Analysis of the Steps

❑ **1.** Have I analyzed each step in the correct order?

❑ **2.** Have I divided a main step into substeps if necessary?

❑ **3.** Have I provided a definition or purpose statement for each step?

❑ **4.** Have I explained the theories that pertain to any one step?

❑ **5.** Have I described any special considerations, requirements, apparatus, preparations, and precautions that apply to any one step only?

❑ **6.** Have I analyzed each step and substep with attention to the reasons *why* each is performed in a specific manner?

Concluding Discussion/Assessment

❑ **1.** Have I discussed, if appropriate, the time and cost involved in the process?

❑ **2.** Have I discussed, if appropriate, the advantages of the process?

❑ **3.** Have I discussed, if appropriate, the disadvantages of the process?

❑ **4.** Have I discussed the effectiveness of the process?

❑ **5.** Have I discussed the importance of the process?

❑ **6.** Have I explained the relationship of this process to a larger process?

❑ **7.** May I expect my reader to be able to judge the reliability, practicality, and efficiency of the process?

   ❑ Will my reader be able to estimate the difficulty of the process?

   ❑ Will my reader be able to foresee the problems that are likely to occur?

   ❑ Will my reader be able to ascertain successful solutions to the preceding problems?

Graphics and Visuals

❑ **1.** Have I provided graphics and visuals to clarify the analysis?

   ❑ A flow chart of steps?

   ❑ A drawing, picture, or photo of required equipment?

   ❑ A schematic?

   ❑ A diagram of the substeps within a step?

   ❑ Tables of costs, time requirements, or advantages/disadvantages?

   ❑ Other?

❑ **2.** Is my document design appropriate?

EXERCISES

1. **Types of Process Analyses.**   Classify the following process analysis subjects as historical (H), scientific (S), mechanical (M), natural (N), or organizational (O).
   a.   How earthquakes occur
   b.   Changing a tire on a hill
   c.   How the liver functions
   d.   Making pennies
   e.   How AIDS is treated medically
   f.   How (your state) became a state
   g.   How intravenous fluids are administered
   h.   How women in the U.S. won the vote
   i.   Evaluating a college teacher

2. **Language.**   The following excerpt is from a process analysis of developing black and white film. Rewrite it to eliminate commands and the pronouns *you* and *your*. Use third person subjects and indicative mood verbs in the active or passive voice.

   > In the first step, loading the film into the roll, you attach the film to the developing reel. Place the end of the film into the notch in the reel. Wind the entire roll onto your reel. Next, you place the reel into the developing tank. Snap the top securely into place. You may now turn on your lights.

3. **Organization and Naming of Steps.**   Using the pattern in this chapter to name and list chronological steps of a process, devise and number the steps in the following organizational processes:
   a.   Balancing my checkbook
   b.   Cooking a favorite dish
   c.   Learning to drive a car
   d.   Scheduling my classes for next term (or this term if you will be graduated next term)
   e.   Planning a spring break vacation

WRITING PROJECTS

1. **Audience.**   Locate a brief process analysis written for a specific purpose and audience. Look in your textbooks, medical or engineering journals, or other sources that publish materials in your field of study. Rewrite the analysis for a different purpose and audience. For instance, change a medical textbook analysis of AIDS progression to a student newspaper account to be included in an article on the perils of the disease.

2. **Group Project.** Collaborate with three other students to write a *historical, scientific, mechanical,* or *natural* process analysis on one of the following subjects. Use the checklist to ensure that all the requirements are covered.

   a. How women in the U.S. won the vote

   b. How the Appalachian Mountains evolved

   c. How abortions became legal

   d. How Jordan and Israel agreed on a peace pact

   e. How gum disease develops

   f. How ulcers develop

   g. How a cellular phone works

   h. How e-mail is transmitted

   i. How the lungs function

   j. How osmosis occurs

   k. How the telephone operates

   l. How sound waves are transmitted

   m. How a president of the United States is elected

   n. How a compact disc is made

   o. How a fire extinguisher works

   p. How a plant propagates

   q. How icicles form

   r. How rust forms

   s. How salmon procreate

3. **Group Project.** Collaborate with three other students to write an **organizational** process analysis. You may select a subject from the following list or analyze a process that is performed in your particular field of study. Use the checklist to ensure that all requirements are covered.

   a. Making clay tiles

   b. Tuning the motor of a particular car

   c. Recording a program on a VCR

   d. Manufacturing tar

   e. Taking X-rays of the hand

   f. Judging a debate

   g. Selecting an automobile

   h. Maintaining a salt water aquarium

   i. Taking fingerprints

   j. Selecting a college or university

   k. Keeping records for income tax purposes

    l. Obtaining a credit card

    m. Booking an airline flight

    n. Interviewing for a job

    o. Planning a campus election

    p. Applying for a transfer to a university

    q. Writing a set of instructions

    r. Selecting a roommate

    s. Renting an apartment

NOTES

PART

three

The Professional Strategies

CHAPTER *10*

Correspondence

DUFFY

by Bruce Hammond

S K I L L S

After studying this chapter, you should be able to

1. Determine whether a conventional letter, electronic mail, or faxed mail is acceptable for a recipient.

2. Understand the advantages of electronic mail and facsimile machines for rapid and cost-efficient correspondence.

3. Recognize a variety of memorandum formats, e-mail headings, and fax cover pages.

4. Implement the International Standard for dating correspondence.

5. Write a brief, focused memorandum in an appropriate format.

6. List, define, and write an example of the regular and special elements of letters.

7. Design letters in full block and modified block formats.

8. Head a second page of a letter correctly.

9. Write effective positive (good news) letters: an inquiry or request, an invitation, an order, a congratulatory or thank-you letter, and a sales or service offer.

10. Write effective negative (bad news) letters: a negative response, a complaint, and a series of collection letters.

INTRODUCTION

All professionals need to perfect the writing strategies required for effective correspondence. Memorandums, electronic mail, and letters are records that promote action, transact business, and maintain continuity.

A one-page letter costs as much as $20. Consider salaries and equipment needed (word processors, computer hardware, computer software, printers, paper, possibly photocopy machines and fax machines, envelopes, and postage). Therefore, mastering the strategies of effective correspondence is crucial for cost-effective transactions.

Memos and letters can be delivered by traditional mail, by computer electronics (e-mail), or by facsimile machine (fax). The latter two technologies permit the writer to correspond throughout the world in just seconds. Business is now conducted daily on a global scale, so an understanding of international practices, such as dating, language differences, and approaches has become essential for the business writer (see Chapter 1, "International/Multicultural Audience Aspects"). As always you must avoid sexist or "politically incorrect" nuances or word usage.

A memorandum, usually called a **memo,** is used primarily for internal (within the company or organization) correspondence. However, memos are being sent externally (outside the company or organization) by e-mail and fax. Aside from the matter of transmission, a memo or an e-mail message is used rather informally to request information, confirm a conversation, announce changes in meetings, congratulate a co-worker, summarize meetings, transmit documents, and report on day-to-day activities. Send your brief messages by e-mail if your correspondents have that computer feature and have indicated that such transactions are acceptable. Why clutter up someone's desk with paper when your e-mail messages can be quickly read, reacted to, saved, or deleted? E-mail is faster and more cost efficient than deliverable interoffice mail and telephone conversations.

Letters are used for external correspondence. A letter may be used to seek employment (see Chapter 11), request information, respond to inquiries, promote sales and services, raise funds, register complaints, offer adjustments, collect payment, provide instructions, offer thanks and appreciation, or report on any other activity.

Standard formats and text may be created by developing templates with a word processing computer program, removing some of the drudgery of setting up a format or repeating text each time for new correspondence, but the writer must guard against too much depersonalization. All types of correspondence are characterized by a *you* perspective as well as brevity, clarity, accuracy, and attractiveness.

DATING CORRESPONDENCE

The dating of material is important to both the sender and receiver of a written communique. It allows both parties to maintain records of correspondence; encourages promptness by discouraging procrastination; and ensures mutually accepted times for meetings, deliveries, and other time-bound events. Currently, there are two widely used dating systems: the American Standard and the International Standard.

The familiar American Standard for dating material is simple, and virtually all North Americans understand it. It is presented as month day, year (e.g., January 11, 2000). In haste, we often use numbers in place of the month, such as 1/11/00. This may present a problem when corresponding with any country outside of the North American continent, because almost all other countries use the International Standard of dating: day month year (with no commas). You should understand that someone in Italy receiving an order to be delivered to New York on 1/10/21 is likely to deliver the merchandise on October 1, 2021, causing quite a delay in the desired January 10 delivery. Thus the U.S. military and all U.S. government agencies have officially adopted the International Standard to avoid confusion. Likewise, the Modern

Language Association, the watchdog of writing practices, suggests that the general public should also adopt the International Standard.

MEMORANDUMS, FAXES, AND E-MAIL

Types of memos may vary from a brief, handwritten message that will probably be discarded by the reader after the appropriate action is taken to a carefully typed message that will be filed as part of the permanent record.

Format

Many companies use a preprinted communications memorandum for telephone and caller messages as shown in Figure 10.1.

To
Date _____ Time _____
WHILE YOU WERE OUT
M
OF
PHONE
Area Code   Number   Extension

TELEPHONED	PLEASE CALL	
CALLED TO SEE YOU	WILL CALL AGAIN	
WANTS TO SEE YOU	URGENT	
RETURNED YOUR CALL		

Message _____

_____

_____

_____

_____
Operator

Campbell 09301

FIGURE 10.1 *Preprinted communication memorandum*

Some organizations also use preprinted memo forms on $8\frac{1}{2}$ x 11 in. or half-page paper. These usually include the company name and logo plus the guide words *TO, FROM, SUBJECT,* and *DATE.* Other organizations use a "speed memo," one with carbon copies that allow the reader to send the message, retain a copy, and solicit a reply, as shown in Figure 10.2:

| **R** **RYDER** | RYDER TRUCK RENTAL, INC. | DATE _____ |

TO _____ A+ _____ SUBJECT: _____

FROM _____ A+ _____ LOCATION CODE: _____

MESSAGE _____
▶
_____
_____
_____
_____
_____
_____

——————————————— Signed ———————————————

DATE _____

REPLY _____
_____
_____
_____
_____
_____

——————————————— Signed ———————————————

S10-01 (1-79)

ADDRESSEE TO RETAIN WHITE COPY—RETURN GREEN COPY TO ORIGINATOR
ORIGINATOR TO RETAIN BLUE COPY—SEND WHITE AND GREEN COPIES WITH
CARBON INTACT TO ADDRESSEE

FIGURE 10.2 *Preprinted speed memorandum* (By permission)

If your company does not have preprinted forms, the model shown in Figure 10.3 (see p. 252) is the most common format. Notice the capitalization, margins, and spacing in Figure 10.3. The message paragraphs are usually single spaced with double spacing between paragraphs. The writer's signature or initials appear after the typed name in the *FROM* line for authorization and for possible legal purposes, too.

MEMORANDUM

TO: John Snyder, President

FROM: Mary O'Neill, Personnel Specialist *M.O.*

DATE: 14 January 199X

SUBJECT: Fire in Reception Area, 13/1/9X

MO:jv

Enclosures

FIGURE 10.3   *Standard memorandum format*

You may design a template for a facsimile cover sheet. Fax cover sheets tend to be designed as memos but include logos or company slogans for easy recognition. Figure 10.4 shows a typical fax cover sheet:

 Making Your Life Easier in the Twenty-First Century

FAX COVER SHEET

TO: John Budge, Racal Electronics
        Media Communications Director
            Fax: 954-201-3700

FROM: Wendy Mills, Motorola
        Technical Writing Specialist
            Fax: 561-323-7890

DATE: 2 April 1999

SUBJ: Line Art for Product 5009D - Completion

FIGURE 10.4   *Sample fax cover sheet on a letterhead*

Addresses and Subject Lines

Although frequently written in haste, a memo or an e-mail message requires careful consideration. The addressee line of a memo should contain the full name and title of the receiver. Most employees wear several hats within an organization. An employee may be the Director of Human Resources, Chairman of the Comptroller Search Committee, and a member of the Ad Hoc Committee on Leave Policies. The receiver of a memo wants to know immediately in which capacity he or she is being addressed. Because e-mail addresses do not include any indication of the recipient's position, the subject line or content should make this clear immediately. Memos and e-mail messages may also be addressed to several people.

Memo Example

TO: Dr. Glen R. Rose, Director of International Sales

Mr. John Bass, Director of Multimedia Development

Dr. Mary Ellen Grasso, Director of Legal Affairs

E-mail Example

Hawk@Icanect.com, LadyvanJ@aol.com, polstein@wsvn.com

The sender should also include his or her own title in a memo, indicating the position from which the memo is written. This may be handled in the subject line entry in e-mail. In all cases the subject line of a memo or an e-mail should contain a precise, descriptive title indicating the topic of the memorandum.

Imprecise

FROM: Susan Long, Committee Member

SUBJECT: Committee Meeting

Precise

FROM: Susan Long, Chair of Technical Writer Search Committee

SUBJECT: Resume Review Committee Meeting—18 May 2000

Because e-mail does not have an area for identifying your full name and position, but only the subject, it is wise to begin the correspondence with an identifying sentence.

Example

Subject: Resume Review Committee Meeting—18 May 2000

I, Susan Long, Chair of the Technical Writer Search Committee, would like you to make every effort to attend . . .

Content

A memo or an e-mail message is a tool, not a piece of literature. One of the most effective memos ever written was by the famous General David Shoup. When Shoup completed his first inspection upon becoming Marine Corps Commandant at Parris Island, he noticed that officers from non-coms to full colonels were carrying swagger sticks, an affectation adopted from the British Royal Marines. Shortly, he posted the following memo:

TO: All Personnel
FROM: Commanding General

_____

If you need one, carry one.

The next day not a swagger stick was to be seen. Shoup had mastered the convention of **brevity.** Most employees read numerous memos and e-mail messages each day and must react to each message quickly. Although I do not advise you, necessarily, to be as brief as Shoup, remember that reading and reacting take time.

Memos and e-mail also get **to the point** quickly. Usually the in-house reader is somewhat informed about the general subject. Lengthy background and tactful motivations to act are usually not necessary. Just state your message and exclude irrelevant detail. With e-mail, your answer can be typed in the appropriate space in the received message and quickly sent back to the sender.

Even though you have several messages to convey to the same person, do not confuse your reader. It is usually wise to cover only **one topic** in each correspondence.

Finally, be **timely.** Announce meetings several days in advance. Respond in writing or e-mail to the correspondence you receive as quickly as possible. Interoffice correspondence is designed for efficiency. Figure 10.5 is a sample memo that not only covers a single subject but also is brief, to the point, and timely.

With word processing software it is easy to write **attractive** messages. Pay attention to all of the document design features (space, headings, margins, list options, emphatic features) discussed in Chapter 4.

E-mail tends to be very informal, short, and staccato. We often write in all capital letters and too often ignore the conventions of punctuation and spacing. In addition, there are any number of typed icons for smiles,

---

MEMORANDUM

TO: Members of the Long-Range Planning Committee

FROM: Kenton Duckham, Chairperson *K.D.*

DATE: 27 April 199X

SUBJECT: Change in May meeting date

    The regularly scheduled 15 May 199X meeting of the Long-Range Planning Committee is rescheduled for 22 May 199X at 3:00 P.M. in the Board Room because President Jean Laird must attend a marketing conference in Atlanta.

    Please notify me at extension 6647 if you will be unable to attend.

KD:jv

FIGURE 10.5 *Sample memorandum*

frowns, and so on. These irregularities may be acceptable for your family, close friends, and chat room messages, but they are definitely not acceptable for professional communications. Also consider how easy it is to misaddress e-mail or how quickly your message may be forwarded to someone else with a couple of clicks. Be guarded about messages that might be misinterpreted in someone else's hands.

LETTERS

Effective letter writing is the key to successful business operation. Letters are the essential link between you and your business connections and between one organization and another. Letter content promotes and confirms most business transactions. Conventions of format, mechanics, and content regularize and simplify these transactions.

Format

Several acceptable letter formats are in use today. Individual companies tend to adopt a uniform format for use on preprinted letterhead stationery.

If letterhead stationery is not used or if you are writing business letters as an individual, select white, unlined, 8½ in x 11 in. bond paper of about 20 lb weight. Cheap, flimsy paper neither feels nor looks important. The quality of the paper may seem like a small point, but it does make a difference in the amount of attention your letter will receive.

Figures 10.6 and 10.7 (see p. 258) show the two most widely adopted letter formats: the **full block** and the **modified block.** Study the spacing and indentation. Side margins are 1 to 1½ in. The top margin is a minimum 1 in. but may be deeper if the letter is short. A letter usually has more white space at the bottom than at the top, and it is conventional to type the body, or text, of the letter over the centerfold of the page.

One thing seldom considered in format selection is the option of a justified or jagged right margin offered by word processors. Although many people feel that the lined-up right margin gives a letter a finished, booklike appearance, some professionals and business psychologists suggest that the jagged edge provides a psychological advantage in that the receiver tends to find it more personalized. Thus, in correspondence of a sensitive nature, it may be wise to use the jagged edge rather than the justified margin.

Heading

409 Northeast Twelfth Avenue
Irving, Texas   75060
12 February 199X

(3–6 lines)

Inside
address

Mr. James W. Nelson
General Manager
Ace Equipment Company, Inc.
1092 East Eleventh Street
Seattle, Washington   98122

Salutation

(2 lines)

Dear Mr. Nelson:

Special
element

(2 lines)

SUBJECT: Your letter of 15 January 199X

(2 lines)

Body

(2 lines)

(2 lines)

Compli-
mentary
closing

(2 lines)

Very truly yours,

Signature

(4 lines)

Carol Hopper

(2 lines)

Notations

CH:cm

(2 lines)

c: David L. Meeks

FIGURE 10.6   *Full block format for business correspondence*

Heading

                                        2107 West New Boulevard
                                        Sterling, VA   22170
                                        3 August 199X

                                                        (3–6 lines)

Inside
address

Mr. Ray Adams, Associate
Sun Air Corporation
State Road 7
Auburn, WA   98002
     (2 lines)

Salutation

Dear Mr. Adams:
     (2 lines)

Body

     (2 lines)

     (2 lines)

Compli
mentary
closing

     (2 lines)

                                        Sincerely,

                                        (4 lines)

                                        Marcia Jordan
                                        Personnel Specialist

Signature

     (2 lines)
MJ:ts
     (2 lines)
Enclosures (2)

Notations

FIGURE 10.7   *Modified block format for business correspondence*

Parts of Letters

All letters have six major parts: the heading, the inside address, the salutation, the body, the complimentary closing, and the signature. In addi-

tion, many letters contain subject, reference, or attention lines; typist's initials; and enclosure and distribution notations.

Headings.   If you are using printed letterhead stationery, add only the date two spaces below the letterhead. If you are using plain, white stationery, type your street address, city, state, zip code, and the date. Decide, too, whether to use the full or modified block format.

Example

1414 Southwest Ninth Street
Florence, South Carolina 29501
12 November 199X

Do not include your name. Each line begins at the same margin. Refrain from abbreviating. Write out *Street, Avenue, Boulevard, East, Northeast,* and so on. Notice also that in street addresses one- and two-digit numbers are written out, but three or more digits are written in Arabic numerals.

Examples

One Landmark Plaza
Twenty-two West Third Avenue
123 Forty-second Street

   but

2134 West 114th Street

Leave two spaces between the state and the zip code. The heading is placed flush left in the block format or flush to the right margin in modified block format.

Inside Address.   The inside address includes the full name, position, company, and address of the recipient of your letter. It is spaced three to six lines below the heading.

Examples

Dr. Mary Jones, President
New Community College
101 South Palm Avenue
Fairmont, West Virginia 26555

Mr. Horacio L. Fernandez
Director of Human Resources
Ace Manufacturing Company, Inc.
Davenport, Iowa 90521

Notice that short titles may be placed on the same line as the name, whereas long titles are placed in a separate, second line. If possible, address your letter to a specific person rather than just to a position within a company. You may abbreviate titles such as *Dr., Mrs., Mr.,* and *Ms,* but do not abbreviate *the Reverend* or *the Honorable* or titles denoting rank, such as *Lieutenant, Captain, Professor.*

The inside address is always placed flush to the left margin.

You may abbreviate the state using the Postal Service two-letter abbreviations, as shown in Figure 10.8:

Alabama	**AL**	Montana	**MT**
Alaska	**AK**	Nebraska	**NE**
Arizona	**AZ**	Nevada	**NV**
Arkansas	**AR**	New Hampshire	**NH**
California	**CA**	New Jersey	**NJ**
Colorado	**CO**	New Mexico	**NM**
Connecticut	**CT**	New York	**NY**
Delaware	**DE**	North Carolina	**NC**
District of Columbia	**DC**	North Dakota	**ND**
Florida	**FL**	Ohio	**OH**
Georgia	**GA**	Oklahoma	**OK**
Guam	**GU**	Oregon	**OR**
Hawaii	**HI**	Pennsylvania	**PA**
Idaho	**ID**	Puerto Rico	**PR**
Illinois	**IL**	Rhode Island	**RI**
Indiana	**IN**	South Carolina	**SC**
Iowa	**IA**	South Dakota	**SD**
Kansas	**KS**	Tennessee	**TN**
Kentucky	**KY**	Texas	**TX**
Louisiana	**LA**	Utah	**UT**
Maine	**ME**	Vermont	**VT**
Maryland	**MD**	Virginia	**VA**
Massachusetts	**MA**	Virgin Islands	**VI**
Michigan	**MI**	Washington	**WA**
Minnesota	**MN**	West Virginia	**WV**
Mississippi	**MS**	Wisconsin	**WI**
Missouri	**MO**	Wyoming	**WY**

FIGURE 10.8   *U.S. Postal Service two-letter abbreviations*

Salutation.   The salutation is your greeting to your reader. It is typed two lines below the inside address and is followed by a colon. Further, it must agree with the addressee of the inside address. We have all seen letters addressed to "Dear Sir or Madam" or "To Whom It May Concern." These vague salutations are useful when the receiver of the letter can be anyone in a certain department or division, as in an order letter. However, in negative situations (such as collection letters or complaint letters), it is easy to have the letter misplaced or even ignored. In such situations people are likely to avoid responsibility by being neither "sirs nor madams," and your letter will probably concern not a single "whom."

To avoid this situation, especially when money is involved, *always* find a name to place on the letter. A simple telephone call to the company to find out the name of the person in charge of complaints or the name of an executive officer will ensure that the letter arrives into the hand of a real person. This person will be more likely to respond in a timely fashion because you have established a personal relationship with him or her.

Further, the salutation must agree with the addressee of the inside address.

Examples

Dr. Susan Clark, Dean
New Community College
101 South Palm Avenue
Miami, FL   30212

Dear Dr. Clark: (agreement with person)

Director of Personnel
Ace Manufacturing Company, Inc.
4012 West Grand Street
Hilo, HI   96720

Dear Sir: (agreement with title)

Ace Manufacturing Company, Inc.
4012 West Grand Street
Augusta, GA   30906

Gentlemen: (agreement with corporate body)

League of Women Voters
Ten Northeast Datepalm Drive
Phoenix, AZ   85062

Ladies: (agreement with gender)

Because many women are joining the corporate ranks of business and industry, it is not unusual to see salutations, such as *Ladies and Gentlemen:, Hello:,* or *Dear Director of Training:.* Unless a woman has expressed a desire for *Miss or Mrs.,* use *Ms* (optional period) whether she is married or unmarried.

Avoid *Sir:* (too formal), *My Dear Sir:* (too pretentious), *Dear Sirs:* (avoid), and *To Whom It May Concern:* (too impersonal).

The salutation is always typed flush to the left margin.

Body.   The body, the text of your letter, begins two lines below the salutation. Notice in Figures 10.6 and 10.7 that paragraphs are not in the block format but are indented five spaces in the modified block. Single space within paragraphs and double space between them. The body should fall over the center of the page.

Complimentary Closing.   The complimentary closing is two lines below the body and is followed by a comma. Only the first word is capitalized.

Examples

Very truly yours,	(formal)
Yours truly,	(less formal)
Sincerely,	(emotional)
Respectfully,	(if addressee outranks you)
Cordially,	(warm)

The complimentary closing is flush to the left margin in block format and at the horizontal midpoint or at the heading margin in modified block format.

Signature.   Your full name is typed four lines under the complimentary closing. Sign your name in black ink between the two. If the addressee is known to you, you may sign less formally than the typed signature.

Example

Very truly yours,

Judy S. VanAlstyne

Judith S. VanAlstyne

Special Elements.   Occasionally, a subject, reference, or attention line is used to alert the reader to the subject, file reference, previous correspondence, account number, or other emphasis.

Examples

Below salutation	Subject: Invoice #20947
Above salutation	ATTENTION: Mr. D. W. Clark
	RE: Your letter of 12 June 199X

Such special elements are typed flush to the left margin two lines below the inside address but above the salutation.

Typist's Initials. If your letter is typed by someone other than you, place your initials in capital letters and the typist's initials in lower-case letters flush to the left margin two lines below the typed signature. Use a colon or a virgule between them. If you type your own letter, your initials after the signature are optional.

Examples

JSV:mt

or

JSV/mt

Enclosure Notations. If you send materials or documents with your letter, add an enclosure notation two lines below the typist's initials.

Examples

Enclosure

or

Enclosures (2)

or

Enc: Copy of Check #1029

Distribution Notation. If you are sending copies of your letter to other readers, add a distribution notation two lines below the last element.

Example

pc: Mr. David Little, Chairman

Since most copies are now photocopies rather than carbon copies, *pc* is favored over the traditional *cc.*

Second Pages

If a letter requires a second page, type the recipient's name, the page number, and the date in a block flush to the left margin or across the page.

Examples

Ms Sally Queen
Page 2
15 July 199X

or

Ms Sally Queen   –2–   15 July 199X

ENVELOPES

Your envelope should be 9½ in x 4½ in. and of the same quality as your stationery. The recipient's name, title, company, address, city, state, and zip code are centered horizontally and vertically. As a guideline, begin 12 lines down from the top. Single space between lines. Your own name, address, city, state, and zip code are placed in the upper left-hand corner. Use the Postal Service abbreviations for states. Figure 10.9 shows a sample envelope:

Dr. Richard Grande
1022 Northeast First Street
La Mesa, CA   92041

                              Ms Julie Maney
                              Director of Personnel
                              Ace Manufacturing Company, Inc.
                              4092 West Grant Street
                              Manchester, CT   06040

FIGURE 10.9   *Sample envelope*

CONTENT

Organization

Organize your message into three parts:

1. A brief introduction that states the purpose of the letter immediately, unless you are conveying "bad news"
2. One or more body paragraphs that contain specific detail
3. A conclusion that establishes goodwill or encourages your reader to act

Introductory Purpose

State your purpose immediately. Do not just fill space until you get around to your purpose.

Vague purpose   It became apparent about five years ago that computerized bookkeeping was to be the answer to the problems which were plaguing our bookkeeping department.

Therefore, we would like to investigate your software program . . .

Clear purpose   I am seeking answers to three questions regarding your software program, Computerized Automotive Reporting Service.

Other clear introductory purpose statements follow:

Examples

Here are the instructions for assembling the Ace 1 Trampoline that you requested by phone on September 20.

Please consider my resume and application for a junior management position at your Pompano resort.

This is in answer to your inquiry about leasing our trucks.

Your vacuum cleaner is repaired and is ready to pick up.

Congratulations on your promotion to Director of Affirmative Action.

You are right. You paid your bill exactly when you said you did.

Second, you may clarify why you need this particular information.

Example

Because I edit our company's in-house newsletter, it is essential that I purchase software that has both WordPerfect and Microsoft Word capabilities.

"You" Perspective Body

Provide the details of your correspondence in the body, using a "you" perspective. Put yourself in your recipient's shoes and consider how that person will respond to your words. Be courteous, direct, and confident. Avoid a slangy, abrasive, pompous, or abrupt tone. Consider the abrasiveness of the following phrases:

Examples

Your department should shape up . . .

I demand that . . .

I am appalled at your slow response . . .

I beg to advise you of my intent to . . .

Rush me information on . . .

Much of your correspondence will be highly repetitive ("Thank you for your order of July 25." "Our records indicate that your payment is past due."), yet trite and cliché expressions should be avoided.

Cliché Expressions	*Plain English*
Having received your letter, we . . .	We received your letter . . .
Pursuant to your request . . .	As you requested . . .
Per your memorandum . . .	As you noted . . .
Enclosed please find my report . . .	Here is my report . . .
It is imperative that you write at once . . .	Please write at once . . .
I am cognizant of the fact that my report is tardy . . .	I know that my report is late . . .
At the earliest possible date . . .	As early as possible . . .
I beg to differ with your . . .	I disagree with your . . .
Please be advised that the new policy . . .	The new policy is . . .
I hereby request that . . .	Please consider . . .
I beg to acknowledge receipt of your check . . .	I received your check . . .
We are in hopes that you succeed . . .	Good luck . . .

Every letter you write should sound fresh and conversational.

Word Processing

A final word about the "you" perspective should be added. Word processing by computers has removed the drudgery from written correspondence by allowing organizations to create standard text for repetitive form letters. If you are already using a word processing computer program, you know that you can delete anything from a character to a paragraph or relocate or insert words, phrases, sentences, and entire paragraphs with ease.

The word processing capability, however, may tend to depersonalize your letters. Great care must be taken to maintain a friendly, "you"-oriented tone.

Purposeful Conclusions

Your conclusion provides an opportunity for you to urge action or establish goodwill. A brief closing should motivate the recipient to follow up on your letter or, at least, to feel favorable to you, as illustrated in these examples.

Examples

May I have your answer by March 12?

If you will call us within the next few days, we can send our sales representative to demonstrate our software capabilities.

I would like an interview and am available weekdays for the rest of this month.

Thank you for pointing out this problem.

I appreciate your services . . .

Do not, however, be obvious or presumptuous. Avoid "I want to thank you in advance for . . ." and "Please feel free to call me if you have further questions." An advance thank-you implies that you may be too lazy to write a proper thank-you when your request is fulfilled. The second closing is unnecessary; the recipient will call you if he or she has questions whether you invite a call or not.

POSITIVE OR GOOD NEWS LETTERS

Recipients of many types of letters are happy to receive them. Some letters offer services or sales, place orders, tender a congratulation or thank-you, or transmit desired information. You will not only receive such letters but will also be called upon to write them in your career field. Good news letters include

- Inquiry and request letters
- Invitations
- Order letters
- Congratulatory letters
- Thank-you letters
- Sales or service offer letters
- Employment application and cover letters
- Transmittal letters

Inquiry and Request Letters

An inquiry or request letter is a good news letter in that the recipient stands to benefit from the writer's interest. Nevertheless, an inquiry or request solicits a response which will ask of the recipient time and, perhaps, effort. Therefore, certain strategies will help to motivate the recipient to respond quickly and accurately.

First, your introductory purpose statement may include the suggestion that you need an immediate response.

Example

I am seeking additional information on your TP-1 Daisy Wheel printer because I plan to purchase one this month.

Second, you may clarify why you need this particular information.

Example

Because I edit our company's in-house newsletter, it is essential that I purchase a printer that performs proportional spacing.

Third, you may subtly compliment the person or the company.

Example

I am interested in obtaining some additional details about your VID-80 Model III. Your company was the first to advertise such a modification more than six months ago, so I believe that you have the most expertise in this field.

Fourth, if you are asking questions, simplify and separate them into numbered items, make each specific, arrange them in a numerical table, and allow sufficient space between each to allow a jot of answers right on your letter.

Example

I have three questions that will affect whether or not I upgrade my system at this time:

1. What type of format does your CP/M use? Is it compatible with the system Radio Shack computers use?
2. Do schematic diagrams come with the unit? If not, are they available for an extra charge?
3. If the added memory is purchased, is there any way to use the additional memory when operating with TRSDOS or similar operating systems?

You may also consider enclosing a self-addressed, stamped envelope. This double strategy of leaving space to answer on your letter and sending a return envelope will allow the recipient to respond right on your letter and place your response in the outgoing mail immediately.

Finally, if it is appropriate, you may offer to share the results of your inquiry.

Example

Because I am compiling information on the technical writing programs of all Florida community colleges, I will be happy to share with you my final report.

Figure 10.10 shows a poor request letter. It exemplifies a "me" rather than a "you" perspective and is poorly organized and rude. Figure 10.11 (see p. 270), however, illustrates a well-written request letter that motivates a quick response.

3072 Southwest Sixth Avenue
Norwich, CT   06360
4 May 199X

Mr. Fred Mandel
Radio-Electronics
200 Park Avenue
New York, NY   70908

Dear Mr. Mandel:

Demanding and insulting

    Please rush me information on how to convert my transistor output voltage to 6VDC usable voltage for my portable radio. I read your article in Radio-Electronics, but it confused me.

Unclear questions

    What I need to know is can I use the same 300 Ohm pot or something else for my 9VAC. Can I use the same number rectifier for my radio? Can you recommend a zenior diode? What else do I need to know?

Rude

    Thank you in advance for answering my questions.

Very truly yours,

Walt

Walter Matthews

WM:jsv

FIGURE 10.10 *A poor request letter*

10111 Kennedy Boulevard
Atlanta, GA   21441
21 February 1993

Mrs. Dawn Michele
Customer Service Representative
Atlanta Bank
100 Peachtree Drive
Atlanta, GA   21445

Dear Mrs. Michele:

I have heard about your excellent loan rates through a friend who banks with
Atlanta Bank. I am considering buying a home, so I wish to obtain some
information about your loan practices.

What are your current interest rates for mortgage loans?

How many points do you currently charge?

Please send me any brochures you have available on your equity loans.

I expect to buy the house within the next month; therefore I would appreciate this
information as quickly as possible. I have enclosed a stamped, self-addressed
envelope for your convenience.

Sincerely,

Suzanne Kelley

Suzanne Kelley

SK:bc

Enclosure

FIGURE 10.11   ***An effective request letter*** (Courtesy of student Susan Kelley)

Invitations

An invitation is a type of request letter, too. You may wish to invite a speaker to address your group or organization. Or you may wish to invite members or guests to attend special functions, such as meetings or luncheons. Invitations should be brief, but detailed. First, extend the invitation including the function, date, time, and place. If necessary, mention guest status, fee, honorarium or the like to be expected. Second, elaborate on the purpose and offer any other particulars that the invitee will need such as the number of expected persons, length of a requested speech, or other program components. Finally, urge a response by enclosing a response card requesting a call, or suggesting you will be calling in a few days for an answer. Figure 10.12 (see p. 272) shows a typical invitation letter.

Order Letters

Another good news letter is the order letter, one that informs a seller that you want to purchase a product or service. Three writing strategies will help you to be clear and accurate about your specific order, shipping instructions, and method of payment. First, accurately describe the product or service by specifying the name, brand, model, stock number, quantity, color, dimensions, unit price, and so on. Include an informal table for multiple product or service orders for easy reader reference. Second, include your shipping instructions, such as first-class or third-class mail, Federal Express, United Parcel, or special mailing address, department, or special attention notation. Third, mention the date needed if this is an issue, and, finally, specify your method of payment: enclosed check or money order, credit card charge number, COD, installment, and so on.

Figure 10.13 (see p. 273) illustrates a clear and accurate order letter.

Congratulatory and Thank-You Letters

Both of these good news letters are characterized by informality and friendliness. The salutation might address the recipient by a first name. The introduction should mention specifically the occasion for congratulations or appreciation. Add detail to underscore your sincerity and end with a warm complimentary close.

Figure 10.14 (see p. 274) and Figure 10.15 (see p. 275) illustrate typical congratulatory and thank-you letters.

Sales and Service Offer Letters

Even though a sales or service offer letter is written to persuade the recipient to purchase a product or service, it may be considered a good

Meet for Good Fellowship

PROVIDENCE JEWELERS CLUB

P.O. BOX 4350
EAST PROVIDENCE, RHODE ISLAND 02914

16 September 199X

Ms. Robin Revell
30 Ormsby Avenue
Warwick, Rhode Island   02886

Dear Ms. Revell:

The particulars

Please accept my invitation to attend as my guest the Providence Jewelers Club "Speakers' Luncheon" on 1 October 199X, 12:00 noon, at the Providence Marriott Hotel.

Elaboration

I would like to introduce you to our members and guests as the recipient of the Providence Jewelers Club 199X-199X scholarship. Your choice of meals should be indicated on the enclosed card and returned as soon as possible.

Urge to action

Should you not be able to attend, please call me at 738-8560 as soon as possible. I look forward to meeting you.

Very truly yours,

Paul J. Austin

Paul J. Austin
President

PJA/eeg

Enclosure

FIGURE 10.12   *An invitation letter* (Courtesy of Paul J. Austin)

2740 Elm Street
Hartford, CT   10031
30 January 1993

J. C. Dime and Co.
8190 W. Presidential Boulevard
Batavia, OH   45103

ATTN. Mail Order Department

Good Morning:

Please send me by United Parcel Service the following items from
your 1999 Catalog:

910	Open Wall Unit	2@$79.95	$159.90
911	2-door Cabinet	1@ 39.95	39.95
912	3-drawer Chest	1@ 49.95	49.95
			$249.80

Label the shipment "Attention Suzanne Kelley" and use the above
address.

I am enclosing a check for the full payment of $249.80.

Very truly yours,

Suzanne Kelley

Suzanne Kelley

SK

Enclosure

FIGURE 10.13 *An effective order letter* (Courtesy of student Susan Kelley)

ACE COMPANY, INCORPORATED
2900 Northeast Seventy-ninth Street
Fort Lauderdale, Florida   33301

5 January 199X

Ms Paula Watkins
Tech Laboratories, Inc.
P.O. Box 37021
Miami, Florida   30321

Informal salutation — Dear Paula:

Source of knowledge —   The good news that you were promoted to Director of Personnel was in the Business News section of this morning's <u>Miami Herald</u>. Tech Labs certainly picked the right person for the job. Congratulations!

Specific message

Underscore of sincerity —   You've earned this advancement, Paula. Your consistent good work and extra efforts, such as your participation in our Broward County Career Festival, set an example for all of us in the field.

Friendly closing —   Warmest wishes,

Ben

Ben Funger

FIGURE 10.14 *An effective congratulatory letter*

news letter in that it offers to enhance the recipient in some manner. Five writings strategies can aid you in obtaining a favorable response.

First, identify and limit your audience. Determine the needs of this group and bear in mind exactly what you want your audience to do after reading your letter. The "you" perspective is critical for your desired response. Keep in mind what you can do for your reader throughout the letter.

Second, begin your letter with an attention-getting statement. You may ask a question, offer a free gift, employ a *how to* statement, or use flattery. In short, hook your reader into reading further.

3502 Southwest Palm Avenue
Athens, GA   30605
7 March 199X

Professor J. John Jenks
Community Service Division
Broward Community College
1 East Las Olas Boulevard
Fort Lauderdale, Florida   33301

Dear Professor Jenks:

To the
point,
informal

    Your generous recommendation of me to Tech
Laboratories, Inc. got me the job. Thank you!

Specific
details of
appreciation

    The word processing skills you taught me plus your
pep talks on organization have opened new doors of self-
confidence for me. I'm truly grateful for your interest.

Friendly
closing

    I'll keep you posted on my advancement.

                              Thanks again,

                              Ralph

                              Ralph Kennery

FIGURE 10.15   *An effective thank-you letter*

Examples

You will receive a free vacation for two, a new car, a dream cottage, or other valuable gift simply by making an appointment to inspect our resort.

Here's how to save $100.00 on your next automobile purchase.

You made a smart choice by enrolling at Broward Community College. Now let us help you make a smart choice in selecting your college wardrobe.

Third, call attention to the product or service's appeal. Persuade the reader that your offer is so desirable that he or she cannot resist it.

Examples

Our time-share condominiums are caressed by gentle ocean breezes and within steps of your very own tennis courts, golf course, and spa. We offer the last word in glamorous vacations.

Ace Motors offers the world-recognized most economical car on the highways—the Ace 400ZT.

Designer jeans, polo shirts, a dazzling array of blouses, dresses, skirts, and accessories are waiting for you.

Fourth, present evidence of your product or service's application. Emphasize its convenience, usefulness, and economy. Endorsements, guarantees, and special features may be highlighted. Present the facts in a manner that emphasizes the attractiveness of your offer.

Examples

A member of Time-Share International, Driftwood Resort offers not only the most reasonable prices on the Gold Coast but also opportunity to vacation in 39 countries of the world. Spacious two and three bedroom plus studio accommodations are available to suit your precise requirements. Fully equipped kitchens and modern hot tubs allow you the casual lifestyle of a truly refreshing holiday.

Fifty-three motoring journalists from 15 European nations voted this newest Ace "Car of the Year." The 400ZT is aerodynamically designed to accelerate from zero to fifty in only 7 seconds. Disc brakes, rack and pinion steering, and a performance-tuned suspension system make this automobile a marvel to drive.

College Corner clothes will make you feel confident and poised. Our College Board representatives make sure we buy the "in" ensembles to fit your classroom, party, and extracurricular needs. We even offer free alterations.

Finally, urge your reader to action. Make it easy to return a postcard to order your product, suggest an appointment next week, invite readers in for a free gift, include a phone number to call, and so on.

Examples

Call 305-455-9000 to arrange a tour of our facilities and to find out what valuable gift is yours. You won't be disappointed.

Stop by this weekend to test drive your next car—the Ace 400ZT.

During Orientation Week we will be open until 9:00 P.M. for you to drop in and browse. Free textbook covers are yours with every purchase.

Figure 10.16 shows a persuasive service offer letter. The audience consists of busy top executives who may be frustrated by the poor quality and time-intensive writing of their employees. The layout is catchy

Jane Hansen & Associates
409 East Seventy-second Street
Suite 9001
Los Angeles, CA   90028
12 May 199X

Mr. Frank Mahoney
Senior Vice-President
City National Bank
100 Northeast First Street
Los Angeles, CA   90066

Dear Mr. Mahoney:

Attention getter

DID YOU KNOW   —it costs $20.00 for each letter your
                      staff types?
                      —your typical employee spends 25% of
                      his or her time at work writing?
OF COURSE YOU KNOW —the ability to write well gets the
                      results you want, and contracted
                      training saves your bank money.

Appeal of offer

To upgrade the writing skills of your employees, you may now contract WRITING SKILLS & STRATEGIES, a training seminar tailored to your employees. Further, we will conduct the seminar at your bank during the times most convenient for your busy staff.

Application and appeal

Endorsement

WRITING SKILLS & STRATEGIES reviews troublesome mechanics, grammar, and usage as well as offers indispensable tips on correspondence, report writing, and much more. An experienced writing instructor will tailor the materials to your specific needs. Over 40 banks in south Florida will attest to the practicality of this training program.

Urge to action

May I make an appointment, Mr. Mahoney, to discuss course content, prices, and times? I will call your office within the next ten days.

Very truly yours,

Jane Hansen

Jane Hansen
Writing Consultant

Application

JH:jsv

P.S. The enclosed brochure highlights features of WRITING SKILLS & STRATEGIES.

FIGURE 10.16 *An effective service offer letter*

and appealing. The service's application and the desired action is effectively covered, yet the letter is brief.

Employment Application and Cover Letters

Because employment application letters often include a resume, a separate writing strategy, they will be discussed separately in Chapter 11.

Transmittal Letters

Letters of transmittal, sometimes called cover or face letters, announce the enclosure of attached material and reports. Content and samples are covered in Chapters 11 and 13.

NEGATIVE OR BAD NEWS LETTERS

Some letters must convey bad news. They may inform the recipients that they are not hired, cannot get a refund, are late with a payment, and so on. Again there are writing strategies to convey your bad news in a positive, result-producing manner. Bad news letters include

- Negative response letters
- Complaints
- Collection letters
- Solicitations

Legal Considerations

"Anything you say can and will be used against you in a court of law." These words from the Miranda Rights are familiar to anyone who has ever seen a film or television show depicting police officers arresting a suspect. They also apply to letter writing.

In the world of business, there is always concern for what one says and how one says it. Political correctness and the avoidance of sexual harassment are probably the most apparent concerns. However, too few people realize that accidentally insulting an applicant's character while refusing credit or a job can be grounds for legal action. Therefore, it is essential that your letters maintain a somewhat neutral tone.

Letters that sound threatening or abusive or appear to solicit bribes or favors can always come back to haunt you. Remember, you are signing your name to every letter you write and are, therefore, personally responsible for every word you write. This is especially true of bad news letters because the receiver is already apprehensive upon receiving the letter.

Negative Response Letters

A letter that must say *no* requires a buffer statement before the bad news. A buffer statement presents a valid reason before the negative response.

Examples

We lease our apartments only through registered real estate brokers; therefore, . . .

Because the criteria for the position of Personnel Specialist require a college degree in business administration, we can consider only those applicants with that credential.

I have referred your letter to Ms Jane Clifford, assistant to the director of computer services, because only that department is authorized to give you the information you request.

A second strategy is to avoid the word *no.*

Too harsh   I'm sorry to say no, I cannot address your engineers next week.

Buffered   Because I will be in Chicago all of July, I cannot speak before your group.

A third strategy in a negative response letter is to avoid an apology. Your reasons are valid, so eliminate *unfortunately, we regret, we are sorry, we wish we could,* and so on.

Fourth, do not leave the opportunity to reopen discussion or consideration. Avoid "If you wish to discuss this further . . ." and "We wish we could . . ."

Finally, conclude by establishing goodwill. Use "We hope we have an opportunity to serve you in the future," "We appreciate your interest in our organization," and the like. Sincerity is an important consideration. If your effort to establish goodwill appears contrived, forced, or formulistic, it will irritate your reader.

Figure 10.17 (see p. 280) illustrates an effective negative response letter to a job applicant. It buffers the bad news, does not apologize, and establishes goodwill.

Complaints

All of us have found it necessary to complain about defective products, delayed orders, billing errors, or inadequate services. Although we usually write complaint letters at a time of anger or frustration, angry tones seldom elicit the action we desire. A complaint letter needs restraint, specificity, and a clear statement of desired action.

ACE COMPANY, INCORPORATED
2900 Northeast Seventy-ninth Street
Columbus, Ohio   43219

5 January 199X

Ms Jane Doe
1107 Northwest Twelfth Street
Dayton, Ohio   33331

Dear Ms Doe:

Buffer statement · Negative response

        We have received your letter and resume, exploring career opportunities at Ace Company. While your education and experience present an interesting background, your qualifications do not fit our particular requirements at this time.

No apology but positive action

        We will place your materials in our potential file. Should a position for which you qualify open, we shall review your file and notify you.

Establishes goodwill

        We appreciate your interest in Ace and wish to extend our wishes for success in the attainment of your career goals.

Sincerely,

Ralph Maran

Ralph Maran

RM:tac

FIGURE 10.17   *An effective negative response to a job applicant*

    The first strategy is to provide a detailed description of the faulty product, service, or suspected error along with the specifics about your purchase or contract.

Example

I am returning for a full refund the U.D.S. Computer Telephone, model 333, for which I sent money order 40920 in the amount of $10.00 on 15 July 199X. I received the defective phone on 20 August 199X.

Second, state precisely what is wrong with the product or service.

Example

The phone malfunctions. The beeper activates on the third dialed digit so that dialing cannot be completed. A persistent, loud buzzing interferes with reception on incoming calls.

Third, consider describing the inconveniences you have experienced. This is not always necessary but may help to underscore the seriousness of your complaint.

Example

Because I conduct a great deal of business from my home telephone, I have lost sales and commissions by not having a properly functioning instrument.

Finally, state clearly what action you desire. You may want a refund, a replacement, copies of all records, or some other consideration.

Example

I am enclosing the warranty and am requesting a full refund for the purchase price. I will not consider a replacement because I have lost confidence in your merchandise. I shall appreciate a prompt refund check.

Figure 10.18 (see p. 282) illustrates a complaint about a credit problem.

Collection Letters

Unfortunately, not all customers pay their bills on time. Therefore, companies must employ several correspondence strategies to urge payment. Frequently, companies send a series of collection letters, each employing a stronger tone than the former.

It is not wise to demand immediate payment in the first collection letter because there may be valid reasons for slow payment, such as misdelivered or misplaced bills or errors in the company's billing. Further, an early threat to begin legal action or collection services may cause the well-intentioned customer to avoid further profitable transactions with the company, or the company may suffer negative word-of-mouth publicity. A tactful "you" perspective is important to maintaining good relationships. Ultimately, you may need to threaten posting with credit agencies, legal action, a lawyer's intervention, or even serving of a summons. Your intent is to avoid those actions.

The first letter should make the customer feel valuable, allow the customer to save face, urge prompt payment, but offer to establish a partial payment schedule if that course of action is appropriate.

1390 Southwest Twentieth Street
Davie, FL   33326
22 September 2000

The Doubleday Store
Customer Service Department
501 Franklin Avenue
Garden City, NJ   07769

Re: Account #96-299-38934

Gentlemen:

Please review my account for a credit. On 12 July 2000 I received the Pierre Cardin canvas luggage set from your company which I ordered on 15 June 2000. When I received the luggage from your company, it was on a trial basis for 60 days. After examining the luggage, I determined that it was not substantial enough for my needs, and I returned the entire set on 12 August 2000.

The charge of $279.95 has continued to be shown on my last two monthly statements. I wrote a note on the statement each time indicating the date and return of luggage and sent the statement back to your company. Copies of these notes are attached to this letter. To date, I have not received an adjusted statement.

Would you please credit my account for $279.95 and send me an adjusted statement?

I will appreciate your prompt attention to this matter.

Very truly yours,

Ruth Burrows

Ruth Burrows

RB

Enclosures: 2 statement photocopies

Details of complaint

Expansion

Desired action

FIGURE 10.18 *An effective complaint letter*

After a reasonable length of time for turn-around mail, send a follow-up letter that refers to the first request for payment and asks for an immediate payment to avoid further action. In your final letter refer to the first requests for payment and alert the customer that the account will be turned over to a collection agency or attorney if partial or full payment is not made immediately. Figures 10.19 through 10.21 show three collection letters illustrating increasing pressure for payment.

Downtown Travel Centre

City Park Mall
140 Southeast First Street
Fort Lauderdale, Florida 33301
(305) 525-1303
Fax: (305) 525-1367

15 January 2000

Ms. Mary Houghton
301 Northeast 109 Street
Plantation, FL 33323

Re: Invoice #47091

Dear Ms. Houghton:

Shows customer value

Downtown Travel Centre values you as a trustworthy client. We appreciate your patronage and are here to serve your future travel needs. Our records, however, indicate that you have not paid the invoice for the balance due on your Travcoa tour to Vietnam and Cambodia issued to you on 15 November 1999. We trust that you enjoyed your holiday and that your nonpayment has been an oversight due to the press of travel and the holidays.

Polite request for payment

Please send your check in the amount of $1,985 as soon as possible. If this is impossible, please call me to establish a schedule of partial payments. If you have already mailed your check, please disregard this letter.

Sincerely,

Jeffrey Duckham

Jeffrey Duckham
Travel Associate

FIGURE 10.19   *An initial collection letter*

Downtown Travel Centre

City Park Mall
140 Southeast First Street
Fort Lauderdale, Florida 33301
(305) 525-1303
Fax: (305) 525-1367

10 February 2000

Ms. Mary Houghton
301 Northeast 109 Street
Plantation, FL 33323

Re: Invoice #47091

Dear Ms. Houghton:

Reminder — May we remind you that this account is overdue for payment and that we notified you of your lateness by our letter of 15 January.

Strong urge for payment — We remind you that our terms are 30 days, and it has now been two months since we issued our invoice for the balance of your Travcoa tour without receiving even a partial payment. In order to avoid further action, please send us a check for $1,985 by return mail.

Very truly yours,

Jeffrey Duckham

Jeffrey Duckham
Travel Agent Associate

Make us your Travel Headquarters . . . you'll be glad you did!

FIGURE 10.20 *A second, follow-up collection letter*

Downtown Travel Centre

City Park Mall
140 Southeast First Street
Fort Lauderdale, Florida 33301
(305) 525-1303
Fax: (305) 525-1367

21 February 2000

Ms. Mary Houghton
301 Northeast 109 Street
Plantation, FL 33323

Re: Invoice #47091

Dear Ms. Houghton:

Curt reminder — According to our records $1,985 is outstanding on your account and is now considerably overdue for payment.

Strongest urge for payment — Since you have not responded to our previous two reminders, I now have no option but to put this matter in the hands of our collection agency, which may affect your credit rating.

If you have any questions concerning your account, please call me at the above number. If we don't hear from you by March 30, we will take the necessary measures to recover this debt.

Very truly yours,

Jeffrey Duckham

Jeffrey Duckham
Travel Agent Associate

Make us your Travel Headquarters. . . . you'll be glad you did!

FIGURE 10.21 *A third, final collection letter*

Solicitations

A solicitation for money or volunteer activities may be classed as bad news because such solicitations offer nothing in return except, perhaps, the opportunity to further a cause that interests the reader. Alumni groups, political candidates, and supporters of various causes frequently solicit money and volunteers. Such letters require inventive, eye-catching, persuasive techniques, including those used in advertising: bandwagon appeal, snob appeal, humor, and endorsements.

The humorous approach is one effective technique, as illustrated in Figure 10.22, a letter soliciting a contribution from a college alumnus.

MIAMI UNIVERSITY   Miami University Fund

SPRING IS
SPRINKLING AGAIN.

Whoever picks the weathermen for southwestern Ohio must assign
the same one here year after year

. . . and he, or she, must like rain.

So up and down High Street, across the Slant Walk, in and out of the
dorms, apartments, classrooms and libraries, thousands of students,
just like when you were here, are going . . .

. . . drip, drip, drip.

Of course, after the "rainy season" the skies brighten. Oxford gets
breathtakingly beautiful. The dogwoods and redbuds, croci and
daffy-dillies all pop out . . . to say nothing of the lifelong friendships
that seem, also, to bloom in the Spring.

Then everything seems better than ever. Remember?

The same kind of cycle repeats each Spring for The Miami University
Fund. Around this time of year we begin to experience a heavy
sprinkle of checks from friends like you who know that Miami's
bright skies next year depend a great deal on how many key
scholarships we can offer and on all the other rays of educational
sunshine the Fund supports.

And the results do show. Miami stays better than other schools;
attracts finer students; offers them a better experience; keeps an
admired faculty; even reflects favorably on the alumni and friends
who make its annual blossoming possible.

So take a moment now, won't you, to check on the weather and see if
this isn't a good time to help "sprinkle" on the Fund. If you'll put a
check in the mail, we'll put it right to work painting the skies sunny,
just as you'd want us to . . .

                              Zip, zip, zip.

                              Douglas M. Wilson

                              Douglas M. Wilson
                              Vice President
                              University Relations

DMW:ehd

FIGURE 10.22 *An effective solicitation letter* (By permission)

CHECKLIST

Correspondence

GENERAL

❏ **1.** Have I chosen the appropriate type of correspondence for the occasion?

   ❏ In-house memorandum?

   ❏ E-mail (usually by permission)?

   ❏ Formal letter?

   ❏ Faxed correspondence (if appropriate)?

❏ **2.** Have I used the International Standard for dating? If not, why?

MEMOS

❏ **1.** Have I used the necessary headings?

   ❏ Full name and title of the addressee?

   ❏ Full name and title of the sender?

   ❏ The date?

   ❏ The subject?

❏ **2.** Is my content to the point, timely, addressed to a single subject, and attractive?

E-MAIL

❏ **1.** Have I included a clear subject line?

❏ **2.** If the receiver is not likely to know me, have I included my full name, business address, phone, and other pertinent information needed by the receiver?

❏ **3.** Is my content to the point, timely, addressed to a single subject, and attractive?

LETTERS

❑ **1.** Have I used a consistent block or modified block format?

❑ **2.** Are all parts of the letter included?

    ❑ A heading, if not using letterhead, which includes your full address and the date?

    ❑ An inside address to a particular person with his or her title and complete address?

    ❑ A salutation to a particular person, if possible?

    ❑ Engaging and clear body paragraphs?

    ❑ A complimentary close, typed name, and written signature?

❑ **3.** If there are two or more pages, are they identified by receiver, page number, and date?

❑ **4.** Is the envelope the same quality paper as the stationery, and its content complete and appropriately placed?

❑ **5.** Does the opening of the letter engage the reader's interest and reveal the purpose of my letter?

❑ **6.** Is my content written with a "you" perspective?

❑ **7.** Is my content free of cliché expressions?

❑ **8.** Is my tone appropriate for a good news letter (an inquiry or request, an invitation, an order, a congratulation, a thank-you, a sales or service letter, a transmittal letter)?

❑ **9.** Is my tone appropriate and does it include a buffer statement if a bad news letter (a negative response, a complaint, a series of collection letters, solicitations)?

❑ **10.** Does my closing urge the reader to act?

❑ **11.** Is my content letter perfect?

❑ **12.** Does my content enhance my image?

EXERCISES

1. **Subject Lines.**   Rewrite the following memo or e-mail subject lines to make them more precise and descriptive:

Test Results	Minutes
Picnic	Training
Hours	Policy Change
Overtime	New Personnel

2. **Letter Formats.**   Correct the errors in the letter elements of the following modified block letter format:

Ms. Julie Wilson
2 N. W. Park Ave.
Chicago, Ill.   33302

Mr. Daniel C. Taylor
four-one-two E. 72nd St.
Dayton, Ohio,   40727

Dear Sir,

Re: Account #407-201 E

_____
_____
_____

_____
_____
_____

                                           Best,

                                           Marcia Morris

                                           Marcia Morris, Treasurer

                                           mm:tc

---

3. **Letter Content.** Rewrite the body of the following complaint letter by dividing it into an introduction, body, and closing, and by eliminating "letterese," a "me" perspective, and an inappropriate tone:

Dear Sir:

In reference to your lousy iron which I purchased recently, I want my hard-earned money back. If you don't refund me the price in full, I beg to inform you that I will take legal action. It spews water all over the clothes I iron, scorches things even on a low setting, will not stand up on its base, and the plug broke the last time I plugged it in. If you have any questions, do not hesitate to call me. I am appalled at the workmanship of this piece of junk. Get with it.
Cordially,

---

4. **Buffer Statements.** Write buffers to the following blunt, negative response statements:

   a. We cannot send you the items you ordered because you did not enclose return postage.

   b. We're sorry to inform you that we cannot hire you at this time.

   c. I must say no to your request for a writing seminar in June.

WRITING PROJECTS

1. **Memo or E-mail.** Write a brief memo to a subordinate to urge that a previously assigned written report be turned in a week earlier than previously scheduled.

2. **Memo or E-mail.** Write a memo to your employer (real or imaginary) to point out a minor problem at your place of work, such as a scheduling mix-up, inadequate lighting, the need for more storage shelves or files, your inability to perform an assigned task, or an error in your paycheck. Be brief and to the point.

3. **Good News Letters.** Write a good news letter based on one of the following suggestions:

   a. Write **an inquiry or request letter** to a company or other organization in response to an advertisement or article in a professional journal in your field. Ask at least four technical questions about a product or service. Review the writing strategies that will motivate a quick response.

    b. Write **an invitation** to a professor or an authority in some field to make a 20-minute address at your annual kickoff meeting of a special-interest organization. Make up all of the details the invitee will need to know.

    c. Write **an order letter** for merchandise from a particular company. Specify the stock or model number, quantity, size, color, unit cost, and total cost as appropriate. Include all details about payment and delivery.

    d. Write **a congratulatory or thank-you letter.** Some suggested topics are congratulations to a friend who has been hired or promoted, won a professional award, completed a degree or other special training, been elected to public or organizational office, or opened up a business. Thank-you letter topics might be a response to a letter of recommendation, a letter of appreciation to a teacher or counselor whose advice you followed, a letter to a hotel or restaurant that hosted a group function or dinner, or to a person who directed some business opportunity your way.

    e. Write **a sales or service offer letter.** Consider your audience's age, occupation, geographical location, needs, and interests. Some suggested topics are the merits of a particular automobile, college, restaurant, new store, personal computer, bank, travel bureau, or flower shop or maintenance service for lawn, pool, or snow removal.

4. **Bad News Letters.** Write **a bad news letter** based on one of the following suggestions:

    a. Write **a negative response letter.** Some suggested topics are turning down an offered job, declining an invitation to speak, declining to serve on a committee or join an organization, refusing to volunteer time to an organization, refusing a job applicant, inability to fulfill a reservation request at a hotel or travel group, declining to refund money for a specific piece of merchandise, or refusing a sales or service offer.

    b. Write **a complaint letter.** Some suggested topics are an error in your credit card or telephone bill, rude service you received at a store or restaurant, late delivery of merchandise, poor quality of some product recently purchased, or damaged goods delivered by a carrier.

    c. Write **a series of three collection letters,** each firmer than the one before. Some suggested topics are late dues payment to a club or organization; late payment for a service you rendered; or late payment for an automobile, credit account, or bank loan.

    d. Write **a solicitation letter.** Select a cause that interests you: save the whales or manatees, abortion legislation, a political candidate, proposed zoning changes, or a halfway house for troubled youth. Solicit donations or an appearance at a civic forum to discuss the issue with top policymakers.

5. Write **an order letter** to be faxed or e-mailed to an overseas company. Be sure to use the International Standard for dating, check multicultural/international language considerations, and convert any currency differences.

NOTES

CHAPTER 11

Resumes, Cover Letters, and Interviews

SHOE   by Jeff MacNelly

S K I L L S

After studying this chapter, you should be able to

1. Define *resume* and *cover letter.*

2. Appreciate the importance of a detailed, letter-perfect resume and cover letter.

3. Compile a file of job descriptions with information on working conditions, necessary qualifications, job outlooks, earnings, potential employers, and employment services.

4. Compile a file of personal education, employment, and other skills and qualifications.

5. Develop a portfolio of outstanding writing assignments, awards, and other accomplishments.

6. Compile a file of prospective references plus actual letters.

7. Understand the distinctions between chronological, functional, narrative, and scannable resumes.

8. Develop and write a personal resume in the appropriate format for your situation.

9. Write a specific and general cover letter with the appropriate inclusions.

10. Understand and practice the *do's* and *do not's* of an employment interview.

11. Prepare yourself for the interchange of questions between an interviewer and yourself.

12. Identify and employ strategies you may use if an interview is going sour.

INTRODUCTION

Résumé is a French word that means "summary." A personal resume is a concise summary of pertinent facts about yourself—your employment objectives, your employment and educational history, personal data, and reference lists. Although most dictionaries include the accent marks (résumé), it is common practice to omit them. A resume and its accompanying cover letter are indispensable job-hunting tools that are submitted to employers to "sell" yourself as a prospective employee and to obtain an interview.

Another term you might encounter is *curriculum vitae,* or *vita* for short. These Latin words, meaning "life's course of events," are used in place of the word *resume* by people holding higher degrees, such as the M.A., Ph.D., Ed.D., D.D.S., M.D., or J.D., and by high-level executives with many years of experience in the business or professional worlds. Although the terms *curriculum vitae* and *resume* are sometimes used interchangeably, you should use the title best suited to your career field and your qualifications, as the misuse of one may be perceived as pretentiousness and the misuse of the other as underselling oneself or a lack of professionalism.

Your resume allows you to organize and amplify your data in a manner that highlights your strengths in order to obtain an interview. A job application form (Figure 11.1) usually is restricted to mere listings of schools and previous employment names and addresses but introduces the basic information concerning a job applicant. Traditionally, resumes and cover letters have been mailed to companies that have advertised for applicants or to those companies where you know an opening exists. You may, in some instances, present a resume during an interview that you have obtained by other means. Posting resumes on the Internet in response to web page and job bank information is increasingly practiced.

Employers receive hundreds of resumes for their positions and must quickly scan them, giving only a few seconds of preliminary attention to each. In order for yours to be put on the review pile, it must be letter perfect and convey immediately that you are qualified, capable, and worthy of an interview.

application for employment

We are an equal opportunity employer, dedicated to a policy of non-discrimination in employment on any basis including race, color, age, sex, religion or national origin.

PERSONAL INFORMATION

Date   Social Security Number

Name

     Last   First   Middle

Present Address

     Street   City   State   Zip

Permanent Address

     Street   City   State   Zip

Phone No.

Referred By

Last   *First*   *Middle*

EMPLOYMENT DESIRED

Position   Date You Can Start   Salary Desired

Are You Employed Now?   If So May We Inquire of Your Present Employer?

Ever Applied to this Company Before?   Where   When

EDUCATION	Name and Location of School	Circle Last Year Completed	Did You Graduate	Subjects Studied and Degree(s) Received
Grammar School			☐ Yes ☐ No	
High School		1 2 3 4 5	☐ Yes ☐ No	
College		1 2 3 4 5	☐ Yes ☐ No	
Trade, Business or Correspondence School		1 2 3 4 5	☐ Yes ☐ No	

Subject of Special Study or Research Work

Activities Other Than Religious (Civic, Athletic, etc.)

EXCLUDE ORGANIZATIONS, THE NAME OR CHARACTER OF WHICH INDICATES THE RACE, AGE, SEX, COLOR OR NATIONAL ORIGIN OF ITS MEMBERS.

Form M660-26NR   Printed in U.S.A.   (Continued on Other Side)   APPLICATION FOR EMPLOYMENT
© 1985 Wilson Jones Company

FIGURE 11.1 *Sample employment application*

FORMER EMPLOYERS List Below Last Four Employers, Starting With Last One First

Date Month and Year	Name and Address of Employer	Salary	Position	Reason for Leaving
From				
To				
From				
To				
From				
To				
From				
To				

REFERENCES Give Below the Names of Three Persons Not Related To You, Whom You Have Known At Least One Year

Name	Address	Business	Years Acquainted
1			
2			
3			

PHYSICAL RECORD Do you have any physical condition which may limit your ability to perform the job applied for? This question is voluntary, and any answers will be kept confidential.

In Case of Emergency Notify

Name	Address	Phone No.

I authorize investigation of all statements contained in this application. I understand that misrepresentation or omission of facts called for is cause for dismissal. Further, I understand and agree that my employment is for no definite period and may, regardless of the date of payment of my wages and salary, be terminated at any time without any previous notice.

Date   Signature

DO NOT WRITE BELOW THIS LINE

Interviewed By   Date

REMARKS:

Neatness		Ability	

Hired	For Dept.	Position	Will Report	Salary Wages

Approved   1.   2.   3.

Employment Manager   Dept. Head   General Manager

FIGURE 11.1 *continued*

RECORD KEEPING

Long before you actively seek initial employment in your chosen career or seek to change jobs, compile two files: one to contain all the information you can gather on future employment possibilities and the other to contain all of your educational and employment background information, materials for a portfolio of personal achievements, and prospective references.

Prospective Positions

In your first file include a list of all prospective employers, their addresses, and phone numbers. Also include a listing of all interesting job descriptions with information on working conditions, qualifications, job outlooks, earnings, and other related information that will help you to define your job search. Some sources for the job search file include:

1. ***Dictionary of Occupational Titles (DOT).***   A good place to start is to check your library, career center, or the Internet for this yearly publication of the U.S. Bureau of Labor, which lists in two volumes 12,000 types of jobs and their required job skills. You may measure yourself against those skills and take courses to obtain new skills. Here is an entry for a technical writer:

    131.267-026 WRITER, TECHNICAL PUBLICATIONS (profess. & kin.)

    Develops, writes, and edits material for reports, manuals, briefs, proposals, instruction books, catalogs, and related technical and administrative publications concerned with work methods and procedures, and installation, operation, and maintenance of machinery and other equipment. Receives assignment from supervisor. Observes production, developmental, and experimental activities to determine operating procedure and detail. Interviews production and engineering personnel and reads journals, reports, and other material to become familiar with product technologies and production methods. Reviews manufacturer's and trade catalogs, drawings, and other data relative to operation, maintenance, and service of equipment. Studies blueprints, sketches, drawings, parts lists, specifications, mock-ups, and product samples to integrate and delineate technology, operating procedure, and production sequence and detail. Organizes material and completes writing assignments according to set standards regarding order, clarity, conciseness, style, and terminology. Reviews published materials and recommends revisions or changes in scope, format, content, and methods of reproduction and binding. May maintain records and files of work and revisions. May select photographs, drawings, sketches, diagrams, and charts to illustrate material. May assist in laying out material for publication. May arrange for typing, duplication, and distribution of material. May write speeches, articles, and public or employee relations releases. May edit, standardize, or make changes to material prepared by other writers or plant personnel and be designated Standard-Practice Analyst (profess. & kin.). May specialize in writing material regarding work methods and procedures and be designated Process-Description Writer (profess. & kin.).

2. ***Occupational Outlook Handbook (OOH).***   Another helpful publication, also put out by the U.S. Bureau of Labor, analyzes 250 jobs by discussing the nature of the work, working conditions, employment statistics in the field, training requirements, job outlook, and salaries. Each of these comprehensive analyses may be five or six pages long so are not reproduced here.

3. ***Journal of Career Planning and Employment.***   Peruse this monthly journal for information on position requirements, resumes, cover letters, and all up-to-date articles on careers.

4. **Libraries.**   Ask the reference librarian to help you locate other occupational handbooks, government publications, and newsletters that contain information on openings and qualification requirements in your field.

5. **Internet sites.**   Search the Internet using keywords CAREERS or RESUMES and explore **www.mon.com, www.careermosaic. com, www.jobweb.com, www.aboutwork.com, http://phoenix. placement.oakland.edu.career/internet.htm,** and **www.yahoo. careers.** These sites have remained stable over time but are subject to change. You will find more than a million resources, including occupational profiles, guides to job hunting, company profiles, places to post resumes, and available job listings all over the world. Also search such sites as **Hoover Company Profiles,** which provides an overview and thumbnail sketches of more than 10,000 businesses detailing products or services, number of employees, and more. **America's Job Bank, Job Bank USA, National Resume Bank,** and the **World Wide Web Employment Office** provide similar information and will post your resume on the Internet. Many of these sites include bulletin boards where you may post questions about companies, positions, and salaries. If your newspaper offers a digital edition, check out all the information on job listings and employment tips. You can also access **The New York Times** classified sections as well as those of other major newspapers. Go exploring. Here is a copy of a position opening that was posted online within www.career.mosaic.com:

Racal-Datacom
Engineering

Title
    SOFTWARE ENGINEERS
Location
    FT. LAUDERDALE, FL 33340, USA
Description
    SOFTWARE ENGINEERS
    Opportunity In Fort Lauderdale, FL!

Join the winning team at Racal-Datacom, a $550 million leader in managed networking systems! We currently seek MTS 3 Software Engineers with a Bachelor's Degree in Electrical or Computer Engineering or equivalent combination of education/training/experience.

You will need at least 5 years related work experience within a high-tech manufacturing/research and development environment to include software design ("C" and assembly, embedded systems 68302/68360) experience within a data communications company.

In-depth knowledge of ISDN networking requirements, an understanding of overall project concepts, as well as project leadership experience are also necessary. If you have a focus toward the future, consider joining a corporation on the cutting edge! We offer highly competitive compensation and benefits including health/dental/life insurance, paid vacation, 401 (K) tuition reimbursement. For consideration, please forward a hard copy of your resume to: Racal-Datacom, Staffing Dept., PO Box 407044, Ft. Lauderdale, FL 33340-7044, FAX: 305/846-5025. Or you may e-mail your resume to us at:

racal_staffing@usa.racal.com

For more information on our products, benefits and vision, please visit our web site within CareerMosaic at:

http://www.careermosaic.com

Racal-Datacom is an Equal Opportunity/Affirmative Action Employer.

Please use our Online Response Form

Learn more about Racal-Datacom. Visit our home page.

6. **Civil service offices.** Call federal, state, county, and city civil service offices. These offices list government positions in such fields as engineering, building and construction, computer technologies, general technology, law enforcement, and parks and recreation. Some offices offer prerecorded recruitment and telephone tapes listing open positions.

7. **Help-wanted columns.** Read your newspaper classified section to determine the openings in your field. Note the number of openings, the job descriptions, the qualifications, and salary ranges. If you are willing to relocate and are able to obtain various out-of-town newspapers, search them for openings. Many major newspapers are online. List the companies and details for future reference.

8. **Classified telephone directories.** Thumb through the Yellow Pages of your city and other desirable cities. List the names of companies or agencies that might be prospective employers.

9. **Trade magazines and newspapers.** Buy or subscribe to magazines or newspapers that specialize in your field. These publications offer help-wanted sections.

10. **College placement offices.** Visit your college or university career service center. Most offices offer interest inventory tests,

career counseling, career guidance software, occupational briefs, recruitment brochures, and job files of local, state, and national full- and part-time positions. Hundreds of companies conduct on-campus interviews for those students who sign up on their postings.

11. **Interviews.** Actively arrange to speak to faculty in your major field and people already employed by companies that interest you. Call to arrange an interview with a top manager or director in a company in your area for career information; many top people will find the time to assist you. Experienced faculty, employed friends, and top personnel in local companies can offer invaluable, practical advice and employment tips.

12. **Employment agencies.** Many companies do not advertise openings but put their employment search tasks in the hands of public employment agencies and private executive search companies. Some agencies specialize in certain types of employment, such as electronics, computers, money and banking, and allied health fields. Search your Yellow Pages and make an appointment to discuss your employment opportunities. Incidentally, employers who hire you often pay the fee to the agency at no cost to you. Some agencies will post your resume online for you. Seek those agencies that can help you.

In summary, be informed and realistic about your job prospects and the companies that might hire you. Learn how your qualifications fit the needs of prospective employers and what you need to learn to make yourself more marketable.

Personal Portfolio

Include in this file information concerning your education and employment records, skills and achievements, and potential references.

1. **Education.** Compile a list of the names and addresses of all educational institutions you have attended: high schools (if you are still in college), community colleges, universities, trade and vocational schools, and military schools. Write down the inclusive dates (months and years) of your attendance. Include degrees, fields of study, major and minor, if any, grade point averages, skills obtained, honors, awards, and outstanding activities and leadership positions in organizations, and other telling achievements. Do not forget scholarships and honor societies.

2. **Employment.** Keep written records of your employment history, including the dates, company names, addresses, your position titles, and summaries of your duties, responsibilities, and skills obtained.

3. **Outside organizations.** List other community or collegiate organizations with which you are associated, officer and chairperson positions, and other pertinent data.

4. **Special achievements.** Each time you obtain a new skill, such as mastering a computer program, earning a certificate for achieving a new skill, earning a new license, or attending a management workshop or seminar, drop a note and the appropriate paperwork into your portfolio.

5. **Presentation portfolio of original work.** Include copies of your best professional writing class assignments or outstanding samples of writing done on the job. If you have published a brochure, newsletter, feature story, proposal, or special report, save it in your portfolio for a later presentation. Save also any stationery, envelopes, business cards, personal logos, and promotion materials for a project you have developed in school or on-the-job. Develop a presentation folder of these materials by placing them into a binder or album. The folder will demonstrate your ability to write, design, and organize your own materials. Plan to take the portfolio to interviews, and be alert to an appropriate opening to present the portfolio to demonstrate your abilities. Wait until you have been asked about your professional abilities. Do not just foist your portfolio on the interviewer.

6. **Prospective references.** List the names, addresses, and phone numbers of prospective references. These may include present and past employers, teachers and professors of courses in your chosen field, recognized leaders in your field with whom you have been in work-oriented contact, clergy, and other professional and experienced men and women who can attest to your skills, abilities, and personal attributes. Ask permission to use them as prospective references. If you are ready to search actively for a position, ask those whom you believe will write the most informed and positive references about you to actually write a letter, leaving the salutation blank. Keep these on hand to send them when requested, or give copies to your career center if you are going to have that center supply your references on request.

THE RESUME

There are several acceptable formats for an effective resume. Traditional resumes organize your information either *chronologically* or *functionally.* Alternatively, sometimes a *narrative* format is appropriate. A new format for *scannable* resumes is emerging so that companies may use computers to store and review resumes.

All resumes are divided into similar sections to provide a complete inventory of your objectives, qualifications, and experience. For easy preparation and reading try to contain your information on only one or two pages. You may use 10, 11, or 12 point type. The sections will include:

- Name, address, and phone number (plus fax number and e-mail address if appropriate)
- Career objective
- Summary of qualifications (optional, but desirable)
- Educational background
- Employment experience
- Special elements (honors, skills, and special activities); optional, may be covered in summary, education, and/or employment data
- References or reference statement

Employers skim resumes, sorting out those that are uninteresting or messy. Therefore, you want to present a letter-perfect resume, which, as a result of good overall document design and emphatic features, clear headings, and white space, not only is readable but also quickly highlights your strengths. You may border traditional resumes. Traditional hardcopy resumes should be prepared on either a word processor or a typewriter and then printed on an ink-jet quality or laser printer. Failing this, it could be quick-printed on heavy bond, white paper rather than photocopied.

Traditional Resumes

CHRONOLOGICAL RESUMES

Traditionally, resume information on education and employment is listed in reverse chronological order; that is, list your present employment and work back, and list your present or latest educational institution and work back.

Figure 11.2 shows a sample resume format. Document design is important to make the professional impact you desire. Use boldface type, bullets, and other emphatic features to make elements stand

Jane C. Doe

Street Address
City, State Zip
Phone: (000)-000-0000
E-mail: JDoe@aol.com

OBJECTIVE

Name of position or broader statement indicating long-term objective. Include statement if willing to relocate.

SUMMARY (Optional)

Number of years experience, overall strengths, outstanding qualifications

EDUCATION

Month, year to Present	**Name of present institution** Street address City, State Zip ■ Phrase on degree sought, major, and expected date of graduation, high GPA ■ [More bulleted listing of honors, activities, organizations, certificates earned, skills obtained, if not included in Summary or Special Skills]
Month, year– Month, year	**Previous schools** Street address City, State Zip ■ Bulleted types of degrees, diplomas, major courses, GPA ■ [Expansion of honors, activities, etc. if not included in Summary of Special Skills]
Dates	Miscellaneous educational experiences, such as company courses, correspondence courses, seminars, home study, computer training

EMPLOYMENT

Month, year to Present	*Name of Company* Street address City, State Zip ■ Position title ■ Bulleted phrases that amplify the duties performed, skills, promotions, awards, and, possibly, the reason for leaving. Stress action verbs.
Month, year– Month, year	*Name of Company* Street Address City, State Zip ■ Position Title ■ Bulleted phrases of amplification

SPECIAL SKILLS, AWARDS, AND ABILITIES (optional)

Dates	In reverse chronological order include scholarship honors and awards, scholarships and grants, languages, professional memberships, skills not otherwise listed, etc.

REFERENCES (include three or four if listing)

Name, Title Company or Institution Street address City, State Zip Phone: (000)-000-0000	Name, Title Company or Institution Street address City, State Zip Phone: (000)-000-0000

or

Available upon request from Career Center, Complete Address and Phone (or from self)

FIGURE 11.2   *Sample resume format*

out. Headings may be centered or placed flush left to the margin. The bracketed items are optional, depending on the position for which you are applying.

Whether you use a traditional or alternative format, do not include unnecessary personal data. **It is illegal** for employers to consider your height, weight, age, religion, sex, sexual preference, or political affiliations. Figure 11.3 shows a traditional chronological resume. As you read through it, notice how it fulfills the following requirements:

Names and addresses.   Decide whether to use a centered, flush left, or right and left page design. Type your full legal name. If fellow workers, teachers, or supervisors address you by a different name, include that in the middle in quotation marks (e.g., Francis "Frank" Jones). Include your street address, city, state, zip code, and phone number with area code. You may include both your home and business phone, or your school phone and home phone if you wish to be contacted during vacation or summer break. Include your fax number and e-mail address if you have them. Review the sample resumes in this chapter to examine a variety of layouts.

Employment objective.   Either center or place flush left the heading "Employment Objective," "Career Objective," or just "Objective." Use boldface type. The heading may be in all capitals or upper- and lowercase. Keep all of your headings consistent. Beneath or beside it write the exact position you are seeking in a brief phrase that spells out your short- and long-term objectives. If you are willing to relocate, include that statement along with your objective. Avoid full sentences.

Examples

Legal Secretary. Willing to relocate.

Position in Customer Service or Sales Field

Position in drafting leading to design responsibilities

Electronic Technician with opportunity for advancement into management

Entry-level position in data processing leading to computer programming. Willing to relocate.

Summary of qualifications.   A summary of your qualifications at this point is optional; however, if you have room in your format, it makes an impact on the reader. Use one of two formats. Either bullet your summary qualifications under the title "Summary" or "Qualifications" or use incomplete sentences like those you may have used in your Employment Objective.

Therese V. Lazzari-Lindo

410 Southwest 10th Street
Boca Raton, FL 37001
Phone: (561)-332-3166
Fax: (561)-332-5444

EMPLOYMENT OBJECTIVE

Associate position in psychological counseling with opportunity to utilize my skills with children, youth, and families. Willing to relocate.

SUMMARY

Eight years experience in children, youth, family counseling plus vocational training including behavioral modification, counseling, crisis intervention, group management, in-service training and supervision

EDUCATION

August 1997–
May 1999
	Florida International University
	900 Glades Road
	Boca Raton, FL 37002
	  ▪ B.A. degree in Social Psychology
	  ▪ Magna Cum Laude, GPA 3.75

May 1995–
July 1997
	Broward Community College, North Campus
	1700 West Cypress Creek Road
	Cypress Creek, FL 33328
	  ▪ A.A. degree in Liberal Arts, 1997
	  ▪ Courses in Psychology, Abnormal Psychology, Sociology, Speech, Technical Writing, Advanced Computers

EMPLOYMENT HISTORY

August 1999
to Present
	Florida Department of Health Rehabilitative Services
	Adolescent Unit
	1403 Northwest 40th Avenue
	Lauderhill, FL 33313
	  ▪ Children, Youth and Family Counselor
	  ▪ Counseling and care management of adolescents; monitoring progress in foster homes; advising foster parents as needed; referrals to community-based services; maintenance of records and working with the judicial system

January 1995–
June 1999
	Covenant House
	733 Breakers Avenue
	Fort Lauderdale, FL 33301
	  ▪ Case Manager, Interim Overnight Shift Manager
	  ▪ Counseling residents and families, resident supervision, crisis intervention, and maintenance of records

June 1993–
January 1995
	Head Start Program
	905 West Atlantic
	Boca Raton, FL 37006
	  ▪ Part-time Counselor
	  ▪ Social work and counseling parents and children

REFERENCES

Immediately available upon request. Contact me.

FIGURE 11.3   *Sample traditional chronological resume* (Courtesy of student Therese V. Lazzari-Lindo)

Sample 1

Summary

- Six years experience in counseling children, youth, and families
- Experienced in behavior modification, counseling, crisis intervention, group management, in-service training, and supervision
- Ability to do statistical operation, SPSS, business systems, computer applications, and office management

Sample 2

SUMMARY   Experience with manual writing and design, speech writing, article writing, interviewing techniques, and advanced computer applications, including online materials and presentations.

Educational background. If your educational preparation is stronger than your employment experience, develop this section first. Type the heading "Education." Beneath it begin with your most recent school and list all other schools in reverse chronological order. List the inclusive dates, including the months, in one column. In another column include the names of the institutions, addresses, cities, states, and zip codes. Cite your major field of study, degrees earned or expected, and graduation dates. Highlight your academic record (if high), awards, special activities, organizations, special skills, and job-related courses. Include only those achievements, however, which are slanted toward the job. A prospective employer in banking might be impressed if you were treasurer of your student government association but would not care if you were on the tennis team.

Grade point average (GPA) may be a concern of employers and may be presented in either of two ways. If you have a better than average GPA, then indicate both your overall average and your average in your major. It is most likely that you did better in courses within your major area of study than in most of your other coursework. Thus, if you believe your overall GPA might hinder your chances at getting an interview, put only your GPA in your major on the resume.

Employment experience. If your previous or present employment is more indicative of your qualifications than your educational background, place this section before your educational data. Type the heading "Employment" or "Work Experience." Present your information in reverse chronological order. List the inclusive dates, including months, in one column. Employers are often looking for gaps in both education and employment histories. A continuous record of employment or education suggests that you are a responsible individual with work-oriented goals. In another column include the names of the companies, addresses, cities, states, and zip codes. List your position titles and whether the work was part- or full-time. Use brief phrases with **strong action verbs**

to highlight your duties, achievements, and awards. Some suggested verbs follow:

accomplished	achieved	analyzed
adapted	collaborated	coordinated
communicated	compiled	conducted
completed	created	directed
established	implemented	invented
increased	initiated	instructed
led	organized	participated
performed	presented	proposed
reorganized	supervised	trained

You may want to list your reason for leaving each position (i.e., "Offered higher-paying position," "Returned to college full time").

Special elements.   In our technological society, employers are seeking applicants with a multitude of skills. Whether or not you have already summarized your qualifications, use an appropriate heading, such as "Awards, Skills, and Activities." Then list your honors and awards; special skills not already mentioned, such as multilingualism; special computer skills (Framemaker software, Desktop Publishing, Corel Presentations, financial spreadsheets); and accomplishments, such as publishing a book or article or mastering Braille. Don't forget teaching or tutoring. If you have not already included organizations (academic, professional, or community) in which you hold office or in which you are active, indicate them here. Remember to include volunteer work. Such participation shows professional involvement and concern. Likewise, professional licenses and certification pertinent to the job applied for should not be overlooked. Think carefully and highlight your positive attributes that have not been included under any other headings.

References.   Under or next to the heading "References," consider listing three or four personal references. These should not be close personal friends, but professionals with whom you have worked or with whom you are acquainted. Past employers, professors, and leading people in your community are best. In the modern job market, the need for quick hiring created by faster and faster job turnover has made immediate referral a necessity. You should have in your files or in a career center file letters of reference written before you distribute your resume. If you do not include the reference names and addresses in your resume, use a statement, such as "Immediate references available; contact me" or "References available upon request from the Career Planning and Placement Center" (or from your college or university), and give the complete address. Take along a file of references to interviews and present them if asked.

FUNCTIONAL RESUMES

If you are short on employment history but strong on skills learned in school or elsewhere, use a functional resume to review your skills in short paragraphs. Essentially, use the same headings as you would in the chronological resume, but move your skills up under your education section and cluster them under such titles as research, leadership, computers, and so on. Figure 11.4 shows a functional resume with a strong skills section.

Alternative Resumes

Occasionally an alternative format will serve you better. The very mature and very experienced person may choose a *narrative* format to cover fully past employment achievements. Many companies now use computers to find, store, and search through resumes. Because computers may not pick up the design elements of other resumes, a *scannable* format is required in this instance.

NARRATIVE FORMAT

Using the same headings as in a traditional resume, you may present your resume in a first-person narrative, much like an expanded cover letter. Use complete sentences to tell the story if your professional life is extensive. Again, cover the material in only one or two pages. This is a good format to use when writing skills are a major requirement of the job for which you are applying.

People in show business, art, photography, or music will present resumes that are quite different from those in most professions. Such persons will submit a narrative and lists of works in which they have performed, a list of professionals with whom they have studied, or a list of their artwork, publications, and the like. This list will often be submitted in a personal portfolio providing actual samples of their work. Persons in these specialized fields will learn more about special inclusions from their peers and from experience.

Figure 11.5 (see p. 312) shows an example of a narrative resume.

SCANNABLE FORMAT

Many large companies use computers to aid their employee search. Such companies may

1. Scan submitted resumes into their computers for future reference.
2. Search out online posted resumes.
3. Establish a database by skills.

PATRICIA D. TAYLOR
11701 Southwest Tenth Place
Fort Lauderdale, Florida 33325
Phone: (305)-475-9644
Fax: (305)-475-9666
E-Mail: TaylorP@icanect.com

OBJECTIVE	Medical laboratory technician at a modern, expanding hospital laboratory. Willing to relocate.
SUMMARY	▪ Three years training in advanced instrumentation, advanced medical laboratory techniques, advanced mycology, and other laboratory procedures
	▪ Experienced laboratory technician
	▪ Advanced computer skills including word processing, desktop publishing, Quicken, Windows 95

EDUCATION   August 1997   **Broward Community College (Central)**
to Present   3501 Southwest Davie Road

Fort Lauderdale, FL 33314

▪ Majoring in Medical Laboratory Technology; will receive Associate in Science degree in May

▪ Grade point average: 3.6 of 4.0

January 1997–   **Sheridan Vocational Center**
July 1998   5400 Sheridan Street

Hollywood, FL 33021

▪ Completed the NAACLS accredited, 12-month Certified Laboratory Assistant program

▪ Grade point average: 4.0 of 4.0

SKILLS   **Laboratory Procedures.**   Extensive practice in hematology with Coulter Model S-Plus, Models, Coulter Diff 3 System, and Coag-A-Mate

Perform urinalyses and serology procedures, whole blood calibration procedures, patient blood-drawing procedures, manual chemistry tests for Glucose, BUN, SGOT, Alkaline Phosphatase, Uric Acid, and Cholesterol

Communication.   Revised laboratory procedures manual for Sheridan Vocational Center. Presented numerous oral training sessions for first-term students. Excel in computer skills (document design, word processing, spreadsheets, Windows 95, and DOS). Top grade in Technical Writing class. Speak fluent Spanish and French.

EMPLOYMENT   August 1999   *Florida Medical Center*
to Present   5000 West Oakland Park Boulevard

Fort Lauderdale, FL 33313

▪ Laboratory Technician in Hematology

▪ Responsible for early morning start-up procedures

▪ Perform routine urinalyses and serology procedures

▪ Experienced in whole blood calibration procedures

SPECIAL AWARDS, ACHIEVEMENTS, ACTIVITIES August 1998—Full-time Brewster Scholarship Recipient, Broward Community College

Dean's List every term.

December 1998—Participant in National Medical Laboratory Technician Seminar, Baltimore, Maryland

June 1993 to Present—Volunteer in County Hospice, 872 Hollywood Boulevard, Hollywood, FL

REFERENCES   Available upon request from the Career Center, Broward Community College,

FIGURE 11.4   *Sample functional resume*

GREGORY WELLBORN
334 South Bonita Avenue
Pasadena, CA 91107
Phone: (202)-596-7081

OBJECTIVE/WORK SUMMARY

Officer in a technology firm. Am willing to relocate. Thirteen years experience in all phases of top management.

EXPERIENCE

NATURAL HOME PRODUCTS COMPANY   **January 1997 to Present**

As **General Manager,** I developed this subsidiary of Prepared Products Company into a full line marketer of all natural consumer packaged goods. My responsibilities included all market and product development, materials sourcing, and establishing independent operating and management control systems. The result was the successful development of an all-natural consumer products line and introduction in targeted metropolitan markets.

ALLIANCE PRODUCTS   **June 1992–December 1996**

General Manager   **June 1994–May 1996**

I undertook the turn-around of the company's unprofitable performance, developed and implemented a strategy to transform the company from a single product distributor to a multiline marketer of synergistic products in key market segments. My accomplishments included successful implementation of the above strategy by broadening the product base to 11 lines, improving average gross margins from 20% to 40%, reducing dependence on distributor sales from 80% to 10%, and maintaining sales volume during the transition period.

Vice-President—Finance and Operations   **January 1993–May 1994**

I established accounting and management reporting systems, negotiated credit lines and equity issues, oversaw production, and managed vendor relations for this new venture. My accomplishments included raising $300,000 in venture capital and developing a flex-time production system which reduced average production time from 2 days to $1/2$ day and eliminated overtime pay averaging 3% of sales.

BANK OF AMERICA   **December 1987–December 1992**

Marketing Director—Eastern L.A. Region   **January 1991–December 1992**

My responsibilities included incorporating 11 autonomous branches into one new region, developing and administering the region's first marketing plan. My accomplishments included obtaining $22 million in new loan commitments, and successful integrating of the branches' management reporting and credit authorization systems.

Assistant Vice-President—Corporate Banking   **March 1989–December 1990**

Commercial Loan Officer   **March 1988–March 1989**

My responsibilities included managing a $15 million portfolio and supervising a team of commercial loan officers.

THE SIGNAL COMPANIES   **June 1986–February 1988**

Acquisition Analyst—Entertainment Division

My responsibilities included all financial and marketing analyses of prospective med acquisitions.

EDUCATION

University of Southern California, 1986, M.B.A. with Finance and Marketing emphasis

SPECIAL ACTIVITIES

I am active in Door of Hope, Harambee Center, and Friendship Corner charities and have served on numerous community boards such as the Chamber of Commerce, Downtown Development Association, and County Industrial Group. Have served as Chairman of the Opera Society and the Philharmonic Symphony.

REFERENCES AVAILABLE UPON REQUEST

FIGURE 11.5 *Sample narrative resume* (By permission)

If you are applying to one of the Fortune 500 or other large companies, call to find out if that company uses computer-related options (scanning, online posted resumes, and/or databases) for filling job openings. If the resume is to be scanned on a left to right scanner, the format of your document must be amenable to scanning. Scanners will not pick up columns, fancy fonts, underlining, italics, or bullets. Shaded or colored paper is likely to decrease the contrast of print and background. If your resume is to be scanned, mail it between cardboard in a large envelope to avoid folds and creases, which will also distort type.

Whereas the stress in the traditional formats and in the narrative format is on strong action verbs to review responsibilities, the emphasis in scannable resumes is on **key nouns** and **adjectives** that reveal positions and skills that can be entered into a database. Some companies will provide you with a disk of listed terms from which to choose. Call the company to determine if such a disk is available.

Some examples of strong adjectives and nouns that you can use to present yourself in a positive and accurate manner follow:

administrator	analytical	fluent in French
capable	communicator	collaboration
consistent	competent	creative
dedicated	diversified	effective
experienced	efficient	exceptional
flexible	global	imaginative
innovative	integrated	motivated
multilingual	a negotiator	reliable
responsible	teamwork	well-traveled

Include a list of skills in noun form. A skilled computer specialist will include such terms as IBM computers, DOS, Windows 95, MacOS, Word, Excel, WordPerfect, FileMaker Pro, Framemaker, Kronos Time and Attendance, WebPage Design, and so on.

Amy Hanson, Texas Tech University technical writing professor, offers these tips for preparing a scannable resume:

- Select a white or light colored paper
- Use a standard size of paper
- Avoid using exotic typefaces, underlining, italics, small type, and graphics
- Fill your resume with keywords and technical skills
- Avoid folding your resume
- Send originals, not copies

Figure 11.6 shows a sample of a scannable resume.

Merrill D. Tritt
880 Northwest 82 Avenue
Plantation, FL 33324

Phone: (954) 370-0733
Pager: (954) 325-1111
E-mail: Hawk@aol.com

Employment Objective: Management in a quality organization that will utilize the knowledge and skills learned in college and those learned in past and present work experience.

Professional Experience:

Incredible Universe, Hollywood, FL, 1996 to present. Human Resource Manager/Payroll Administrator. Implementation, coordination, and execution of human resource programs and functions. Maintain an effective employment and recruitment process. Direct all training activities. Maintain performance appraisal and merit raise system. Administer and process weekly payroll and monitor payroll budget. Administer and coordinate corporate benefits programs. Supervise other human resource personnel. Participate in resolving employee relations issues.

Broward Community College, Downtown, Fort Lauderdale, FL, 1996 to present. Computer Instructor. Develop lesson plans and instruct students on IBM-based microcomputer applications. Compose quizzes and examinations.

Incredible Universe, Hollywood, FL, 1994 to 1995. Sales Associate. Responsible for sales of computer hardware and related peripheral devices. Designed and created department signs and advertisements. Effectively scheduled employees. Maintained item prices on IBM AS-400 mainframe computer system. Authorized and overrode changes to item prices. Developed and taught a semi-weekly training class on Internet Services.

Chase Federal Bank, Miami, FL, 1987 to 1994. Customer Service Representative. Established new accounts and authorized data entry requests. Completed monthly branch audits. Reconciled branch totals at end of each day. Developed lending opportunities in the community. Processed Home Equity Loan applications. Oversaw branch operations in the absence of other supervisors. Supervised tellers and authorized transactions. Trained newly hired tellers and customer service representatives.

Education:

Florida International University, Bachelor of Arts in Business Administration. Majors: Management and Personnel Management. Graduated Magna Cum Laude. Overall GPA: 3.758.

Scholastic Honors and Awards:

Florida International University: Recognized for having the highest GPA in the School of Management for the 1996 graduating class.

Member of Golden Key National Honor Society, The Honor Society of Phi Kappa Phi, and The Honor Society of Beta Gamma Sigma.

Miami-Dade Community College, Miami, FL. Certificate of Outstanding Academic Achievement.

Skills:

Strong organizational, leadership, and interpersonal skills. Technical knowledge of IBM-based microcomputer hardware. In-depth knowledge of DOS, Windows 3.x, Windows 95, and MacOs. Quick adaptation to software with proficiency in Word, Excel, WordPerfect, FileMaker Pro, Kronos Time and Attendance, and various Internet client software applications. Typing speed: 65 wpm.

References: Immediately available upon request.

FIGURE 11.6 *Sample scannable resume* (By permission)

COVER LETTER

A cover letter, also called a *letter of application* or a *face letter,* accompanies your resume and serves both as an introduction of yourself and as a strategy to interest employers sufficiently to read your resume and to grant you an interview. The letter should reflect your personality and emphasize your potential. While a model letter may be developed, each letter should be tailored to a particular prospective employer.

Each letter should be individually typed on bond paper of the same quality as your resume. Address your letter to a specific name, if possible.

Your opening paragraph must attract the reader's attention without being overly aggressive or "cute." You must indicate the position or type of employment you are seeking and refer to your enclosed resume. You may want to mention how you learned of the position and subtly praise the company. Consider these opening paragraphs:

> I am an ideal candidate for the electronics technician position that you advertised in the July 2 edition of <u>The Miami Herald</u>. My two years of experience in the field and an Associate of Science degree in Electronic Technology qualify me for employment with your progressive corporation. My enclosed resume amplifies my background.

> Mr. Jack Smith, Chair of the Data Processing Department at Port Washington Community College, has alerted me that there will be a computer technician opening in your computer maintenance group. As my resume details, my specific qualifications prove that I would be an asset to your organization.

Your body paragraph(s) should summarize and emphasize your specific qualifications and personal qualities as they relate to the position. Use an enthusiastic tone and project self-confidence. Mention your outstanding personal attributes. Consider this sample:

> I can offer you four years of experience in employment and education. I will earn my Associate of Science degree in radiology technology this May and have been employed as a medical assistant to Dr. James E. Perry, radiologist, here in Fort Lauderdale for two years. My personal attributes include determination, courtesy, and attention to detail.

The body material should be short, direct, and persuasive.

Your closing paragraphs should urge action on the part of the reader. Ask for an interview or suggest that you will phone shortly to arrange for an interview. If it is appropriate, seek additional information, applications, and so on. Mention your flexibility and availability. Consider these closings:

> May I have a personal interview to discuss your position and my qualifications? I will call your office on Monday to arrange a convenient date.

I would welcome your consideration for an entry-level position and can start work any time next month. Any information you may have regarding my prospects with your company will be greatly appreciated. You may reach me after 3:00 P.M. weekdays at the number listed on my resume to arrange an interview.

Ultimately, remember that your cover letter is an overall view of your personality, your attention to detail, your communication skills, your enthusiasm, your intellect, and your qualifications. Your cover letter and resume are usually all a prospective employer has to decide whether or not you will reach the next phase in your employment search process— the interview.

Figures 11.7 through 11.9 show the complete texts for three sample cover letters. Figure 11.7 is for an entry-level position; Figure 11.8 (see p. 318) is for a position requiring some experience; and Figure 11.9 (see p. 319) is for a position requiring extensive experience.

THE EMPLOYMENT INTERVIEW

The final step in the employment search is the interview. You may be interviewed by one or more people individually or in a group. Your first interview may be considered preliminary and is often called a **screening interview.** In this interview, numerous applicants are invited to speak with a number of interviewers, perhaps a search committee, as well as the officers and directors of the organization. The objective is to narrow the pool of candidates for more intensive follow-up interviews. If you are asked back for a second interview, the indication is that you are a good candidate for the job. This will most probably entail a **line interview,** during which you will be more intensively interviewed by fewer person-nel. Here you may meet other supervisors and the people with whom you will work if you are hired so that they can offer the interviewer opinions. It is essential that you are prepared, informed, gracious, and self-contained.

The "Do's"

It is at the interview that your prospective employer evaluates you—your appearance, your personality, and your abilities. Chapter 17 discusses a number of verbal strategies to employ in an information-gathering inter-view, other verbal transactions, and nonverbal messages that you may want to review quickly. To prepare yourself for the employment interview, here are a few tips:

1. **Be on time.**   Know the exact time and location of your interview. Arrive a few minutes early and state clearly your name and purpose,

7661 Hood Street
Troy, NY 12181
6 July 199X

Mr. Jerry Diener
Personnel Director
American Express Company
777 American Expressway
Plantation, FL 33324

Dear Mr. Diener:

I am an applicant for an entry-level position in your data processing
department. Dr. Ted Smith, head of the Data Processing Department
at Queens Community College, South campus, has told me about your
progressive practices, and I believe I can offer a substantial
contribution to American Express Company. My resume highlights
my qualifications.

My main experience is in the accounting field. To further my
understanding of the complete accounting cycle, I enrolled in 198X at
Hudson Valley Community College in Troy, New York. At that time I
received my first exposure to computers. I found that I have a
natural ability for the logical thinking that is required for a good
computer programmer. Because a career in computer programming
seems the ideal goal for me, I am now working towards my A.S.
degree in data processing and will be graduated this August.

I will call you on Wednesday, 13 July to arrange an appointment for
an interview.

Yours truly,

Sandra J. Burnette

Sandra J. Burnette
SJB
Enc: 1

FIGURE 11.7 *Sample cover letter for entry-level position* (Courtesy of student
Sandra J. Burnette)

5000 Griffin Road
Hickory, NC 28603
2 May 199X

Mr. J.W. Duran
Attorney-at-law
346 North Andrews Avenue
Hickory, NC 28603

Dear Mr. Duran:

Please consider my application for the position of legal secretary which you advertised in the 1 May edition of the Hickory Sentinel. I am the "trained professional who can perform a variety of office duties" you seek. My training in legal techniques, business law, and legal secretary practices plus considerable experience as a general secretary for Ace Construction Company definitely meet your qualifications. My complete resume is attached.

I will be graduated on 9 June 199X, from Catawa Valley Technical College with an associate degree in Legal Secretarial Science. In addition to the required courses of this program, I have studied word processing software techniques. My shorthand rate is 120 words per minute, and I type at 65 words a minute.

In addition to specialized training, I can offer you three years of experience in general office work, bookkeeping, and salesmanship. You will find me to be reliable, efficient, and personable.

I trust you will consider me for your position. May I have a personal interview at your convenience? I may be reached by telephone between noon and 5:00 P.M. at 583-4771.

Very truly yours,

Sharon Cates

Sharon Cates

SC

Enclosure

FIGURE 11.8 *Sample cover letter for a position requiring some experience*

410 Southwest Tenth Street
Fort Lauderdale, FL 33315

14 May 199X

Dr. Peter Parrado, Executive Director
Juvenile Services Program
3435 First Avenue South
St. Petersburg, FL 33711

Dear Dr. Parrado:

I am a Children, Youth, and Family counselor with the Foster Care Division of the
Florida Department of Health and Rehabilitative Services, and I was recently informed
by Tom Piz about your new six-month program for girls being opened in the Oakland
Park area. As I have a great deal of experience in dealing with the problems specific to
girls in the adolescent years, I am most interested in a position with your program. My
resume highlights my qualifications.

I have earned a Bachelor of Arts degree in Social Psychology and have a strong background in behavior modification and positive reinforcement techniques. My experience
with Covenant House was extremely positive, and working with Broward County's
troubled youths provided me with a strong insight and feeling for their very distinct
needs. My work with adolescents in foster care further broadened my affinity for working with youth and allowed me to develop further my skills as a counselor.

My nemesis, however, has always been that neither of these programs provided enough
time and depth to result in a sufficiently positive outcome for most adolescents.
Furthermore, limited resources and resulting high case loads greatly restrict the proper
use of behavior modification. This, along with the fact that I am very concerned with
properly helping these troubled youths, is the primary reason that I find your program
so positive and exciting.

I look forward to discussing employment opportunities with you as well as hearing more
about the details of your program.

Very truly yours,

Therese V. Lazzari-Lindo

Therese V. Lazzari-Lindo

Enclosure: Resume

FIGURE 11.9 *Sample cover letter for a position requiring extensive experience*
(Courtesy of student Therese V. Lazzari-Lindo)

as well as the name of the person who has established the interview, to the secretary, receptionist, or security desk guard. You have some control here. Although the time, place, and type of interview will be set by the interviewer, you have alternatives such as suggesting alternate times for the interview that may alleviate rushing, unpreparedness, or any diversion you encounter.

2. **Dress appropriately.**   Dress conservatively and neatly in a suit, white shirt and tie, polished leather shoes or a dark dress or suit, closed heels, and hose. You might wear just one article that is eye-catching and memorable, such as a striped tie, a lapel rose, or an unusual, but not showy, piece of jewelry.

3. **Be prepared.**   In advance, learn as much as you can about the company so that you can project your interest and ask informed questions. Take a copy of your resume along to refresh your memory in response to direct questions about your background. Be ready to answer questions about your reasons for wanting to work at this company, what you see as your strongest assets and your biggest weaknesses, what you would like to be doing in ten years. Take your personal portfolio along with you in case it becomes appropriate to show examples of your work.

4. **Modulate your voice.**   The acoustics of the room, the distance from the interviewer, and your own anxiety should be noted in order to modulate your voice to create a friendly, yet assertive, impression. If you talk too loudly or your voice is too high pitched, you will be offensive and boorish. If you talk too softly, you may appear "wishy-washy" or uninterested. Accept or ask for water or coffee if your voice becomes raspy. A few sips will relax your vocal chords.

5. **Show your personality.**   Be enthusiastic and excited about the job. Use your interviewer's name. Smile and shake hands firmly. If you are a woman, offer your hand first, for many males are still uncertain if it is proper etiquette to initiate a handshake with a woman. A handshake establishes confidence and warmth. Wait to be asked to be seated. Listen well. Know at least five things about yourself that make you unique in the marketplace, but do not brag. Ask open-ended questions about company policies, working conditions, advancement possibilities, and salary range. Let the interviewer know you want the job. Ask if your prospects of being hired are good.

5. **Be courteous.**   At the end of the interview thank the interviewer for the time given you. Do not dawdle; your interviewer is a busy person. Be prepared to shake hands again. Even if you are disappointed with the outcome, use your best manners. Leave decisively.

6. **Follow-up letter.** Write a follow-up letter to your interviewer expressing your appreciation for the time given to you and reiterating your interest and availability.

The "Do Nots"

Conversely, guard against making the wrong impression. Consider the following items to avoid:

1. **"Winging" it.** Remember your competition is coming to the interview prepared.

2. **Familiarity.** Avoid being too familiar with the interviewer. Ask how he or she would like to be addressed, and do not ask personal questions. Do not drop names; the name you drop may be a political or personal enemy of the interviewer.

3. **Nervousness.** Do not fiddle with your hands or rings. Do not smoke, chew gum, or suck on a mint. Keep your hands and pencils away from your face. Sit still instead of fidgeting. Breathe normally rather than in gasps or sighs. Sit in a comfortable, but not slouchy, position, and keep both feet on the floor with one foot slightly more forward than the other.

4. **Language.** Do not tell off-color jokes or use even mild profanity or overused slang. Avoid "ums" and "you know." Avoid pretentious words. Be as articulate as you are able.

5. **Questions.** Do not fail to answer questions, or at least acknowledge that you do not know the answer but would be interested in finding out.

6. **Endings.** Let your interviewer indicate when the interview is over. Do not mention next appointments, nor look at your watch.

Questioning Skills

Your resume and application will have presented basic information about you, but both your interviewer and you will ask questions. The interviewer will expect you to summarize your qualifications. In addition, the employer wants to learn about your potential, your expectations, your attitude toward the organization, and your personality and character. By thinking about the following questions that you may be asked, you can prepare oral responses that are complete and confident.

Interviewer's Questions

1. Can you tell me about yourself? Expand on your resume? (Give a summary of your employment, education, and personal background.)

2. Why do you believe you are qualified for this position? (Stress skills.)

3. Why did you select us in your job search? (Demonstrate that you have done your homework about the organization.)

4. What are your long-term career goals? What do you want to achieve in five years? Ten years? (Demonstrate that you have goals and expectations that are realistic.)

5. What are your strongest (weakest) personal qualities? (Name two or three each; negatives can be balanced against positives or recognized with a goal for improvement.)

6. How do you handle stress? How would you handle conflict between yourself and a subordinate (supervisor, peer)? (Summarize how you cope with personal stress, such as physical exercise, reading a book, gardening, and the like. Be prepared to answer how you approach problems.)

7. What have been your most satisfying and most disappointing school or work experiences? (Recite your accomplishments and stress how you overcame disappointments.)

8. What are your attitudes toward unions (absenteeism, punctuality, geographical transfers, shift work, weekend assignments, travel)? (Know in advance what the organization expects.)

9. Do you have any home or personal problems that may bear on your performance? (Be honest while stressing how such situations are being modified.)

10. What salary do you have in mind? Would you accept x amount? (Be prepared to give a minimum or a range, emphasizing need and goals. Accept or ask for clarification about work hours, promotions, and raise potential.)

Throughout the interview you should feel free to ask questions yourself. Seek clarification of any vague questions asked of you. Let your own line of questioning reflect that you are interested and serious about the organization. Decide which of these typical questions you want answered.

Interviewee's Questions

1. What are the organization's long- and short-term plans? Is the organization expanding or shrinking?

2. What is the potential for advancement in this position?

3. How does your organization support employee advancement? Are there educational opportunities? Grants? Training programs? Performance reviews?

4. What is the leadership structure? Will I have any decision-making authority?

5. What are the organization's relations with the community? Consumers? Related organizations?

6. What are the grievance procedures? Layoff policies? Leave policies?

7. What are your hours of operation? How much travel is expected? Do employees normally work many hours of overtime?

8. Are there any benefit provisions? (Insurance, pension plans, stock options, cars or housing provided, travel per diem, employee discounts, and the like.)

9. How is the housing market in this area? What cultural, recreational, and social opportunities are available in this city? How would you rate public transportation, schools, municipal services?

10. What are the wages and salaries? Are increases based on performance or indexes? What are the means of advancement?

Salvaging the Sour Interview

Sometimes, despite all of your preparation, the interviewer may seem uninterested in what you have to say, and you detect that you are not coming across well. Don't give up. It *is* possible to salvage a job interview that is not going well. Robert Half of Robert Half International, a corporate recruiting company, offers this advice:

1. Analyze the problem. Are you stressing the wrong information? Try to change the focus of discussion.

2. If the interviewer has missed your strongest points, don't be subtle; tell about them confidently.

3. If the interviewer seems rushed or preoccupied, offer to return when things are calmer.

4. Return to a previous question that you feel you may have fumbled and add any more necessary information.

5. Find a trigger of common interest in the office such as a book, art piece, or other item that you can discuss as an ice-breaker.

6. Be honest and ask the interviewer what you're doing wrong. Example: "I'm not sure I've impressed you enough today. I know I can do this job. Would you tell me where I'm going wrong so I can address your fears?"

7. Ask for the job. The most successful salespeople know the best way to get an order is to come out and ask for it. Example: "I like this

company and I would like to work for you. And if you hire me, I won't let you down."

A realistic examination of your employment prospects, a letter-perfect resume, a self-confident cover letter, and a thoughtful interview are going to get you the position you seek.

CHECKLIST

Resumes, Cover Letters, and Interviews

RESUME

❑ **1.** Have I compiled a record of my prospective positions by examining sources?

   ❑ The *Dictionary of Occupational Titles?*

   ❑ The *Occupational Outlook Handbook?*

   ❑ The *Journal of Career Planning and Employment?*

   ❑ Career handbooks, government publications, and newsletters in my field?

   ❑ Internet sites?

   ❑ Civil service offices?

   ❑ Help-wanted columns?

   ❑ The Yellow Pages?

   ❑ Trade magazines and newsletters?

   ❑ College placement office materials?

   ❑ Interviews with professionals?

   ❑ Employment agencies?

❑ **2.** Have I compiled a personal portfolio demonstrating my background?

❏ Education details?

❏ Employment details?

❏ Community involvement?

❏ Special achievements?

❏ Bound presentation portfolio of my best and original work?

❏ Prospective references?

❏ **3.** Have I decided on the appropriate format for my needs?

   ❏ Chronological?

   ❏ Functional?

   ❏ Narrative?

   ❏ Scannable?

❏ **4.** Have I included all of the proper sections with the necessary details?

   ❏ Name, address, and phone/fax/e-mail address?

   ❏ Career objective?

   ❏ Summary of qualifications?

   ❏ Educational background?

   ❏ Employment background?

   ❏ Special elements?

   ❏ References or reference statement?

❏ **5.** Have I used a pleasing document design on white bond paper with proper placement of sections, consistent headings and indentations, and appropriate boldface/italics?

COVER LETTER

❏ **1.** Have I prepared a letter with a consistent block or modified block format?

   ❏ Heading with date?

   ❏ Inside address?

❑ Proper salutation to a specific person?

❑ Body paragraphs?

❑ Complimentary closing?

❑ Signature?

❑ Enclosure notation?

❑ **2.** Does my first paragraph attract favorable attention by indicating the position, my source of learning about the position, subtle company praise, and reference to my resume?

❑ **3.** Do my body paragraphs summarize and emphasize my specific personal and professional qualifications?

❑ **4.** Does my closing urge action on the part of the reader?

❑ **5.** Have I reviewed my letter for an enthusiastic tone and projection of self-confidence?

INTERVIEW

❑ **1.** Is there a specific time and place scheduled; do I know exactly how to get there and how much time is required?

❑ **2.** Do I have the suitable clothing for a professional interview?

❑ **3.** Have I reviewed what I know about the company so that I can engage in intelligent discussion?

❑ **4.** Am I familiar with my resume and my strengths and weaknesses?

❑ **5.** Do I have a resume on hand as well as my personal portfolio?

❑ **6.** Am I prepared to ask significant questions?

❑ **7.** Am I prepared to be enthusiastic, mannerly, courteous, and calm?

❑ **8.** Do I know how to behave at the end of an interview?

EXERCISES

1. **Prospective Positions.** Compile an employment opportunities file for yourself. Include photocopies from the DOT, OOH, and other position requirements from journals, government publications, newsletters, Internet sites, civil service offices, help-wanted columns, Yellow Pages, trade magazines and newspapers, your college placement office, interview notes, and employment agency telephone or in-person interviews. Submit this file to your professor for analysis.

2. **Employment and Educational History.** Compile a file of your employment and educational background including all of the institutional or company names, complete addresses, degrees and achievements, position titles, responsibilities, extracurricular activities, honors and awards, GPAs, and three to five greatest character and skill strengths. Include military records, community organization data, and other pertinent records. Be prepared to submit this file in a folder to your professor for analysis.

3. **Presentation Portfolio.** Place in a binder or album your outstanding class or on-the-job writing projects plus any brochures, newsletters, feature stories, proposals, and special reports. Also include professional items, such as letterheads, fax cover sheets, business cards, banners, or other items that you have designed. Submit the portfolio to your professor for a personal critique.

4. **References.** Obtain permission and actual letters of reference from outstanding past employers, professors, and other professionals to submit to your teacher for analysis.

5. **Development of an Employment Section of a Resume.** Using the following information, develop an employment section of a chronological resume. Use headings.

    The applicant is seeking a position as an electronics technician with the opportunity for advancement. His first job was as an auto mechanic with Bird Ford Agency, 101 North Garfield Avenue, Davenport, IA, where he serviced cars and repaired radios for two years, January to January. (Add zip codes to addresses. Provide months and years.) Next he worked from February 199X to May 199X as an automotive electronics supervisor at Borg Corporation, 6025 South 50 Street, Philadelphia, PA. Third, the applicant is currently employed as a service manager for Ace Auto Company at 1102 Broad Way, Delray Beach, Florida, where he began work in June of 199X.

6. **Cover Letter.** Revise the following cover letter content to make it more assertive, specific, and persuasive. The applicant is seeking a position as a draftsperson with a large architectural firm.

    Pursuant to your recent ad, I am applying for a job with your company. I don't have much experience but am willing to learn. I will be graduated from college this June, and although I changed my major three times, I will earn a degree in drafting technology.

I have studied some very pertinent courses and held a number of part-time jobs with local architects, learning about materials, office procedures, and on-site supervision. I think I have a pretty good design ability and am a careful draftsperson.

I realize you want a more experienced person, but I would like to talk to you about myself.

WRITING PROJECTS

1. **Personal Traditional Resume—Chronological or Functional.** Write either a chronological or functional resume—whichever would be most suitable for you if you were looking for a particular position that you could fill immediately while continuing your college education. If you have solid employment experience in the field in which you seek employment, the traditional chronological format will probably be your choice. If you are short on experience but long on the skills that this job will require, select the functional. If you are devising your resume in response to a real job listing, copy the listing and attach it to your resume.

2. **Group Project—Narrative or Scannable Resumes.** Find two or three classmates who are pursuing the same field of study as you are. Assume that you all have extensive experience in your field. Review the duties of a collaborative team as discussed in Chapter 1.

   Step 1. Parcel out the following tasks: one member of the team should obtain copies of the DOT and OOH job descriptions and any other library materials that define the training requirements, job outlook, and salaries in your field. Another team member should obtain job postings from the Internet, local or out-of-town classified ads, and trade magazines or newspapers. The third member should check the college Career Center and call one or two employment agencies to obtain additional position postings.

   Step 2. Collaboratively, devise realistic details for a narrative resume and design it carefully. If you are responding to an actual position, attach it to the resume.

   Step 3. Using that same information, convert the data into a scannable format for possible posting on the Internet.

3. **Group Project—Cover Letter.** Each member of your three- or four-member team is to use his or her traditional format resume and write a cover letter to a specific person at a specific company, agency, or institution. Be brief, persuasive, and self-confident. Exchange letters with each member of the group and critique each other's work. Be specific about possible additions, deletions, rewording, and any other possible improvements. Peer evaluation is one of the most valuable tools there is to writing an effective cover letter.

VERBAL PROJECTS

1. **Group Project—Interviews.** Divide the class into two groups. Each member of each group will be prepared to submit his or her resume and cover letter to a member of the other group for a mock interview. Allow the interviewer to review the resume and cover letter. Arrange the desk and chairs so that class members can see clearly the interviewer and the interviewee. Then conduct the mock interview. Then change groups and have the interviewers be the interviewees. After each interview, critique the work. Remember these will not be perfect but an exercise toward honing your skills.

2. **Mock Interviews.** Invite a professional interviewer to your classroom or to your television recording studio to conduct mock interviews. If the studio is not available, arrange for someone to videotape the proceedings in the classroom. Usually someone from the Human Resources (Personnel) department of your college or someone from your Career Center, or any other experienced interviewer (a member of a faculty selection committee, perhaps) will be willing to give the time for the interviews. Have your professor present three or four resume/cover letter combinations to this volunteer so that he or she may become familiar with the material. Following the interviews, play back the videotape of each mock interview in order to critique the pros and cons of the interviewee. Be honest but tactful in your criticism.

NOTES

CHAPTER *12*

Brief Reports

DUFFY   by Bruce Hammond

S K I L L S

After studying this chapter, you should be able to

1. Appreciate the diversity and purpose of a wide variety of standard brief reports.
2. Explain the differences between analytical, evaluation, and recommendation reports.
3. List and define five common types of analytical reports.
4. List and define three common types of evaluation reports.
5. Incorporate the appropriate language, organization, and content into brief reports.
6. Devise appropriate formats including headings, visuals and graphics, and emphatic features for clarifying the content of brief reports.
7. Write a periodic or progress report.
8. Write a laboratory or test report.
9. Write an incident report.
10. Write a field report.
11. Write a feasibility report.

INTRODUCTION

Engineers, scientists, technicians—in short, all employees in an organization—must assume the everyday task of informal and formal report writing. Reports are the written record of the events and progress of work. They assist professionals and management in making decisions, or they may become the working documents to help all employees carry out a program.

Informal reports are usually short, tightly focused on one subject, and addressed to a single receiver or small audience. Formal reports, on the other hand, cover more complicated topics; contain title pages, tables of contents, extensive body content, appendixes, and the like; and are frequently addressed to overlapping audiences. This chapter addresses the briefer types of typical reports, and the following chapter covers the longer, formal report particulars.

RECEIVER/AUDIENCE

Professional and technical reports are characterized as intensely factual and objective and are written in an authoritative and serious tone. You should avoid opinionated and judgmental language even though your goal may be to persuade your audience to certain conclusions or actions. Consider this biased paragraph from an accident report:

> A minor accident occurred in the Data Storage Department recently. One of the departmental technicians carelessly overturned a cup of coffee on a microfilm machine, which resulted in a massive short circuit when the machine was turned on. Luckily, there were no personal injuries.

The emphasis here is on blame rather than on the facts of the incident itself. An improved version would emphasize the facts in a detailed and objective account:

> An explosion occurred on the second floor, east wing of Building C, Avtec Corporation, on 13 June 199X, at 3:42 P.M.
>
> The cause of the explosion was an overturned cup of coffee spilled by Technician Seth Deutsch onto an Acme 710 microfilm projector (Inventory #703-11201-8). The liquid penetrated the felt light seals and dripped onto the main power supply, resulting in an explosive short circuit when the machine was turned on by Technician Brenda Bailey.
>
> Although there were no personal injuries, damages to the machine and surrounding area total $1,480.00.

The revised version contains essential facts and presents them in an unbiased manner.

FORMAT/GRAPHICS

Brief reports are often presented on preprinted forms, in a memo, or in a letter format. If you are devising your own format use emphatic features, such as topical headings, capital letters, underlining, boldface print, variable spacing, numbers, and bullets to enhance and emphasize the data of even the briefest report.

Consider visuals and graphics to organize your data and to convey a clear message. Study your message to determine if informal tables, formal tables, pie diagrams, maps, charts, and other graphics will present your information more clearly than lengthy paragraphs of dry material.

An incident report detailing a work-related accident might include a brief location diagram and an informal table of repair or replacement costs. A progress report may call for a formal table of materials, costs, or scheduling dates. A lab or test report will best display testing results in tables or performance curves. A feasibility report may include the results of surveys presented in circle graphs, bar charts, or tables. Further, tables comparing features of optional equipment may be appropriate. Be alert to graphic possibilities that quickly convey vital information and are appropriate to technical writing.

TYPES OF REPORTS

You may classify typical reports in several manners. One approach is to group them as analytical, evaluation, or recommendation reports. The first two types are generally brief, while recommendation reports may entail much more detail, persuasive tactics, and supporting evidence.

ANALYTICAL REPORTS

Analytical reports, in contrast to evaluation or recommendation reports, emphasize findings without expressing much opinion. They are confined to facts and a few logical inferences. Further, they are objective and informative rather than persuasive.

Here is a list of typical analytical reports required in business or industry:

- **Work logs:** Control cards or sheets indicating work assignments, locations, and work performed accounts
- **Expense reports:** An itemized accounting of all expenses incurred while performing duties, such as promoting sales, attending conferences, or inspecting field progress
- **Request for leave reports:** Requests for vacation, work-related travel, sick leave, or other absences
- **Periodic or progress reports:** The written information on the status of a project
- **Laboratory and test reports:** The written results of laboratory or testing experimentation

Work Logs/Expense Reports/Requests for Leave

Many companies preprint forms for the types of reports that are requested most often, such as work logs, travel expense reports, and leave requests. Figures 12.1 through 12.3 (see pp. 334–337) show typical preprinted report forms. Considerations necessary for periodic/progress reports and laboratory/test reports follow.

Periodic Progress Reports

PURPOSE

A periodic or progress report provides a record of the status of a project over a specific period of time. The report or reports are issued at regular intervals throughout the life of a project to allow management and workers to keep projects running smoothly. A periodic or progress report

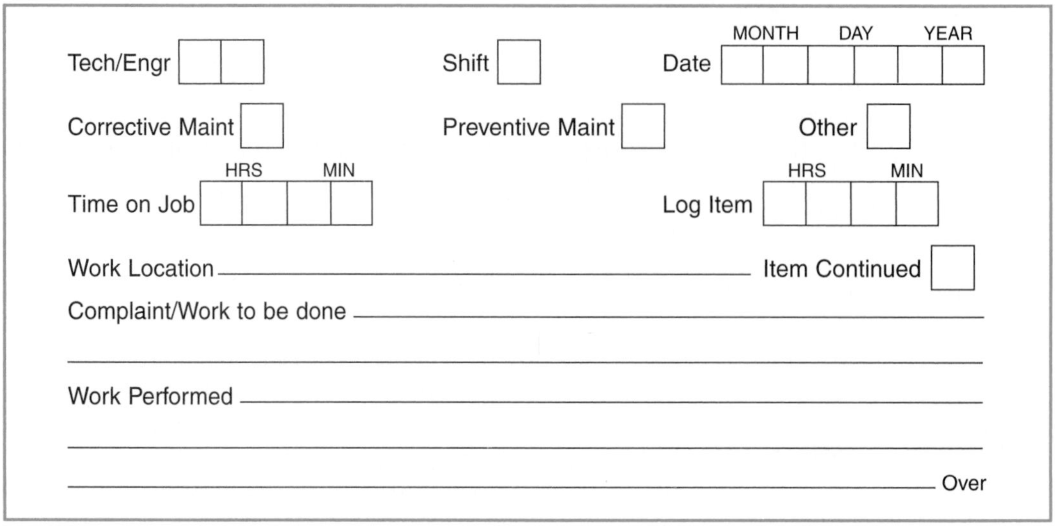

Tech/Engr ☐☐ Shift ☐ Date MONTH DAY YEAR ☐☐☐☐☐☐

Corrective Maint ☐ Preventive Maint ☐ Other ☐

Time on Job HRS MIN ☐☐☐☐ Log Item HRS MIN ☐☐☐☐

Work Location ______ Item Continued ☐

Complaint/Work to be done ______

Work Performed ______

______ Over

FIGURE 12.1 ***Typical preprinted work log report***

states what has been done, what is being done, and what remains to be done. The report usually reviews the expenditures of time, money, and materials; therefore, decisions can be made to adjust schedules, allocate budget, or schedule supplies and equipment. Often a company employs preprinted forms to report routine progress. Employees simply fill in the blanks at the completion of tasks. Figure 12.4 (see p. 338) shows a fire department's weekly and monthly Apparatus Report form.

Similar periodic reports may be required daily, weekly, quarterly, semiannually, or annually. They ensure uniformity and completeness of data.

ORGANIZATION

All narrative reports in a series of progress reports should be uniform in organization and format. A progress report covers

- A review of the aims of the project highlighting accomplishments or problems
- A summary or explanation of the work completed
- A summary or explanation of the work in progress
- A summary or explanation of future work
- An assessment of the progress

The accounting of work in progress may be chronological, topological, or even a combination of the two. A chronological organization covers

RYDER TRUCK RENTAL, INC.

TRAVEL EXPENSE REPORT

NAME _____

TITLE _____

WEEK ENDING _____

VENDOR NUMBER	REFERENCE NUMBER	LOC CODE	MO REC.

VEHICLE NUMBER	ACCOUNT NUMBER	AMOUNT
W/E DATE		TOTAL AMOUNT

AIR TRAVEL CHARGED TO RYDER
(FOR H.Q. AND REGION MANAGER USE ONLY)

TICKET NUMBER (LAST 3 DIGITS)	TICKET DATE	TICKET AMOUNT
		$

BUSINESS PURPOSE OF EACH TRIP

DATE OF DEPART.	RETURN	EXPLANATION (IF NOT CHECKED BELOW)	
			☐ VISIT TO COMPANY LOCATION ☐ SALES CALL
			☐ VISIT TO COMPANY LOCATION ☐ SALES CALL
			☐ VISIT TO COMPANY LOCATION ☐ SALES CALL

	DATE	SUN	MON	TUES	WED	THURS	FRI	SAT		TOTAL
DAILY ITINERARY	FROM									
	TO									
	TO									
AIR TRAVEL PAID BY EMPLOYEE										
CAR RENTAL*										
PERSONAL CAR EXPENSE (DETAIL ON REVERSE SIDE)										
ROOM*										
MEALS PLUS TIPS**										
COMPANY CAR EXPENSE (DETAIL ON REVERSE SIDE)*										
MICELLANEOUS (DETAIL ON REVERSE SIDE)*										
ENTERTAINMENT (DETAIL ON REVERSE SIDE)*										
TOTAL										

ACCOUNTING FOR ADVANCES

* ATTACH RECEIPTS
** INCLUDE ENTERTAINMENT MEALS ON REVERSE

IN ADDITION TO THE REQUIRED RECEIPTS AS SPECIFIED ABOVE, RECEIPTS FOR EACH EXPENDITURE OF $25.00 OR MORE MUST BE ATTACHED.

ADVANCE RECEIVED	
EXPENSES THIS VOUCHER	
BALANCE DUE COMPANY	
OR BALANCE DUE TRAVELER	

TRAVELER'S SIGNATURE

APPROVED BY (SIGNATURE)

6-22 (11/80)   SIDE ONE   10395   Litho By R In U.S.A.

FIGURE 12.2 *Typical preprinted travel expense report* (Courtesy of Ryder Truck Rental, Inc.)

PERSONAL CAR EXPENSE									
DATE	SUN	MON	TUES	WED	THURS	FRI	SAT	TOTAL	
ODOMETER READING-ENDING									
ODOMETER READING-BEGINNING									
MILEAGE									
LESS PERSONAL MILEAGE									
NET COMPANY MILEAGE									

AMOUNT AT _____ ¢ PER MILE

DATE	AMOUNT	VEHICLE NO.	COMPANY CAR EXPENSE (DETAIL BELOW)

DATE	AMOUNT	MISCELLANEOUS (DETAIL BELOW)

ENTERTAINMENT EXPENSE					
DATE	AMOUNT	TYPE OF ENTERTAIN-MENT	PLACE NAME, ADDRESS, OR LOCATION	BUSINESS RELATIONSHIP OF INDIVIDUALS OR GROUP ENTERTAINED (Give Name, Title, Etc., Include Names of Co. Employees.)	BUSINESS PURPOSE Date, Duration, Place, and Nature of Association Business Discussion or how otherwise related to active conduct of the business.

FIGURE 12.2 *continued*

REQUEST FOR LEAVE OF ABSENCE*
FIRE DEPARTMENT

NAME _____
Last   First

DATE _____

*1 Leave requests for vacation and anticipated leaves must be submitted two weeks prior to requested starting date.

*2 Leave requests for sick leave and emergency leave must be submitted by 9:00 A.M. the day before your shift works.

_____
Signature

LEAVE CODES

Department _____

V	—Vacation
S	—Sick with pay
FS	—Family Sick with pay
I	—Job Injury with pay
Z	—Sick without pay
W	—Personal Absence w/o pay
A	—Absence w/o leave
PG	—Maternity Leave w/o pay
FL	—Funeral Leave with pay (next of kin)
M	—Military Leave (calendar days)
CL	—Conference Leave
JD	—Jury Duty
SV	—Vacation from Sick
MV	—Management Vacation
CT	—Comp Time

_____ Days _____ from _____ to
Code

_____ (inclusive dates).

_____ Days pay in advance requested on

_____ (last shift worked).
Date

Regular pay checks of: _____

_____

Explanation _____

_____

STATION OFFICER'S SIGNATURE NECESSARY FOR SICK LEAVE

DISAPPROVED
APPROVED _____
Station Officer

DISAPPROVED   DISAPPROVED
APPROVED _____   APPROVED _____
Commander   Chief Officer

FORM AA-108 Rev. 4/82

FIGURE 12.3 *Typical preprinted leave request form* (Courtesy of Fort Lauderdale, Florida, Fire Department)

APPAR. NO. ___ RADIO NO. ___ MO. ___ YEAR ___ CO. ___ MAKE ___

WEEKLY APPARATUS REPORT

	1st Monday	2nd Monday	3rd Monday	4th Monday	5th Monday
Check pumps—Record vacuum test					
Check pumps—Record pressure test					
Were all drains flushed?					
Did connections or packing glands leak?					
Does relief valve or governor operate satisfactorily?					
Change Hurst Tool Fuel					
Check and flush foam pick-up					

	1st Tuesday	2nd Tuesday	3rd Tuesday	4th Tuesday	5th Tuesday
Resuscitator—Blood pressure of lowest cylinder					
Air Chisel—Record pressure of lowest oxygen cylinder					
Demand Regulator Mask—Record pressure					
Portable Spotlight—Running test					
Wench & Cord—Operating check					
First Aid Kits—Inventory check					
Hurst Tool & Compressor—Running test					
Tires—Record pressure					

Form AA-122 Rev. 8/77

FIGURE 12.4 *Sample periodic report* (Courtesy of Fort Lauderdale, Florida, Fire Department)

the work done by time (first A was done, then B, then C; work in progress includes first A, then B, then C; and remaining work includes first A, then B, then C). A topological organization covers the work done by task (Task 1, Task 2, Task 3, etc.). A combined organization covers the tasks within each time period (first time period: Tasks 1, 2, and 3; second time period: Tasks 1, 2, and 3, and so on).

Headings for the sections, however brief, are usually used. Consider these topical headings:

Introduction		Overall Goal
Work Completed		Work Completed
Work in Progress	*or*	Expenses Incurred
Work Remaining		Present Work
Appraisal		Future Work
		Conclusions

The major factor of a progress report is time. Other features, such as costs, materials, and personnel, may be incorporated into the appropriate sections or attached as support materials. If the reports are being prepared frequently during the course of a project, they are characterized by brevity—phrases rather than sentences and paragraphs. Figures 12.5 and 12.6 (see pp. 340 and 341) show progress reports in memo and letter formats.

Laboratory and Test Reports

PURPOSE

Students and employees in the fields of chemistry, data processing, fire science, electronics, nursing, and other allied health areas must frequently write laboratory or test reports. The reports present the results of research or testing.

ORGANIZATION

Typically the report includes

- Statement of purpose
- Review of method or procedure of testing
- Results
- Conclusions and recommendations

Purpose Statement. Here you clarify what is being tested and for what purpose (durability, colorfastness, safety factors, customer preference, and so forth).

MEMORANDUM

TO: Susan Niles, Law Department

FROM: Catherine Robertson, Executive Secretary *C.R.*

DATE: 12 May 199X

SUBJECT: Ace Company, Inc. Acquisition—Progress Report

INTRODUCTION

Here is the requested progress report on the Ace Company, Inc. acquisition.

WORK COMPLETED

- The stock purchase and Noncompete Agreements have been typed in final draft.
- Copies of the above have been sent to the Seller and acknowledged as received.
- Hotel and plane reservations have been confirmed by Downtown Travel Centre for arrival in Detroit on 25 June at 3:30 P.M. and a suite at the Downtown Hilton.

WORK IN PROGRESS

- Limousine has been requested from Detroit Limousine Service for transfer of all parties to the Hilton Hotel; awaiting confirmation.

WORK REMAINING

- Meeting with the Seller and John Galt Associates (law firm) to take place 26 June.
- Stock Purchase and Noncompete Agreements to be executed by all parties.
- Upon return, file on acquisition to be completed, and all documents to be filed as previously discussed.

APPRAISAL

Although this acquisition was originally scheduled for a 1 May closing, all work is up to date, all parties have been notified of the delays, and all parties are scheduled for the 25–26 June closing. With the final execution of all documents, this acquisition will be concluded.

CR:js

FIGURE 12.5   *Sample progress report in memo format*

Acme Advertising, Inc.
206 Lexington Avenue
New York, New York 10112
24 September 199X

Mr. Bert Campbell
Widget, Inc.
2552 Washington Street
Youngstown, Ohio 37602

Dear Mr. Campbell:
As you requested the following is a progress report on the West Coast
advertising campaign of your new product, Super Widget.

WORK COMPLETED

On 23 September 199X, thirty-second radio commercials began
appearing in the following markets: Portland, Seattle, San Francisco,
Los Angeles, and San Diego. As we agreed, the commercials aired on
local sports shows on one station in each of those cities.

WORK IN PROGRESS

On 1 October 199X, full page ads will appear in the Sunday morning
sports sections of newspapers in the thirty largest West Coast
markets. Regent Stores has agreed to special promotions for Super
Widget at all their retail outlets on 5 October 199X.

WORK REMAINING

Depending on the success of the first few weeks of sales of Super
Widget on the West Coast, we will determine our best approach for
national advertising. A target date for national television advertising
is 15 November.

APPRAISAL

So far our surveys reveal that the public is enthusiastic about the
new product, and if the West Coast is any indication, Super Widget
will be a financial success.

Very truly yours,

Nancy R. Merrell

Nancy R. Merrell
Account Executive

NRM:eq

FIGURE 12.6 *Sample progress report in letter format* (By permission)

Method/Procedure. Explain the particulars of your testing method.

Results. Present the test results. Use graphics for quick comprehension or comparison to preset standards.

Conclusions. Explain the implications of the test results and make recommendations.

A test report is usually less formal than a true laboratory report. The test report may be transmitted in a memorandum or business letter. Figure 12.7 shows a test report from an avionics engineering firm to determine the efficiency and cost effectiveness of a dimmer design for a light behind a liquid crystal color display on an airplane flight instrument panel. Although a layperson could understand it, the test report is written to an informed audience and includes many technical terms. It is clear that the *purpose* of the tests was to determine the better of two designs by comparison testing. The *procedures* used to conduct the tests are explained only to the extent that the reader needs to know. The *results* of the tests are reviewed and presented in text and graphics for clear comprehension, and the *conclusion* is presented in the final paragraph of the report.

Laboratory reports generally cover the same four considerations but are more formally structured. Figure 12.8 (see pp. 349–350) shows a report on the preparation of aspirin. It includes flowcharts of the main chemical reactions, potential side reaction, and the separation scheme. The procedure section of the report details the method used to prepare aspirin with attention to the equipment used. The conclusions are covered under separate headings.

Memorandum

Alltec Aerospace Company

Aircraft Avionics Division
3100 S.W. 22nd Avenue
P.O. Box 4582
Melbourne, CA 35821
Telephone (217) 338-5900

**Allied
Signal**

DATE: 9 June 199X

TO: Distribution

FROM: Robert Johns *R.J.*

SUBJECT: Evaluation of GAAD Dimmer Driven Circuitry

A parallel effort to devise a cost-effective and efficient dimmer design was undertaken by GAAD. The data which they presented to us was not relevant to the Boeing specifications. Moreover, the data could not be directly compared with previous experimental data relating to our design. Consequently, an experiment was designed to provide a direct comparison of the two designs so that we could determine the better design for our needs.

The following aspects of the design were compared:

- Efficiency
- Circuit performance
  —Operation over temperature
  —Dimming ratio
  —Light level controllability
- Lamp life
- Reliability of the circuit itself
  —Breakdown mechanisms
  —Part count
- Manufacturing costs
  —Repeatability
  —Part cost

Efficiency

Initially, it appeared that the efficiency of the resonant circuit was considerably less than the efficiency of the flyback design. This was due to the difficulty of measuring the voltage and current waveforms at the lamp itself. The determining factor, however, is the amount of light out for a given value of power in. The power dissipated in the circuit is not that important since most of the power in the lamp is reflected back into the LRU anyway. Figure 1 shows the light out versus power in for each circuit.

FIGURE 12.7 *Sample test report* (Courtesy Robert Johns, Senior Design Engineer)

Circuit Performance

Figure 2 shows the power needed to light the lamp at full power over a wide temperature range. Clearly, the resonant circuit produces the light more efficiently over most of the temperature range.

A comparison of the dimming ratio between the two circuits is illustrated in Figure 3. It should be noted that the quality of the light output associated with the flyback design was superior to that of the resonant design. That is, the resonant design shows some perturbations (barber-polling) at low light levels. Some of these perturbations may be tolerable in that they cannot be noticed after the light travels through the LCD. Both circuits require some modification of the reflector backplate to even out the capacitance between the lamp and the backplate. This is required to maximize the dimming ratio of the lamp from a 100 foot-lambert reference. Figure 3 indicates that the resonant supply is superior in dimming capability. However, this data is somewhat misleading because the flyback design has a discrete dimming control mechanism. (At low duty cycles, the duty cycle control does not have the resolution necessary.)

Lamp Life

It is our opinion, based on experimental data, that the lamp life/reliability is clearly a function of the peak to R.M.S. current through the lamp. Figure 4 presents the current peak to R.M.S. value for each light level at room temperature. In an effort to reduce the ratio, the resonant circuit has a dual method of dimming the lamp controls. The bursting is common to both circuits, but the resonant method operates at 100 percent duty cycle over much of its high brightness range. The light is controlled by varying the high voltage DC into the bridge circuit. This results in a current ratio near unity. At the lower duty cycles, it should be noted that the peak currents in resonant circuit are only one cycle in duration and much smaller in absolute magnitude (13 to 15 mA peak), whereas in the flyback design the high peak currents (200 + mA peak) last the entire duration of each burst on time. It is believed that this is more detrimental to the cathodes.

Circuit Reliability

No real time tests using the resonant circuit have been started. However, nine lamps are undergoing life test using the flyback design. Of these nine, one failed completely in less than 120 hours, and a second lamp is indicating deterioration of the cathodes.

FIGURE 12.7 *continued*

The greatest virtue of the flyback design is its simplicity with a total of fifteen parts. Although it presently does not meet the dimming requirements (poor resolution in dimming control), it could probably be modified to do so with relatively little added cost.

Manufacturing Costs

The primary cost in the circuit is the magnetics associated with the flyback transformer. I speculate that the cost of the parts for the flyback circuit will be about $60.00 per circuit. However, the circuit requires some improvements in the design. That is, it is very sensitive to any fluctuations in the 16-volt power input: the transformer fails catastrophically. Also, there is no protection circuitry in the event that the lamp fails.

The resonant circuit is more complex with approximately 100 parts costing about $110.00 per unit. The corresponding reliability is, therefore, reduced. However, the reliability of the lamp is the critical factor. Consequently, the increased complexity of the circuit was a tradeoff for increased lamp life. The circuit includes protection circuitry for all modes of lamp failures. Planer magnetics were used to lower costs through increased reliability and repeatability. Additionally, a phase locked loop was incorporated to compensate for component tolerances—both part to part and over time. The circuit requires no tuning or setting during manufacturing.

Conclusion

The overall conclusion regarding the preferred circuit is very much dependent on the final application of the unit being designed. In most applications, however, the life of the lamp is critical to the overall MTBF of the equipment. I, therefore, recommend that the resonant driver approach be used. While the cost and simplicity of the flyback design are beneficial, those factors are outweighed by the expected disadvantage of the limited lamp life.

Robert Johns

RJ/rj

cc: W. Antle, O. Tezucar, V. Frazier, G. Catsimpiris, L. Evans, S. Hammack, T. Mickie—GAAD, W. Tombs—BGCS

FIGURE 12.7 *continued*

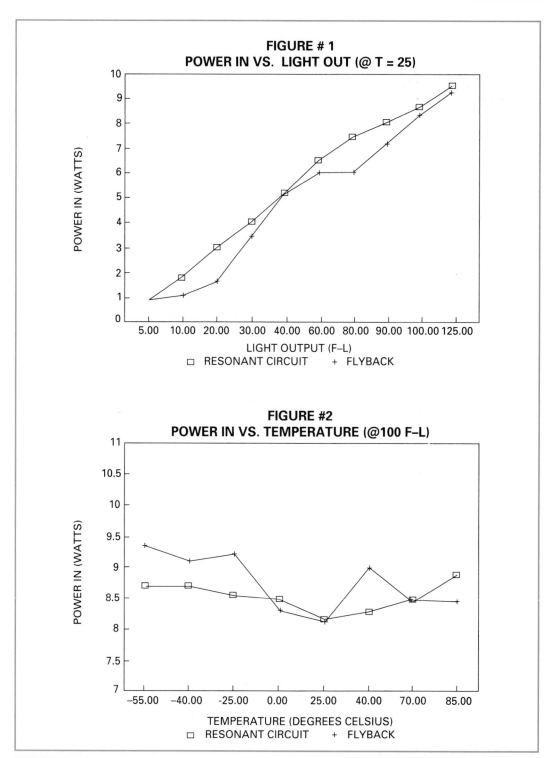

FIGURE 12.7   *continued*

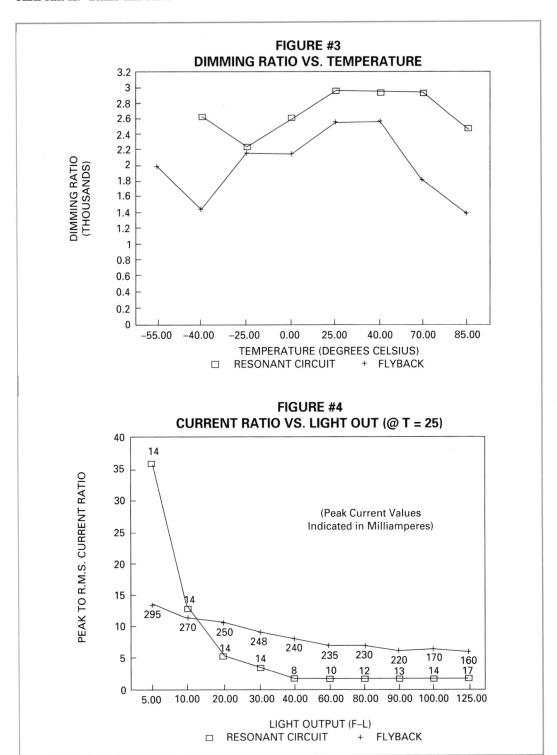

FIGURE 12.7   *continued*

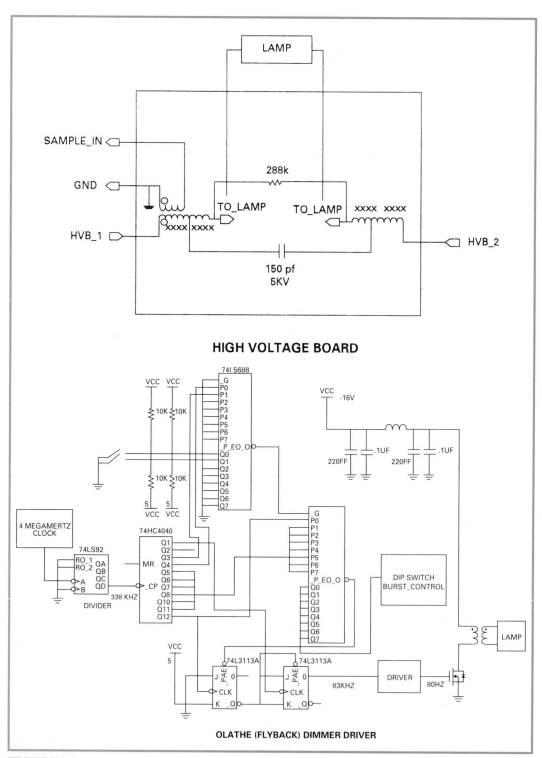

FIGURE 12.7   *continued*

THE PREPARATION OF ASPIRIN

Main Reaction

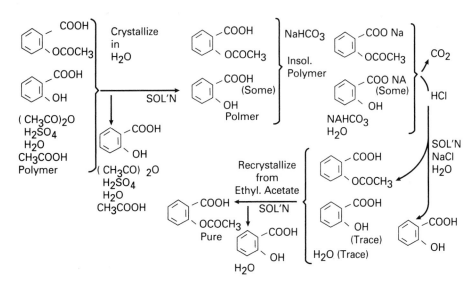

From Literature:

Salicylic Acid   M.P. ‖157@159®

Aspirin (Acetylsalicylic Acid)   M.P. ‖135@136®

Aspirin ‖ $C_9 H_8 O_4$ M.W. ‖180

Calculations $C_7H_6O_3$ M.W. 138

$$\left(\frac{1 \text{ mole Salicylic Acid}}{138\text{g. Salicylic Acid}} \right) = \frac{0014 \text{ Moles}}{\text{Salicylic Acid}}$$

(2og. Salicylic Acid)

$(CH_3CO)_2O$ $C+H_6O_3$ MW   102   Density = 108 g./ML.

The limiting reagant is <u>salicylic acid.</u>

Side Reaction

Separation Scheme

(CH_3CO)_2O
H_2SO_4
H_2O
CH_3COOH
Polymer

Procedure

Salicylic acid: Sample + Paper 3.12g.   2.00g. = 0.014 moles
                        Paper 1.12g.   138g./mole
                        Sample 2.00g.   Salicylic acid

The salicylic acid was placed in a 125 ml Erlenmeyer flask. Acetic anhydride (5 ml) was added along with 5 drops of conc. H_2SO_4. The flask was swirled until the salicylic acid was dissolved. The solution was heated on the steambath for 10 minutes. The flask was allowed to cool to room temperature, and some crystals appeared. Water (50 ml) was added, and the mixture was cooled in an ice bath. The crystals were collected by suction filtration, rinsed three times with cold H_2O, and dried.

Crude yield: Product + paper 3.07g.
                            Paper 1.15g.
                                1.92g. aspirin
Theoretical yield =(0.014 moles) (180g. aspirin/mole) = 2.52g.
Actual yield   = 1.92g.
Percentage yield = 76%

The crude product gave a faint color with $FeCl_3$. Phenol and salicylic acid gave strong positive tests.

The crude product was placed in a 150 ml beaker, and 25 ml of saturated NaHCO was added. When the reaction had ceased, the solution was filtered by suction. The beaker and funnel were washed with CA. Two ml of H_2O dilute HCl was prepared by mixing 3.5 ml conc. HCl and 10 ml H_2O in a 150 ml beaker. The filtrate was poured into the dilute acid, and a precipitate formed immediately. The mixture was cooled in an ice bath. The solid was collected by suction filtration, washed three times with cold H_2O, and placed on a watch glass to dry overnight.

Yield: Paper and product: 2.78g.
                    paper: 1.08g.
                  product: 1.70g.
Theoretical yield = 2.52g.
    Actual yield = 1.70g.
Percentage yield = 67% mp. 133-135°C.

The solid did not give a positive $FeCl_3$ test. The final product (CA. 075g.) was dissolved in a minimum amount of hot ethyl acetate. Crystals appeared. The crystals were collected by suction and dried.
MP. 135-136°C

FIGURE 12.8   *continued*

EVALUATION REPORTS

Evaluation reports are investigative in nature and emphasize strictly logical conclusions. They include an overview of a problem, an evaluation of findings, and a few conclusive evaluations or reasonable recommendations tightly based on the evidence. Evaluation reports may be spatial, chronological, or topical in organization. Some typical evaluation reports include:

- **Incident reports:** The written record of an unforeseen occurrence, such as accidents, machine breakdowns, delivery delays, cost overruns, production slowdowns, or personnel problems
- **Field reports:** The written evaluation of data to determine appropriate action, such as estimating real estate value, determining services costs, or establishing claims for damage
- **Feasibility reports:** The written evaluation of data to determine the practicality of future products, land development and environmental impact, expansion programs, new equipment, or services

Incident Reports

PURPOSE

No matter where you work, the unexpected frequently occurs. Such digression from normal operating procedure generally requires an incident report to supervisors or others to prevent the incident from recurring. The incident report is a written investigation of accidents, machine breakdowns, delivery delays, cost overruns, production slowdowns, or personnel problems.

The incident report may be reviewed when the next budget is planned if your evaluations involve finances. The report may constitute the basis of a longer proposal to improve procedures. It may even be used as legal evidence in a follow-up investigation. A carefully detailed report becomes part of the written record of what goes on in your place of work.

ORGANIZATION

The incident report adheres to fairly conventional organization. Its parts cover

- What happened (factual, not opinionated)
- What caused it (detailed and chronological)
- What were the results (injuries, losses, delays, costs)
- What can be done to prevent recurrences (evaluations)

In a lengthy incident report, it is a good idea to include topical headings, such as

Incident   Accident Description
Cause   Analysis of Causes
                    or
Results   Corrective Action
Evaluations   Evaluations

Your reader will be able to cull the appropriate information quickly by glancing at the headings.

Incident Description. In your introductory material, write a concrete statement detailing what happened. Include the exact date, time, and location. Personnel details, such as employees' names, titles, and departments, should be included. If personal injuries occurred, include the name(s) of victim(s), titles, and departments, or in the case of victims who are not employees, include home addresses, phone numbers, and places of employment. Describe the actual injury. If equipment is involved, identify it by including brand names, serial numbers, inventory numbers, or other pertinent descriptive detail.

Analysis of Causes. In this section write a chronological review of what caused the incident. Include what was happening prior to the incident and each step that caused the incident.

Results. In this section, explain what happened as a result of the incident, such as the action that was taken immediately. If anyone was injured, describe the extent of the injury and how, when, and where the person was treated. Explain who was immediately involved. This may include paramedics, police, repair experts, or extra workers. In the case of equipment failure, explain how it was required or replaced, how late deliveries were speeded, or how high costs were curtailed. Detail what was done to settle a personnel problem or to satisfy a customer demand.

The results section may require an actual or estimated expense review. Include a precise breakdown of medical expenses, equipment replacement, repair costs, profit loss, or other applicable costs.

Evaluations. This section should include concrete suggestions to prevent the incident from recurring. Consider what should be done, who should do it, and when it should be done. Include as much detail as your position authorizes.

Figure 12.9 shows a brief incident report illustrating logical organization and detail. The incident, a fall and supplies breakage, requires a replacement breakdown, which is presented in an easy-to-read continuation table.

Figure 12.10 (see p. 354) shows an incident report concerning an environmental problem.

MEMORANDUM

TO: James Friedman, Supervisor
FROM: Cathy Venci, Night Manager *CV*
DATE: 6 October 199X
SUBJECT: Busboy Fall Near Kitchen Entrance, 6/10/9X

On Sunday, 6 October 199X, at 8:15 P.M. Mike Sullivan, busboy, fell at the kitchen entrance located in the rear of the dining room. Mr. Sullivan was not injured, but there was considerable breakage of dishes and glassware. Figure 1 shows the exact location of the fall.

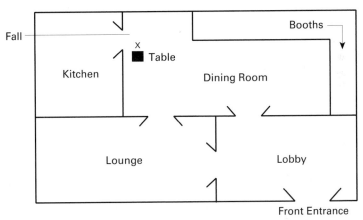

Figure 1   Exact location of incident

Prior to the fall, Mr. Sullivan had cleared dishes from three tables, as is customary. He was returning to the kitchen when his foot caught on the leg of the table next to the kitchen. The weight of the dishes he was carrying caused him to lose his balance and to fall forward. As he fell, the tray dropped, and the dishes broke.

As a result, the replacement of the dishes and glassware will cost $155.00, which includes:

> $42.00 for Anchor Hocking stemware (6 @ $7.00 ea.)
> 25.00 for Whitehall water glasses (5 @ $5.00 ea.)
> 48.00 for Sango dinner plates (6 @ $8.00 ea.)
> 30.00 for Sango coffee cups (6 @ $5.00 ea.)

In order to prevent such a fall from recurring, I recommend that the table next to the kitchen entrance be removed. This will give the busboys and waiters more clearance when entering and exiting the kitchen area.

CV:jv

FIGURE 12.9   *Sample brief incident report* (Courtesy of student Cathy Venci)

MEMORANDUM

TO: Enforcement Administration

FROM: Gerald Mills, Marine Resources *G.M.*

DATE: 28 April 199X

SUBJECT: NOV Request to Redress Complaint

On 10 February 199X, one Herbert Johnson, a resident of Beautiful Waters Condominium, Longboat Key, Florida, called to complain that his condominium's Board of Directors had ordered a contractor to bulldoze several small sand dunes and their associated vegetation on the beach fronting the property. Mr. Johnson indicated that the dunes had been flattened and the vegetation destroyed on 6 February 199X.

On 20 February 199X I received a visit from Mr. Johnson to present documentation that included the accompanying photographs and a copy of the contract between the Board President of Beautiful Waters Condominium and Coastal Construction. A visit to the site by Marine Resources Staff on 13 April 199X confirmed that no dunes or vegetation were present.

Following consultations with Board President Lee Santini and the Zoning Board, it was determined that the dune destruction was a violation of Section Z145773 and that Coastal Construction operated without a permit.

It is respectfully requested that NOV procedures be initiated in this matter. I will attend to this personally with your approval. Please contact me at (561)-542-3455 if you have questions or need additional information.

GM:acd

FIGURE 12.10 *Sample incident report* (By permission)

Field Reports

PURPOSE

A field report presents an analysis of a location, site, or situation to record and determine appropriate action. Realtors prepare field reports on undeveloped, commercial, and industrial properties to determine the value and prospects of such properties. Service people from every field inspect property to determine costs and plans for building on, improving, or repairing the property. Firefighters, health inspectors, and others report their work in the field. Insurance adjustors inspect sites to establish claims for damages.

ORGANIZATION

Often preprinted forms with space for narrative reporting are used. Many companies devise organization formats in outline form to be followed by their reporting personnel. These regulating outlines list specific topical headings along with brief instructions about the data to be included under each.

Despite the diversity of headings required for any one specific field report, all such reports include

- Essential background data
- Account of the field inspection
- Analysis of findings
- Conclusions and recommendations

Background.   Explain what is being investigated and for what purpose (safety or health factors, cost determination, need for improvement, validity of claims, etc.).

Account of Inspection.   Clarify what the field investigation found.

Findings Analysis.   Interpret the significance of the findings as related to the purpose of the investigation.

Evaluations/Conclusions.   Detail what action should be taken.

Figure 12.11 (see p. 356) shows a partial fire department company run report. This second page contains a narrative report about what happened in the field. Figure 12.12 (see pp. 357–358) is a field report concerning a site inspection to determine a potential violation. Figure 12.13 (see pp. 359–361) is a field report by an insurance field claim adjuster.

OPERATIONS AND COMPANY AND STORY OF FIRE

Give a complete account of the work of the company. The sequence of operations and where the company worked (floor of building, for example) should be clearly recorded. To the best of your recollection show where each man worked and what he did. Describe where fire apparently started and from what cause; what material burned; progress of fire upon arrival of company; and, any dangerous condition of the building. List all company personnel responding. List all injury and/or fatality information.

Condition of doors or windows (locked or open, for example); any unusual presence of flammable or combustibles; or, obstacles placed to hamper the fire department deliberately should be noted. This information should so far as possible be coupled with the name of the man who encountered the condition so that he can be witness to it if necessary in connection with further investigation of the fire.

Also, utilizing Form 901-J, make sketch to show the relative positions of pumper, hydrant, ladder or other apparatus with respect to the building afire. Show hose lines manned by this company and their approximate lengths.

R-5 Crew: Lt. Sicliri and Ferranti
Man Min: 140
Operation:

This unit was dispatched to a reported structure fire at topic address. Upon arrival on the scene District Commander One told me to bring a second line to the fire floor. We pulled a 1½ in. line free from engine one S and advanced it to the fire floor. This line was not needed at this time. Driver Ferranti brought smoke ejectors and a pike pole to the apartment involved. DC-1 told me to check for fire extension. I entered the attic and crawled the length of the building checking for fire. Finding no fire in the area, I reported back to the Commander. Commander Zettek then had my unit stand by in reserve. When the area was safe, we cleared the area and returned to quarters.

FIGURE 12.11   *Second page of fire field report* (Courtesy of Fort Lauderdale, Florida, Fire Department)

MEMORANDUM

TO: Ken Vordermeier, Vice-president
FROM: Jane Ellyson, Realtor *J.E.*
DATE: 22 September 199X
SUBJECT: Unimproved Land Inspection

On Friday, 20 September 199X, I inspected 3.4+ acres of unimproved land fronting on Powerline Road and Northwest Twentieth Street, Boca Raton, Broward County, Florida, to determine its potential and to obtain the seller's listing.

DESCRIPTION

Legal—Hillmont Middle River Vista, Replat of a portion of Plat Book 59, Page 188, Parcel B less pt., Desc. in CRS 3197/876, 3195/834, 433/711, and 4638/335, less approximately 1.049 acres conveyed to McDonald's Corporation and less approximately 0.358 acres to be conveyed as Northwest Twentieth Street.

Style code: 97
Area code: 45
Tax number: 9228-13-002

Price—Listing price is $600,000.00. The property is free and clear, and the seller desires cash. The price per acre is $174,240.00 or $4.00 per square foot. Sale lease back is not available.

INSPECTION

This vacant property is 3.4+ acres, 150,000+ square feet, with 450 to 500 feet of frontage and irregular depth. The property is waterfront with the property line following river bank.

It is zoned for commercial land use and platted. No rezoning is required. A survey is available in the listing office. No improvements have been made.

Minor clearing and/or grubbing, and fill will be required. The need for easements and restrictions are unknown at this time.

FIGURE 12.12 *Sample field report* (Courtesy Vordermeier Company, Realtors; adaptation)

The property is accessible to Port Everglades waterport, Fort Lauderdale/Hollywood airport, and the main highways: I-95, Florida Turnpike, and Oakland Park Boulevard. No railroad access is available. The property has access to electricity, city water, and sewers.

ANALYSIS

The surrounding area is developing rapidly. New Lake Pompano Park adjoins the property at the north end, and a fast-food restaurant is being built immediately south of the subject property.

The land offers an excellent prospect for a retail or service business.

RECOMMENDATIONS

1. Obtain the listing.

2. List on Realtron Computer service.

    Style code: 97 Commercial
    General Search codes: V2 Electricity
                             V3 Public Water
                             V4 Sewers or services
                             V5 Paved streets or sidewalks
                             V8 Major Road Frontage

3. Verify financing at time of contract.

FIGURE 12.12 *continued*

Feasibility Reports

PURPOSE

You may be assigned to look into a new project—a new product, the development of a new program, a relocation, the purchase of new equipment—to determine the practicality of the project. The feasibility report presents the evidence of your investigation and analysis plus your conclusions and evaluations.

Typically, feasibility reports analyze data to answer specific questions:

- Will a given product, program, service, procedure, or policy work for a specific purpose?
- Is one option better than another option for a specific purpose?

MEMORANDUM

TO: Robert Austin, Superintendent, Ace Business Services, Inc.

FROM: Gloria Dunn, Field Claim Adjuster *G.D.*

DATE: 26 March 199X

SUBJECT: Fancy Fast Food Corporation -#43101-08459
               Claimant Adele Clarke - Claim #172579

SUGGESTED RESERVE

Bodily injury - Adele Clarke - $10,000

FACTS

On 3/12/9X the claimant was a customer at the Fancy Fast Food
premises located at 1688 South Lake Drive, Erie, Pennsylvania. The
claimant was apparently bouncing on a chair at station 7 nearest
the kitchen when the left rear leg of the chair broke, collapsing the
chair and causing the claimant to strike her chin on the edge of the
table in the ensuing fall.

LEGAL VIOLATIONS

There are no known legal or code violations.

DIAGRAM

Enclosed is a diagram of the restaurant interior and station 7 showing the location of the table and chair at the time of the incident.

PHOTOGRAPHS

Enclosed is a photograph of the table and broken chair.

INSURED

The insured is Fancy Fast Food's licensee, 1688 South Ocean
Boulevard, Miami, Florida. The manager's name is Raymond Krick.
We have met with Mr. Krick, and he has indicated that this is the
first incident involving a collapsed chair.

FIGURE 12.13 *Sample field report*

The chairs and other furniture were purchased and installed in February 198X by Restaurant Suppliers, Inc. of Erie, Pennsylvania. The rest of the tables, banquettes, and chairs appear to be in good condition, and no defects are noted. His attached statement is self-explanatory.

BODILY INJURY - Adele Clarke

The claimant is ten (10) years old and lives with her parents, Mary and Jerry Clarke, at 7210 West Seventh Street, Johnstown, Pennyslvania.

Injuries

The claimant received two damaged teeth that were pushed completely up into her gums. The claimant was transported to Erie General Hospital emergency room where she was treated and then referred to an oral surgeon. The claimant will be having oral surgery including a bone transplant to correct the damage. The claimant's oral surgeon is Dr. M. C. Popper, located at 1400 South Belvedere Street, Erie, Pennsylvania. Since the claimant is represented by an attorney, we were not able to obtain a statement or medical authorization from the claimant's parents.

Damages

The claimant's damages are unknown at this time; however, we expect the medicals and specials to be several thousand dollars.

CLAIMANT ATTORNEY

The claimant is represented by Attorney Julie King located at 609 North Andrews Avenue, Johnstown, Pennsylvania. We have spoken to Ms King, and she is very cooperative and has indicated to us that the claimant seeks medical coverage because the Clarkes do not have any medical or dental insurance.

Although Attorney King did not make a definite statement about liability, she insinuated that a liability claim may be forthcoming.

WITNESSES

Enclosed is the statement of waitress Paula Kline, an employee of Fancy Fast Food, and it is self-explanatory.

FIGURE 12.13 *continued*

The claimant's mother, Mary, was seated in the booth when the accident occurred, but since she is represented, we have not obtained her statement. There were no other witnesses to the occurrence.

EVALUATION

At this time there appears to be a valid liability claim on behalf of Fancy Fast Food Corporation as well as claim to medical bills.

Once we have been able to obtain the medicals and specials from the claimant's attorney, I recommend we pay the medicals and attempt to obtain a Parent/Guardian Release for the amount of the medicals. Should a liability claim be pressed, I recommend we attempt compromise.

UNFINISHED ITEMS

1. Obtain medical and special costs

2. Determine if liability claim is forthcoming

3. Next report 4/30/9X

ENCLOSURES

1. F2-205 Form

2. Diagram

3. Photograph

4. Transcribed statement of Raymond Krick

5. Transcribed statement of Paula Kline

GD:jv

FIGURE 12.13   *continued*

- How can a problem be solved?
- Is an option practical in a given situation?

ORGANIZATION

The feasibility report usually includes

- Explanation of the problem
- Preset standards or criteria
- Description of the item(s) or subject(s) to be analyzed
- An examination of the scope of the analysis
- Presentation of the data
- Interpretation of the data
- Conclusions and evaluations

A brief feasibility report does not require headings; however, for a longer report consider these headings:

Background		Introduction
Standards		Problem
Options	*or*	Criteria
Method		Options
Data		Limitations
Conclusions		Evaluations

Background.   This section includes all introductory material, such as the purpose of the report, a description or definition of the question, issue, problem, or item(s). You may discuss the scope or extent of the report.

Standards.   Here you present a detailed explanation of the established criteria, aims, or goals of the question being investigated.

Options.   If applicable, present each alternative according to your established criteria. Consider costs, capabilities, procedures, personnel involved, required training, or other appropriate features of each option.

Method.   In this section explain how each option was analyzed to determine its practicality. This may include description of testing methods, survey instruments, research source material, qualifications of consultants, and discussion of limitations to your investigation.

Data.   In this section present the test results, survey results, or research findings.

Conclusions. Finally, the feasibility report summarizes the investigation, drawing logical conclusions. Discuss the limitations, if any, of your study. Build in a time schedule for action and a review of the results. In short, interpret the data and offer your recommendations.

Figure 12.14 (see pp. 364–365) shows a brief feasibility report. Figure 12.15 (see pp. 366–368) shows a longer feasibility report. This latter report addresses itself to solving a problem of abuses in a present bonus program. The organization remains much the same; the report presents the *background* and *problem,* the *standards* for an equitable bonus plan, the details of an *optional plan* along with its possible *limitations,* and the *recommendations* for adopting the new bonus program.

RECOMMENDATION REPORTS

Recommendation reports, in contrast to analytical reports (which emphasize findings) and evaluation reports (which emphasize conclusions), emphasize recommendations. Recommendation reports are strongly persuasive and urge action. Because they usually require extensive evaluation of the present situation or problem plus complex supportive data on the included recommendations, we consider them in Chapter 13, "Longer Reports, Proposals."

MEMORANDUM

TO: J. D. Big, Director, Accounts Department
FROM: Jane White, Secretary, Accounts Department *J. W.*
DATE: 9 May 199X
SUBJECT: Replacing Copy Machine

Inasmuch as we have been experiencing difficulties with our Atlas copy machine, I have investigated the feasibility of our renting a new copier. The Atlas costs $311.00 per month to rent. It does not make two-sided copies. In the past 30 days we have needed seven service calls costing a minimum of $30.00 a call. In April we were without copy capability for three days while a part was being located by the service representative.

Our criteria for a new machine include

- low cost (below $400 a month)
- two-side print capability
- various paper quality capability
- reduction capability
- immediate delivery

Only two copy machines are available within our price range. Table 1 compares the costs, capabilities, and limitations of Brand X and Brand Y:

Table 1   Comparative Features of Brand X and Brand Y Copy Machines

Brand	Rental Cost ($) Per Month	Capabilities	Limitations
X	316.00	• One step operation for two-sided print • Any paper • Reductions • Immediate delivery	• $35.00 base service fee • Reputation for frequent breakdowns
Y	389.00	• One step operation for two-sided print	• $75.00 base service fee • Reputation for slow delivery

FIGURE 12.14   *Sample brief feasibility report*

- Any paper
- Immediate delivery
- One-size reductions

Both copiers meet our criteria; the Brand Y one-size reductions are suitable. In order to assess the disadvantages, I surveyed five departments that use the X or Y copiers to determine the number of service calls required in a one year period. Table 2 shows the departments and number of service calls required for each:

Table 2   Service Calls for Brand X and Y Copy Machines in One-Year Period

Department	Brand	# of calls (one year)
Personnel	X	11
Payroll	Y	2
Data Processing	X	25
Records	X	7
Purchasing	Y	0

Both Brand X and Brand Y appear to be more reliable than our Atlas copy machine. Although Brand Y charges more ($75.00) for a base service fee than does Brand X ($35.00), the survey data suggests that Brand Y is the more reliable machine.

Therefore, I recommend that we rent a Brand Y copy machine, which will give us a two-side print capability that we do not presently have as well as meet all other criteria.

May I have your authorization by Friday to negotiate a Brand Y rental?

JW:eg

FIGURE 12.14   *continued*

NORTHEAST REGION MCS

BONUS PROGRAM

Feasibility Report

I. INTRODUCTION

The Northeast Region MCS Partners are considering the feasibility of discontinuing annual bonuses and of paying bonuses to professional staff on an individual basis throughout the year immediately following bonus-worthy events or conditions. In this way the bonus would be kept separate from salary considerations, and the reward would more closely relate to the event or condition.

II. BACKGROUND

For many years the Northeast Region MCS has had a bonus program for professional staff. Each staff member was eligible for an annual bonus payable 30 September. Whether or not he or she was paid a bonus and what the amount would be were determined at the time of the June performance evaluation. When paid, the bonuses ranged up to 15% of annual salary.

The bonus was meant to recognize and reward unusual contribution or difficult conditions during the previous year. It also reflected the economic conditions of the practice. In good times, total bonus payments were higher than they were in bad times. In one recent year, when operations were showing a loss, there were no bonuses.

The program was abused in two ways. To some extent partners and staff began to consider bonuses as regular recurring payments with only the amount subject to annual determination, and they sometimes were used as a substitute for salary increases that were not as permanent as a salary increase. Also, during wage controls, some payments that would have been salary increase in other times were awarded as bonuses.

FIGURE 12.15   *Sample feasibility report* (Courtesy of Charles E. Smith)

III. STANDARDS

An equitable bonus plan must

1. Motivate staff toward desirable activities
2. Reward unusual contribution
3. Compensate for difficult conditions
4. Make salary adjustments more representative of a staff member's overall performance

IV. OPTIONAL BONUS PLAN

Under the new program, bonuses will be awarded for the same reasons as before—unusual, meritorious contribution and undue hardship. Some examples follow.

A. Meritorious Contribution

- For any professional, a contract or other development work leading to securing a new audit client.
- For a manager, a self-conceived practice development program of self-initiated contact leading to a consulting engagement. (Bonuses should not be awarded to managers for an excellent sales record per se.)
- For any professional, the conception and eventual use of a unique approach to an engagement-related work task.
- For a staff consultant, completion of a clearly defined engagement work task in significantly less time than had been budgeted by a partner or manager.
- For a staff consultant, a contact or other development work leading to securing a consulting engagement.

B. Undue Hardship

- An extended period of travel away from home, defined as spending more than 75 percent of weekday nights away from home over a six-month period or spending one-half of the weekends in a four-month period away from home.
- A close working relationship for a month or more with intransigent, unreasonable, abusive, or otherwise difficult client personnel.

FIGURE 12.15 *continued*

- Uncomfortable physical working conditions for a month or more, such as at a remote, isolated community or in a noisy, dirty, hot, or cold facility.
- An extended period of weekend or evening work where total hours worked are more than twice normal hours for a month or more.

V. PROBLEMS

This bonus plan option may present some problems. Some members of the staff, both managers and consultants, are apt to direct their efforts toward earning bonuses, and thereby pay less attention to their regular professional work. Furthermore, they may embarrass the firm in doing so.

For example, a consultant may devote too much time on an engagement looking for a unique approach rather than following the work plan as laid out by the manager or partner. It is also possible that a manager or staff consultant may actively seek new audit clients at the expense of MCS engagements, or his activity in developing an audit client may conflict with plans of the general practice or infringe on professional ethics. There is also a remote possibility that members of the professional staff might try to encourage rather than allay intransigence in a client.

VI. EVALUATION

1. The optional bonus plan should be adopted for a period of two years because it provides for more equity than the previous bonus plan.
2. The Northeast Region MSC Partners should develop preliminary guidelines listing events and conditions and the amounts of bonus merited by each.
3. To provide initial equity any partner may propose a bonus for any staff member who may warrant bonus within the next six months. The bonus proposal will be discussed at each scheduled partners' meeting and approved or disapproved by the Regional Director. In this way the guidelines may be further refined, and all of the partners will develop a consistent point of view.
4. Subsequently, bonuses will be approved or disapproved directly by the Regional Director without joint discussion, but the justification for the amounts of all bonuses paid in the preceding period will be presented for informational purposes at each partners' meeting.

CHS:e

FIGURE 12.15   *continued*

CHECKLIST

Brief Reports

ANALYTICAL REPORTS

Periodic or Progress Reports

❑ **1.** Have I selected a subject that requires a record of the status of a project over a specific period of time?

❑ **2.** Have I reviewed the aims of the project highlighting accomplishments or problems?

❑ **3.** Have I provided a summary or explanation of the work completed?

❑ **4.** Have I provided a summary or explanation of the work in progress?

❑ **5.** Have I provided a summary or explanation of future work on the project?

❑ **6.** Have I provided an assessment of the progress?

❑ **7.** Have I designed a document with headings, emphatic features, and graphics, where applicable?

Laboratory or Test Reports

❑ **1.** Have I selected a subject that presents the results of research or testing?

❑ **2.** Have I provided a statement of purpose?

❑ **3.** Have I reviewed the method or procedure of research or testing?

❑ **4.** Have I presented the results?

❑ **5.** Have I presented conclusions and recommendations?

❑ **6.** Have I designed a document with headings, emphatic features, and graphics, where applicable?

EVALUATION REPORTS

Incident Report

❑ **1.** Have I selected a subject, such as an accident, machine breakdown, delivery delay, cost overrun, production slowdown, or personnel problem, suitable for a written report?

❑ **2.** Have I described completely what happened in a factual, not opinionated, tone?

❑ **3.** Have I explained what caused the incident with sufficient, chronological detail?

❑ **4.** Have I explained the results, such as injuries, losses, delays, costs, repairs, paramedic or other auxiliary personnel involvement, problem solving, and satisfaction of customer demands?

❑ **5.** Have I suggested exact measures to be taken to prevent recurrences of the same nature?

❑ **6.** Have I designed a document with headings, emphatic features, and graphics, where applicable?

Field Report

❑ **1.** Have I selected a subject that involves an analysis of a location or situation requiring action, costs, plans, improvements, repairs, and so on?

❑ **2.** Have I presented all of the essential background data, including what is being investigated and for what purpose?

❑ **3.** Have I clarified exactly what was discovered by the field investigation?

❑ **4.** Have I interpreted the significance of the findings as related to the purpose of the investigation?

❏ **5.** Have I presented a logical and complete evaluation and other conclusions?

❏ **6.** Have I designed a document with headings, emphatic features, and graphics, where applicable?

Feasibility Report

❏ **1.** Have I selected a subject that requires an analysis of data to determine if a given new approach is attainable, sound, and practical?

  ❏ Am I investigating a new product, program, service, procedure, or policy to suggest improvement?

  ❏ Am I evaluating one option over another for a specific purpose?

  ❏ Am I recommending how a problem may be solved?

  ❏ Am I investigating if an option is practical in a given situation?

❏ **2.** Have I provided a complete explanation of the problem?

❏ **3.** Have I established preset standards or criteria?

❏ **4.** Have I described the item or subject to be analyzed?

❏ **5.** Have I presented test results, survey results, or research findings to support my recommendations?

❏ **6.** Have I adequately interpreted the supporting data?

❏ **7.** Have I presented conclusions and evaluations for implementation, including limitations, if any, a time schedule for action, and related costs?

❏ **8.** Have I designed a document with headings, emphatic features, and graphics, where applicable?

❏ **9.** If a test, survey, or research was involved, have I presented the instrument(s)?

EXERCISES

1. **Progress Report.** Use the following information to write a progress report in letter format. Arrange the data in chronological, topographical, or combined order. Use headings and any other emphatic features and graphics that make your report clearer. Elaborate on the tasks as you see fit. Make up names, company name, and so on.

   Position: You are a training specialist who works for New World Communications. You are reporting to writing consultant Paula Kline, whose services have been contracted to conduct on-premises technical writing instruction at your company.

   Project: Technical Writing seminar to upgrade employee skills—scheduled 5, 6, and 7 March 2000.

   Data: a. Consultant evaluation forms to be devised.

   b. Negotiating luncheon menus with George Kelly, cafeteria manager; all three dates.

   c. Twenty registrants from 15 departments confirmed.

   d. Writing samples from each registrant being solicited to forward to consultant for evaluation before the instruction begins.

   e. Training Room C scheduled from 9:00 to 5:00, 5, 6, and 7 March.

   f. Certificates of completion to be designed and printed.

   g. Seminar announcements sent to 54 departments on 1 February.

   h. Printing and binding of 20 sets of instructional materials in progress.

   i. Overhead projector to be ordered for 6 March, 9:00–5:00.

   j. Wayne Salsbury to be notified to prepare introduction of Consultant for 5 March.

   k. Memos to supervisors explaining cost center billing sent on 15 February.

   l. Tables to be arranged in U for registrants on 5, 6, 7 March.

   m. Lectern for consultant to be placed in front of blackboard at the opening of the U arrangement.

2. **Incident Report.** Use the following information to write an incident report in memorandum format. You are a work supervisor reporting to the head of the department. Use the appropriate format, emphatic features, and graphics to make your analytical report clear and readable.

   Incident: Water pipe in ceiling burst; water damage to office equipment, carpet, and ceiling panels.

Where: Work Station A, Engineering Department

When: 12:40 P.M., Friday, 18 April 199X

Damage: IBM laptop computer soaked and short-circuited; requires overhaul maintenance and troubleshooting test.

IBM laser printer (serial no. 809-71-2654) exploded due to short circuit; requires replacement.

Carpet (20′ x 16′) needs replacement.

Three ceiling panels soaked; need replacement.

Water pipe repairs required by a plumber.

Costs: Computer overhaul: $1,000

Printer replacement: $2,500

Carpet replacement: $900

Ceiling panels replacement: $45

Plumber: $150

Further results: Estimated 1-week delay in preparation of proposal project in progress by engineer Thomas Waters.

Analytical analysis: Position engineer Thomas Waters into Workstation 5 for one week. The workstation does not include a printer, but he can reprocess information that was not on back-up or zip disks at the time of the incident.

3. **Test Report.** Write a three or four paragraph test report based on the following information. Indicate the test results in a table as well as in words. Include your conclusions and recommendations. Use other document design features as appropriate.

You are an engineer at Universal Testing Laboratories. You have been given four electronic pool alarms to compare for costs, attachment method and ease, alarm length, battery information, remote receiver features, connectors to home security system, and warranty information. It is known that some 500 people drown in backyard swimming pools each year. More than half are children under five. These pool alarms are triggered by waves from a fall into the pool. Each begins beeping within 30 seconds. Your testing reveals the following:

a. **Unit #1**—Cost: $160. Warranty: One year. Attaches to pool edge with Velcro-style pads. Can be turned off and left in place when using the pool. Alarm sounds for eight minutes and resets automatically. Can be set off by gusty wind and rain. Requires a nine-volt battery. Has a low-battery indicator. Has a remote receiver to sound inside the house. Can be connected to home security system. Not very prone to false alarms due to gusty winds and rain.

b. Unit #2—Cost: $260. Warranty: one year. Rests on the edge of the pool. Must be removed before people can use the pool. Alarm sounds for seven minutes and resets automatically. Requires nine-volt battery. Does not have a low battery indicator, but a weak or irregular alarm signals time for a new battery. Has remote receiver to sound alarm indoors. Can be kicked into the water accidentally. Not prone to false alarms due to gusty wind and rain.

c. Unit #3—Cost: $150. Warranty: one year. Floats on the water with a 16-foot tether to attach it the side of the pool. Must be removed before pool use. Alarm sounds for three minutes and resets automatically. Requires a nine-volt battery. Has a low-battery indicator. Has remote receiver to sound alarm indoors. Prone to false alarms due to gusty wind and rain.

d. Unit #4—Cost: $150. Warranty: one year. Floats on the water. Has a 26-inch lanyard to attach to a pool ladder. Must be removed before using the pool. Alarm sounds for two minutes and resets automatically. Requires a nine-volt battery. Sounds a weak alarm when time for a new battery. Has remote receiver for indoor alarm. Only one that requires a square-head (Robertson) screwdriver to install battery. Prone to false alarm in gusty winds and rain.

4. **Feasibility Report.** Use the following information to write a feasibility report in memorandum format. You are the Grounds Committee Chairperson reporting to the Golden Lakes Condominium Association. Rearrange the data into a logical organization. Present the data in appropriate graphics. Based upon the information, draw logical recommendations in your conclusion.

Problem: Remodeling of Golden Lakes Condominium recreation building has resulted in grass damage in common areas.

Fact: Rainy season begins 15 June.

Data: Luxury Landscape will require 3 days to resod at bid of $4,839. Guarantee includes six inspections in 4-month period with necessary sod replacement at no extra charge.

Landscaping Professionals will require 2 days to resod at a bid of $3,984. Guarantee includes six inspections in 6-month period with necessary sod replacement at no extra charge.

Green Company will require 3 days to resod at a bid of $3,707. No guarantee offered.

K-Mart Professional Crew will require 3 days to resod at a bid of $4,000. Guarantee includes six inspections in 12-month period with necessary sod replacement at no extra charge.

Criteria: Twenty-square-foot area needs resodding. Budget allows $4,500 expenditure. Guarantee required.

WRITING PROJECTS

Any of the following five projects may be handled by a group.

1. **Periodic or Progress Reports.** Select one option.

    a. In letter form, write a periodic report on your monthly expenses to your parents or spouse. Include a circle graph or bar chart on the percentage and the actual dollar expense of each category. Include, or comment on the lack of, the following categories:

Housing	School Supplies
Utilities	Insurance
Food	Charge Accounts and/or Car Payments
Transportation	Leisure
Clothes	Miscellaneous

    Include other categories as they are appropriate to your expenses. Conclude with an appraisal of your expenses.

    b. In memo form, write a progress report to your academic advisor showing your progress toward completing a degree or receiving a certificate or license. Begin with a statement of your overall goal and its requirements. Include tables of your courses, credits, grades, and grade point averages for courses completed, courses in progress, and courses remaining. Conclude with a discussion of your career plans.

    c. Write a progress report on a project in which you are involved either in school or on the job. Include an introduction and sections on completed work, current work, future work, and an appraisal of progress. Be alert to graphic possibilities.

2. **Lab or Test Reports.** Select two simple products (ballpoint pens, glues, paints, stepladders, brooms, car waxes, toothbrushes, garlic presses, etc.). With the purpose of determining which is the better product, devise a method or procedure to test each. Carry out your testing and then write a test report that states the object or purpose of the experiment, the explanation of the test method, a step-by-step analysis of the test and the results, and your analysis of which is the better product. The results section should offer the opportunity to present data in a table or performance curve.

3. **Incident Reports.** Write an incident report on a real or imaginary business/industrial accident for your "employer." Suitable subjects are damaged equipment, brief fire, broken merchandise, minor burns or sprains, collapsed shelving, broken windows or doors, and so forth. Write a concise description of the incident or accident. Next, write a sequential analysis of the cause. Follow this with a review of the results, and, finally, present your recommendations to prevent the

incident from recurring. Include graphics of the location and cost breakdowns.

4. **Field Reports.** On your campus or at your place of employment, select a location to conduct a field examination that will include conclusions and recommendations based on your evaluation of the efficiency or safety of the location. Suggested fields to investigate are

On the Campus	*At Work*
parking lot layout	room furnishings
cafeteria food arrangements	office layout
classroom layout	restroom facilities
registration procedures	lounge or coffee room
recreational areas	locker space
library study carrel layout	emergency exit doors/stairwells
campus bookstore displays	storeroom arrangements
a piece of equipment	a piece of equipment
a small structure	security arrangements

Write a field report structured to include the purpose of your inspection, the methods of gathering data, the facts and results of your investigation, and your analysis, which will make the chosen field more efficient or safer. Use appropriate headings.

5. **Feasibility Reports.**   Select one option.

   a. Write a feasibility report that analyzes a new purchase. State your purpose and the requirements. Then describe two or more probable alternatives (cars, office equipment, water beds, appliances, etc.). Next explain a method for evaluating the products (survey, testing, research). Present your findings. Evaluate the data, and conclude with logical recommendations for purchasing one of the optional items.

   b. Write a feasibility report that analyzes a procedure or problem in your place of employment. Identify the problem (the present means of advertising a product, scheduling personnel, awarding salary increases, providing in-service training, promoting personnel, handling tasks, or other similar procedures). State the standards that should be set to remedy the problem. Devise two solutions to the problem and analyze the merits and limitations of each. Interpret the feasible solutions to draw logical recommendations for implementing one or the other.

CHAPTER 13

Longer Reports, Proposals

HI & LOIS   by Chance Brown

Reprinted with special permission of King Features Syndicate.

S K I L L S

After studying this chapter, you should be able to

1. Identify the usual types of longer reports.
2. Identify the special features of longer reports.
3. Design title pages, tables of contents, and lists of illustrations.
4. Write transmittal correspondence.
5. Distinguish between and write several types of abstracts and summaries.
6. Attach appropriate back matter—supplements, appendixes, exhibits, and the like.
7. Understand the terms *internal, external, solicited,* and *unsolicited* as they refer to proposals.
8. Design traditional and streamlined formats for proposals.
9. Write a lengthy proposal with all appropriate features
10. Design long reports and proposals using all document design features.

INTRODUCTION

Although the need for conciseness is always present, a lengthy report is occasionally necessary. Quarterly and annual reports, long-range planning programs, system evaluations, and proposals are a few typical long reports that most organizations produce. Frequently, these reports fall into the recommendation report category in that they emphasize recommendations, are strongly persuasive, and include considerable supportive data. This chapter covers the special features of longer reports and reviews the particulars for a proposal, a widely used recommendation report.

PRESENTING THE LONGER REPORT

To avoid "gray material"—pages of dull, gray type—a longer report is distinguished by special features to make the information contained in it more accessible. These features may include

- Title page
- Transmittal correspondence
- Table of contents
- List of illustrations (tables and figures)
- Abstract or summary

- Body of report with topical headings
- Back matter

Title Page

Include an attractive and clarifying title page. This page should include

- A precise title
- The name, title, and company of the person(s) to whom the report is directed
- The name, title, and company of the writer(s)
- The date

A precise title, such as

**PROPOSAL FOR PURCHASE
AND INSTALLATION OF IONIZATION
AND PHOTOELECTRIC FIRE ALARM SYSTEMS
IN OCEANVIEW CONDOMINIUM UNITS**

is more effective than a vague title, such as

PROPOSAL TO DECREASE FIRE HAZARD

You have an opportunity here to use landscaped placement, multisize type, and a variety of fonts. Occasionally, modest art work is incorporated on to the title page. The first pages of Figures 13.7 and 13.8 (see pp. 397 and 407) illustrate basic title pages to longer reports.

Transmittal Correspondence

A letter or memorandum of transmittal accompanies most longer reports. The purpose of the transmittal correspondence is to orient the receiver to the long report or proposal in a suitable explanatory manner. It is usually very brief—three or four short paragraphs. It contains the following information:

1. The title and purpose of the report.
2. A statement of when and by whom it was requested or why it is being submitted.
3. Comments on any problems encountered (limited scope, unavailable data, deliberate omissions).
4. Acknowledgment of other people who assisted in assembling the report.
5. A statement eliciting feedback.

Do not include repetitions of the data from the actual report. These details will be covered in the **abstract** or **summary.** Figures 13.7 and 13.8 (see pp. 399 and 408) at the end of this chapter illustrate typical transmittal correspondence.

TABLE OF CONTENTS

Your reader(s) will want to be able to refer to sections quickly. A table of contents not only helps the reader(s) to turn rapidly to a particular section of the report, but also gives an initial indication of the organization, content, and emphasis of the report. A table of contents should accompany every written report that exceeds eight or ten pages and may be helpful in some shorter reports. Title the page *Table of Contents*. All headings used in the report are included in the table, and the subordination of sections is indicated by indentation. The starting page of each section is included as shown in the sample proposals in Figures 13.7 and 13.8 (see pp. 398 and 409).

List of Illustrations

If graphics (tables and figures) are used throughout your report, include their table and figure numbers, titles, and page references on the same or a separate page from the Table of Contents. Center the title *List of Illustrations*. List tables separately from figures. Figure 13.7 (see p. 399) contains a list of illustrations.

Back Matter

The back matter, or supplements to the body of the document, contains glossaries, tabulated survey results, financial projections, job descriptions and resumes of new personnel, equipment brochures, lists of references, and the like. These supportive materials are attached as exhibits or appendixes and are grouped in the back of the binder of the parent document. Should you include back matter, follow these guidelines:

- Number and title each supplement

  Examples: Exhibit 1   Interview Topic Outline

  Exhibit 2   Sample Policy Matrix

  Appendix A   Employee Safety Survey

  Appendix B   Resume of Donald C. French

- Refer to the exhibits or appendixes in the body of your text.

  Examples: Although I have interpreted the survey data here, the survey instrument and numerical tabulations are in Appendix C.

- Include the exhibit or appendix titles in your Table of Contents.

Figures 13.7 and 13.8 (see pp. 405–406 and 418–428) include supplemental back matter following the recommendations section of the sample proposals. Note their numbers and titles, their listing in the appropriate Table of Contents, and references to them in the text.

ABSTRACTS AND SUMMARIES

Abstracts and summaries, short and concise reviews of a parent document, describe basic information contained within the document and are vital parts of longer reports. There are two kinds of abstracts and two kinds of summaries: descriptive and informative abstracts and executive and concluding summaries. In practice, the terms *abstract* and *summary* are often used interchangeably but should not be.

Abstracts and executive summaries are written to orient the reader to the material to follow, yet they are written after the main document has been written. All are reduced representations of the parent document and reflect the purpose and content, scope, methodology, and sometimes the conclusions, of the parent document. Both kinds of abstracts and executive summaries are located in the prefatory section of a long report; concluding summaries are placed at the end of sections within the document.

Abstracts accompany scientific and technical reports and articles, whereas executive summaries accompany business and industrial reports written for executives concerned with resource management. In short, abstracts and summaries share a number of similarities but serve different purposes and require different writing strategies. The potential audience of a long report expects an abstract or executive summary to accompany it. After reading the abstract or summary, the busy professional or executive can decide whether to read the entire report or article. Because the reader may not possess the technical knowledge and language that your full report embraces, be careful to avoid technical terminology in your abstract.

Abstracts

An abstract for scientific and technical reports and articles is a brief *descriptive* or *informative* preview of a longer report. Abstracts, written in an objective, detached tone, present an overview of the complete, longer report by identifying the organization of the paper to come, the scope, the methodology, and sometimes the conclusions and evaluations contained in the parent document. The information is usually presented chronologically.

Libraries subscribe to abstracting journals that provide brief descriptive or informative abstracts of articles published in a number of journals in a specific field. Some of these are written by the author of the article, although others are written by professional abstract writers. In a research environment these abstracts allow the researcher to obtain an overview of content without a lengthy search for the article and thorough reading of it to determine its usefulness. The reader can review the abstract and then reject or pursue the original article.

DESCRIPTIVE ABSTRACTS

The *descriptive abstract* (also called the *indicative abstract*) represents the complete document and can be thought of as functioning separately from it. The reader of a descriptive/indicative abstract is usually an engineer, scientist, technician, or other expert. The abstract serves as an extended statement of the purpose, scope, and methodology used to arrive at the findings. A descriptive/indicative abstract is useful for a very extensive report because it indicates the organization of the report, although usually not the conclusions, results, or recommendations, which are covered in the parent document. The descriptive or indicative abstract is usually short (perhaps one to five sentences totalling 150–300 words). Figures 13.1 and 13.2 show two descriptive abstracts for extended technical reports:

Abstract

This paper addresses the ship, system and equipment design features, operational doctrine and training that has been developed to provide effective shipboard damage control. Both the ship and the sailor are addressed, since both are integral and interdependent components of the damage control "system." Also the enhancements afforded to protection of personnel from the effects of conventional and nonconventional weapons are discussed. Finally, a brief look into the future is presented. With the shrinking defense budget and corresponding reduction in fleet size and ship manning, novel system designs and automated decision aids will be required to do the job.

FIGURE 13.1 *Descriptive abstract for a 17-page technical report* (From *Naval Engineers Journal*, January 1992 [63])

Abstract

A review is given of the uses of programs written in languages that enable machines to respond to unrestricted conversational natural language input, and the special requirements are addressed. The operation and applications of one particular system, the Computer Assisted Socratic Instruction Program, or CASIP, are described. CASIP is a dedicated authoring language, comprising over 7000 lines of code. It is generally used as an interpreter that uses external data files of words, phrases, responses, and run-time instructions.

FIGURE 13.2   *Descriptive abstract for four-page article on voice-driven testing and instruction computer applications* (From *IEEE Journal,* March 1992 [57])

INFORMATIVE ABSTRACTS

The informative abstract distills the essential information from the parent document. An informative abstract is frequently confused with an executive summary, but careful consideration elicits the differences. Like the descriptive abstract, the informative abstract is intended for the technical reader or expert who is interested in the methodology and validity of the results covered in the longer document. In contrast, the executive summary is intended for management personnel interested in the definition of a problem or a proposal and the best solutions recommended to solve the problem or realize the proposal.

An informative abstract will accompany research-related documents, such as laboratory reports or systems evaluations serving not as an introduction but as a complete description of the information in the parent article or report, the scope of the research or experimentation, the methods of procedures used as well as the results, conclusions, and recommendations. The informative abstract is longer and more detailed than the descriptive abstract. It is typically 300 to 500 words in length. Figure 13.3 shows an informative abstract that includes the results and conclusions of a 10-page report:

Abstract

As part of a blind longitudinal study, 5,465 job applicants were tested for use of illicit drugs, and the relationships between these drug-test results and absenteeism,
turnover, injuries, and accidents on the job were evaluated.
After an average 1.3 years of employment, employees who
had tested positive for illicit drugs had an absenteeism rate
59.3% higher than employees who had tested negative
(6.6% vs 4.16% of scheduled work hours, respectively).
Employees who had tested positive also had a 47% higher
rate of involuntary turnover than employees who had
tested negative (15.41% vs 10.51%, respectively). No significant associations were detected between drug-test
results and measure of injury and accident occurrence. The
practical implications of these results, in terms of economic
utility and prediction errors, are discussed.

FIGURE 13.3 *Informative abstract containing essential information from the
parent, 10-page report* (From *Journal of Applied Psychology*, December 1990 [629])

Summaries

Summaries are also of two principal kinds: *concluding summaries* and
executive summaries. They differ from abstracts in that they are designed
for management rather than for scientists and technicians. In addition to
summaries to reports, busy executives often request assistants to present
summaries of newspaper and other articles so that the executive may
have access to the content of pertinent reading material in brief form.

CONCLUDING SUMMARIES

A concluding summary is most often found following a section or chapter
of a report or textbook. It is also used in instruction manuals. A concluding
summary about brief report writing that could be placed at the end of the
preceding chapter of this textbook is illustrated in Figure 13.4.

EXECUTIVE SUMMARIES

An executive summary is similar to an abstract in that both are prefatory
sections of a document and discuss what is to follow in the full text.
Executive summaries often deal with resource allocation and preface

Engineers, scientists, technicians, and other employees of an organization will often write formal reports. Brief reports are the written record of events and the progress of the company's activities and assist all personnel in making decisions. They usually focus on just one subject and are addressed to a single receiver or small audience. Their tone is factual, objective, authoritative, and serious. Some types of reports are written in specific formats, whereas others may be presented as memorandums or titled documents. Graphics and visuals often characterize the content.

Brief reports tend to be analytical or evaluative. Analytical reports include work logs, expense reports, requests for leave or transfer, periodic or progress reports, and laboratory and test reports. Evaluation reports include incident reports, field reports, and feasibility reports. Each type of report, analytical or evaluative, requires a statement of purpose, background information, specific organization of material, and, frequently, results, conclusions, and calls to action. An analytical or evaluation report differs from the recommendation report in that the latter are strongly persuasive in tone, tend to be considerably lengthier, and call for implementation of the final recommendations.

FIGURE 13.4 *Sample concluding summary*

long feasibility reports, proposals, and the like. Although they can be distributed separately, they usually precede the text. They serve to highlight the problem or proposal described, discuss its present implications, and offer solutions or recommendations to implement a program of action.

The **tone** of an executive summary is frequently persuasive; that is, it may be more personal and emotional than straightforwardly objective. Consequences resulting from inaction may be stressed. Appeals to forward thinking and fair decision making may be incorporated. Word choice may be more connotative.

The **organization** of material may be more causal than the chronological or procedural organization in an abstract. That is, the summary may consider first the most persuasive details and then the lesser ones.

The executive summary is typically from one to five pages in **length,** never exceeding more than 10 percent of the length of the original document. Figure 13.5 shows an executive summary from a 54-page home products business plan:

EXECUTIVE SUMMARY

Natural Home Products Corporation (NHP) was started in July 1990 by Mr. Thomas W. Lehmer and Mr. Gregory J. Welborn to develop unique nonfood or general merchandise products for distribution through Prepared Products Company's broker/distributor network in the United States.

Our specific focus is to take advantage of two predominant trends in consumer purchasing patterns: the demonstrated demand for greater convenience and better performance in consumer products, and the predominant interest in healthier and environmentally safer products.

It has been taken previously as an article of faith that product attributes that provide greater consumer appeal are incompatible with those features or components that are healthier or environmentally sound. We believe these attributes can be very compatible, that products can both provide greater convenience or performance and be health conscious and environmentally safe.

Over the last year, NHP has researched and identified three market segments in which these demands overlap, the growth rate is substantial, and the potential exists for greater-than-category-average profitability. The market segments are grilling products, home fragrances, and home hearth products. Furthermore, we have developed two products for immediate introduction for the grilling market and three additional products for the home fragrance and home hearth markets. At the end of 199X, our projections indicate net income of $3,515,586 (11.8%) on revenues of $29,675,071.

NHP has assembled a strong and experienced management team and an affiliate relationship with Prepared Products Company (PREPCO) allowing us to draw upon PREPCO's successful record in innovative packaging and its distribution strength in the grocery, convenience, and club store trade channels, as well as in food service. NHP is operated as an autonomous company because of its unique and specific focus and its need to cultivate favorable media attention. We are seeking a $750,000 investment for promotion, advertising, and working capital needs.

FIGURE 13.5   *Executive summary from a 54-page report*

We believe we offer a unique investment opportunity. NHP is at the forefront of two converging trends in consumer markets, has already developed five strong products, and has ready access to established distribution channels. NHP's conservative financial projections demonstrate substantial sales growth and profitability. The investment risk is minimized since utilization of the $750,000 investment will be dedicated to controllable expenditures, not to fixed assets or unpredictable Research and Development. Finally, NHP will be well positioned in three to five years for sale or an initial public offering.

FIGURE 13.5 *continued*

This summary includes background information on the company, a statement of focus, persuasive elements, the actual proposal to introduce three new products, costs and projected net income, management information, and projected consequences.

Figure 13.6 (see p. 388) shows another executive summary from a 45-page proposal to obtain venture capital (funds necessary for a new business undertaking) to establish an Internet advertising agency in a specific locale. Notice the persuasive terminology.

Writing Abstracts and Summaries

To write an abstract or executive summary, follow these steps:

1. Read the entire report to grasp its full content.

2. Estimate the number of words and plan an abstract or summary that does not exceed 10 percent of the original length.

3. If you are writing an abstract, decide whether it should be descriptive or informative. In either case identify the organization, scope, and methodology.

SUMMARY

Purpose—AdNet, an Internet Advertising Agency

The following three-year business plan has been created to obtain $60,000 in venture capital to establish AdNet, the first Internet advertising agency in the South Florida area. AdNet is designed to be a distinguished advertising agency offering a new type of advertisement that may utilize full-color pictures, CD-quality stereo sound, three-dimensional graphics, floor plans, and even customized movies, as well as ordinary text. It will be registered on Internet search utilities allowing prospective customers and viewers to find us quickly and easily. AdNet will also feature a "What's Happening, South Florida?" page that will highlight special events for that week. Additionally, links will be provided to other South Florida event pages. Internet subscribers from all over the world will be able to view AdNet's advertisements anyplace they have Internet access—at home or work, in specialized cafes and coffee houses, or even in their automobiles—24 hours a day, 7 days a week.

Capitalization

This three-year initial plan proposes a limited partnership business structure because of its ease of formation, flexibility, freedom from many governmental controls and regulations, and tax advantages. Three limited partnership units (LPUs), each priced at $10,000, will be offered yielding a $30,000 capital fund. The six general partners of AdNet will between them initially contribute $12,000 in capital; they will also contribute their salaries for the first six months of operation as reinvestment into the venture. Total capital funds will thus be $90,732. Each LPU will realize an 85 percent ROI (return on investment) before taxes in the three-year period. The financial forecasts included in this proposal, though conservative, show that this investment has significant promise.

Marketing Strategies

Marketing strategies are based on the fact that the Internet "population" doubles every nine months and the growth shows no sign of slowing. Each month approximately one million new

FIGURE 13.6   *An executive summary from a 45-page proposal* (Courtesy of Merrill Tritt)

subscribers sign on to this world wide system. There are currently more than 10 million subscribers. The target market is small- and medium-sized business owners in South Florida plus individuals seeking classified advertisement placement. AdNet will utilize direct sales by its management team and a customer referral system with a five percent commission. AdNet will initially offer each customer one free month of advertising. In addition to the sales and promotions, AdNet will also wage an aggressive advertising campaign through print media, radio, handbills, and on-line banners.

Financial Estimates

A three-fiscal-year sales projection indicates $415,565 in gross sales by the end of year three—a 31 percent increase over year one. Profits are projected at $102,417 over the same three-year period after a first-year loss. The initial $30,000 of capital raised by offering the LPUs will be used for advertising in *PC World Magazine* and *Forbes Magazine,* Internet advertisements, and radio advertisements.

Management

AdNet's general partners will consist of the six original founders and employees plus the three outside limited partners. The general partners have significant experience in management, international business, personnel management, accounting, and marketing, including Internet marketing.

Risks

Although some possible risks—delays in raising the required capital, faulty initial design of AdNet's World Wide Web home page, slow public acceptance, possible new competition, and lack of Internet regulations—are to be anticipated, our able and experienced personnel will help minimize and resolve them. The proposal offers a dynamic and original business opportunity, provides a vital service to South Florida residents, and ensures investors and management significant profit realization.

FIGURE 13.6   *continued*

4. If you are writing an informative abstract, include also the key facts, statistics, and conclusions.

5. If you are writing a concluding summary, reread the material and summarize the content, omitting specific detail, examples, graphics, and the like.

6. If you are writing an executive summary, condense the detail of the parent document and consider persuasive words and phrases that would predispose the reader to think favorably toward your report or proposal. Figure 13.5 (see p. 386) appealed to the public's desire for convenient, high performance, and environmentally safe products. Such phrases as "our focus . . . is to take advantage of . . . ," "growth rate is substantial," and "strong and experienced management team" are appealing and persuasive terms.

7. If you are writing an informative abstract or an executive summary, explain your key findings in a very condensed form. Omit or condense lengthy explanations, tabulated material, and other supporting explanation.

8. Edit for completeness and accuracy.

PROPOSALS

A proposal is an action-oriented report. While most reports include recommendations for ongoing accomplishments, a proposal suggests a future task and includes a complete plan of how to accomplish this task. That is, a proposal contains procedure or equipment analysis, cost analysis, the capabilities of existing facilities, information on involved personnel, and usually a timetable for accomplishing the work. A brief report may recommend that a new policy be devised. A proposal provides exactly what the policy must cover, a schedule for adoption, a procedure for implementing the policy, and the personnel who should be in charge. A proposal's purpose is to persuade the reader.

To be persuasive a proposal must emphasize the advantages to the organization. You need to convince the decision makers to take action. As you develop your data and organize your material, stress one or more of the following advantages that your proposal will effect:

- Money savings or increased profits in the short and/or long term
- More efficient time applications
- Improved employee and client safety and comfort
- Compliance with laws or ethics
- Enriched employee morale

Proposals persuade people into action by appealing to their sense of responsibility and, perhaps, even playing to their fear of failure as leaders.

Types of Proposals

Proposals are classified as *internal* and *external, solicited* and *unsolicited.* Government and industry often solicit external proposals from independent agencies to solve problems or to develop services prior to awarding contracts or grants. A county commission may advertise for competing firms to submit proposals to develop a county-wide transportation system. A national airline may solicit proposals to develop a larger and faster jet. The federal government solicits grant proposals for services in education, environment, energy, science, medicine, rural development, and other areas. A university may hire a consulting firm to develop a long-range expansion program. The agency that solicits external proposals spells out general requirements, cost ceilings, deadlines, and criteria for evaluation.

In business and industry the internal proposal, one written by a member of the organization, may be solicited or unsolicited. Management may appoint an individual or a committee to devise a program to change or improve some existing procedure or practice. A cafeteria manager may ask employees to submit proposals to increase sales in the fast-food line. An office manager may solicit proposals for policy and procedure to avoid charges of sexual harassment. Or any employee or group of employees may initiate an unsolicited proposal to management to purchase new equipment, improve working schedules, or alter procedures.

Writing a good unsolicited proposal for your employer is an excellent way not only to improve working conditions but also to demonstrate your interest in and commitment to your company.

Organization and Format

A proposal explains an existing problem and proposes the concrete measures, procedures, or steps for its rectification, along with an explanation of costs, equipment, personnel needs, and a time schedule. A proposal usually involves

- A clear statement of what is being proposed and why
- An explanation of the background or problem
- A presentation of the actual proposal, including methods, costs, personnel, and action schedules
- A discussion of the advantages and disadvantages
- The conclusions, recommendations, and/or an action schedule

All formal reports are more readable if they contain headings. The following topical headings should be considered for organizational purposes for a proposal, although each actual proposal will suggest additional major and minor headings.

Traditional Format	*Streamlined Format*
Introduction	Subject
Purpose	Objective
Scope	Problem
Background	Proposal
Investigative procedure	Advantages
Findings	Disadvantages
Proposal	Action
Equipment	
Capabilities	
Costs	
Personnel	
Timetable	
Consequences	
Advantages	
Disadvantages	
Conclusion	

The streamlined format (SOPPADA) is usually used only for *brief* proposals.

Writing the Proposal

For purposes of instruction we concentrate on an unsolicited proposal, the kind you may originate in an entry-level career position or devise for consideration at your college or university. It is important to remember your audience. You essentially are writing a persuasive report; therefore, you must justify your proposal by presenting compelling reasons for its adoption. Although you may address your proposal to your immediate supervisor, a proposal is often reviewed by superiors further up the chain of authority. Your explanations, data, and language must be clear to those people who may not have any familiarity with the situation you describe. You must be objective and diplomatic.

    Introduction. The introduction usually includes a *statement of purpose* or an *objective* and comments on the *scope* of the report. State briefly exactly what you propose along with a general statement of why the proposal should be given serious consideration. The scope statement will orient your reader(s) to the material to follow. Examples follow:

Example 1—Purpose and Scope

Purpose. This proposal, to purchase and install bicycle supports and gate locks in the ABC Elementary School bicycle compound, is designed to eliminate prevalent vandalism and to decrease personal injuries.

Scope. In this report an examination of the problem, a tabulation of survey results, and a description of the proposed concrete bicycle supports and gate locks are presented, followed by the suggested layout, costs, product availability, and conclusions.

Example 2—Objective Heading Only

This report proposes the purchase of a table saw to increase production in our store fixture manufacturing plant and to increase profits. This proposal will document the problem, examine the capabilities of the proposed equipment, detail the costs, recommend the location, discuss the advantages, and present conclusions.

Problem/Background. The problem/background section details the existing problem, such as high costs, inefficiency, dangers or abuses, or low morale among employees. The solution you intend to propose will probably cost money or involve personnel in new responsibilities, so you must spell out that a very real and perhaps costly problem presently does exist. If it is not obvious how you researched the problem, you may need to include an explanation of your investigation techniques. If appropriate, research the operational costs of the present system. Project these costs over a week, a month, a year, or other appropriate time frames. Present data in tabular form.

Your problem may be that hazards or inconveniences exist under the present system. Document accidents, work slowdowns, late production schedules, or other related evidence. If the problem is causing low morale, research the turnover rate of personnel or incidents of friction.

The **deductive** organization pattern is easier to read than the **inductive** (see Chapter 2). In a proposal to purchase a computer for two floral shops the problem was documented as follows:

Problem:

Bookkeeper transportation and telephone costs are excessively high. Because both of our floral shops maintain separate inventory control of customer account information, bookkeeper Mary Jacobs must frequently travel between the two shops and telephone for customer account information. In addition, this system requires twice the amount of time necessary to review the total daily sales. Table 1 shows the transportation and communication expenses of the bookkeeper under the present system:

TABLE 1 *Current Bookkeeper Transportation and Telephone Expenses*

Time Frame	Travel Time (hr)	Gas ($)	Telephone Time (hr)	Cost @ $6.00 per hr wage ($)
Day	1	5	1	12
Week	5	25	5	60
Month	20	100	20	240
6 Mo	120	600	120	1,440
Year	240	1200	240	2,880
		TOTAL COST.........$2,880		

The $2,880.00 wage and gas expenditure can be put to more productive use. The emphasis here is on inefficient costs. The following is from a proposal to establish a regulating committee to end sex discrimination at a community college. The problem does not entail costs, but details concrete evidence of unethical discrimination:

Problem:

Some sex discrimination facts present themselves:

1. Of the 215 faculty members 79 or 36.7 percent are women, a percentage that is not reflected in either administrative positions or standing committee membership.
2. Of the 36 administrators only three (3) are women; of the ten (10) division chairpersons none are women; of the twelve (12) department heads only two (2) are women; of the twenty (20) area leaders only five (5) are women; of the total 78 positions only ten (10) or 12.8 percent are women.
3. Of the 207 faculty and administrators serving on standing committees only 52 or 25 percent are women. One committee has no women members.
4. Women staff have voiced concern that the inequitable number of women in administrative and standing committee positions is a "negative incentive" for innovative teaching, volunteer assignments, and requests for advancement.
5. Women students have voiced concern through the agency of the student government association that the college does not actively counsel and provide for women students to excel nor to set goals commensurate with the expanding opportunities for entry into the previously male-dominated professions.

Be thorough and exacting in your documentation of the problem. The information in this section will be referred to when you detail the consequences of adopting your proposal.

Proposal.   Present the solution to the problem by providing all of the particulars of your proposal. As already suggested, you may wish to consider the following subheadings:

Equipment/Procedure
Capabilities
Costs
Personnel
Timetable

If the proposal involves the purchase of new equipment, describe it accurately and explain its function and capabilities. If the proposal involves a new procedure or policy, explain exactly how it will work. Under costs include initial purchase price, financing, installation, labor, and training costs. If new or transferred personnel are involved, include the qualifications for the position or the qualifications of the employee. Summarize duties and salaries. If you use the traditional format, include a timetable or work plan for making your proposal operable in this section. If you use the streamlined format, place your schedule of tasks to be accomplished in the final Action section. An effective proposal not only seeks action but also becomes the blueprint by which the action shall be performed. Because the actual Proposal section of your report is often lengthy, samples are not given here but may be reviewed in Figures 13.7 and 13.8 (see pp. 403–404 and 414–416).

Consequences. This section may be divided into *Advantages* and *Disadvantages* or *Strengths* and *Limitations*. Here you refer to the details in your Problem section and explain how each problem will be solved or alleviated.

If your proposal costs money, show in graphics how implementation will save the company money over projected periods of time. Often a large expenditure will ultimately save money over a few years by increasing efficiency or production. If you propose a new system, number and list all of the benefits of it. Employee morale is an important concern of management; if adoption of your proposal will improve employee morale, detail these benefits.

Do not ignore disadvantages (temporary work stoppage, layoffs of employees, or limited capabilities). Discuss these disadvantages or limitations in a positive manner.

Following are brief sections from proposals outlining consequences. The first example is from a proposal to pave an American Legion Hall parking lot:

Consequences:

By paving the existing parking lot we will be able to accommodate 30 or more cars in addition to the present 140. By eliminating the problems of dust and erosion, the physical appearance and value of the property will be enhanced. During the two weeks of construction Sunset City officials have agreed in writing to allow our 57 employees and approximately 100 daytime visitors to park in the Sunset City Hall west parking lot.

The second example is from a proposal to adopt new displacement and furlough rules for an airline pilots' association:

Advantages:

Adoption of the system-wide seniority rules in the pilot displacement and furlough process results in the following advantages:

- Senior pilots are able to displace at any base.
- The total number of displacements arising out of a curtailment situation is small, and the displacements are mostly confined to the base where the curtailment situation occurs. Thereafter, significantly fewer pilots are affected by curtailment.
- A maximum of two base moves exists for every pilot curtailed.
- Protection of captaincy is guaranteed for all but the junior captain at base. The junior captain has base protection.
- The indicated cost saving is 30 percent over the current operating agreement.

Disadvantages:

These benefits are achieved at the cost of limiting pilots' freedom of choice in the following manner:

- A pilot's displacement choices are limited.
- If a captain desires to revert to first officer status to displace in a particular equipment, he or she may not be able to do so.

Conclusions.   The body of your proposal has examined the problem and presented a blueprint for action. You have anticipated the questions and possible objections to your proposal and objectively, but persuasively, responded to them. Your Closing section is essentially an urge to action. In the traditional format summarize your proposal features and reemphasize the advantages in numbered statements. Close with a persuasive statement, such as

> I urge you to give serious consideration to this proposal and am available to discuss its particulars with you.

In the streamlined format, title your section *Action* and present your schedule, work plan, or timetable. Figures 13.7 and 13.8 (see pp. 404 and 416–417) illustrate the optional Closing sections for proposals.

Your signature, position, and date are appended to the final page in the right-hand corner of the paper:

Janet Gorky
—————————————————————————————
Janet Gorky, Assembler   11/15/9X

PROPOSAL TO IMPROVE

EMPLOYEE WORK SCHEDULES

IN THE

CUSTOMER INQUIRY DEPARTMENT

Prepared for

Andrea Brooks

Assistant Vice-president

Operations Division

State Trust Charter Bank

Bigtown, New Hampshire

by

Monica Ferschke

Administrative Assistant

19 October 2000

FIGURE 13.7 *Sample proposal with formal report features* (Courtesy of Monica Ferschke)

MEMORANDUM

TO: Andrea Brooks, Assistant Vice-president

FROM: Monica Ferschke, Administrative Assistant *M.F.*

DATE: 19 October 199X

SUBJECT: Proposal to Improve Employee Work Schedules

Enclosed is a proposal to improve the employee work hours in the Customer Inquiry Department. Implementation of this proposal will increase the department's productivity and efficiency, raise employee morale, and, ultimately, improve the level of service to customers.

I am grateful for the cooperation that the department employees gave me by completing the survey and providing thoughtful comments.

May we discuss this proposal when I return from vacation on Monday, 26 October?

MF/mb

Enclosure

FIGURE 13.7   *continued*

TABLE OF CONTENTS

ILLUSTRATION

FIGURE 13.7 *continued*

SUMMARY

The schedule of employee work hours in the Customer Inquiry Department needs revision. Staff experience difficulties in completing work by deadlines, customers wait while staff work on assignments, and morale is low because of long hours and stressful pressures. The low volume of calls on Fridays after 4:00 P.M. doesn't warrant that a full staff be on hand. I propose the creation of two alternating shifts:

Week A	Monday–Thursday	8:00 A.M. to 4:30 P.M.
	Friday	8:00 A.M. to 4:00 P.M.
Week B	Monday–Thursday	8:00 A.M. to 4:00 P.M.
	Friday	8:00 A.M. to 6:00 P.M.

An equal number of employees will work in each shift, but employees will not have to work for ten hours every Friday. Adoption of the proposal will improve the department's productivity, efficiency, morale, and customer service.

iv

FIGURE 13.7 *continued*

PROPOSAL TO IMPROVE EMPLOYEE WORK SCHEDULES

IN THE CUSTOMER INQUIRY DEPARTMENT

INTRODUCTION

Purpose

This proposal to change employee work hours in the Customer Inquiry Department will provide a schedule to raise the levels of productivity and efficiency, to improve morale within the department, and to provide more attentive customer service.

Scope

This report documents the department's current operations, presents employee survey results, proposes a new work schedule, and reviews the advantages to the bank and its customers.

BACKGROUND

Problem

Currently, employee work hours in the department are 8:00 A.M. to 4:00 P.M. Monday through Thursday and from 8:00 A.M. to 6:00 P.M. every Friday, a 10-hour work day. Training sessions are often conducted between 8:00 A.M. and 9:00 A.M. Monday through Friday, making it difficult for staff to complete paperwork and assignments by deadlines. To accomplish assignments, certain staff members must stop accepting customer calls before 4:00 P.M. to complete the work. This creates added pressures for staff members still accepting calls, and customers must wait longer for service due to a smaller available staff.

Employees have voiced their concerns to their supervisor, Ms. Annabelle Higgins, who, in turn, recognizes that staff productivity and attention to detail decreases after 4:00 P.M. on Fridays.

1

FIGURE 13.7 *continued*

2

Procedure

Initially, I examined the volume of calls after 4:00 P.M. on Fridays during September this year because that month has proven to be the busiest over the past three years. Appendix A contains a tabulation of calls per hour. Additionally, I conducted a survey among the staff regarding a change of their current work hours and asking for their comments on subsequent benefits to their department and the bank if hours were changed. Appendix B contains the survey instrument and numerical tabulations.

Findings

A.   Volume of Calls

After examining the volume of calls after 4:00 P.M. on Fridays during September, I find that the volume does not warrant having a full staff available. The departmental average number of calls per employee per hour is 50 (based on the annual number of calls divided by the hours of operation and the number of employees). Table 1 shows the actual number of calls handled by staff on Fridays in September after 4:00 P.M. and the number of employees needed to handle those calls.

Table 1   Number of Employees Needed for Calls After 4:00 P.M.

September Fridays	Hours	Calls Taken (#)	Employees Needed (#)
1st Week	4:00–5:00	45	1
	5:00–6:00	21	1
2nd Week	4:00–5:00	42	1
	5:00–6:00	16	1
3rd Week	4:00–5:00	53	2
	5:00–6:00	25	1
4th Week	4:00–5:00	46	1
	5:00–6:00	37	1

As illustrated in Table 1, only two employees are actually needed from 4:00 P.M. to 6:00 P.M. to answer customer calls. Assignments and paperwork are, in most cases, expected to be completed prior to 4:00 P.M. on Fridays; therefore, to require all ten employees to work after 4:00 P.M. is inefficient.

FIGURE 13.7   *continued*

3

B. Employee Survey

All ten staff members in the Customer Inquiry Department responded to my survey requesting their opinions regarding the implementation of a new schedule and commenting on the benefits of the change. Of the ten employees who responded,

- 100 percent favor the new schedule,
- 60 percent think that the proposed change will improve morale within the department,
- 70 percent feel the change will allow them to finish their work by deadline, and
- 80 percent feel customer services will be improved.

Employee opinion, long working hours on Fridays, and inadequate employee scheduling lead me to conclude that a change should be made regarding the work hours within the department.

PROPOSAL

Therefore I submit the following proposal.

Schedule

Employees will alternate work week schedules; there will be five employees on each schedule.

Schedule A (5)	Monday–Thursday	8:00 A.M. to 4:30 P.M.
	Friday	8:00 A.M. to 4:00 P.M.
Schedule B (5)	Monday–Thursday	8:00 A.M. to 4:00 P.M.
	Friday	8:00 A.M. to 6:00 P.M.

The number of hours worked by any one employee in a week will not change, but the assignment of long and short hours on Fridays will alternate.

This schedule will ensure that a sufficient number of employees will be on station to answer calls during additional operating hours on Fridays. Duties involving mid-week deadlines could be assigned to employees working the late (4:30) shift, allowing the extra half hour after calls to complete paperwork and assignments.

FIGURE 13.7 *continued*

4

Date

I propose that this new schedule be implemented by November 1.

CONSEQUENCES

By implementing the proposed hours, these advantages should accrue:
- Employees on the 4:30 P.M. shift will be able to complete more work, and accuracy should improve as interruptions will be minimal.
- Employee morale should improve. Acceptance of this proposal will demonstrate that management values employee opinions and that employees offer sincere efforts to increase operational efficiency.
- Customer service will improve because the entire staff will be available to answer calls until 4:00 P.M.

CONCLUSION

I strongly urge you to implement this proposal so that the Customer Inquiry Department can become a more productive and efficient work unit of the bank. Although Ms. Annabelle Higgins will have the final responsibility for monitoring the schedule, I will be happy to assist her in posting the schedule and clarifying its points to the staff. Please let me know your decision.

Monica Ferschke
_____
Monica Ferschke   10/19/199X

FIGURE 13.7   *continued*

APPENDIX A

CUSTOMER INQUIRY DEPARTMENT

CUSTOMER CALL LOG TABULATIONS

DATE	4–5 P.M. CALLS (#)	5–6 P.M. CALLS (#)
Friday 9/4	45	21
Friday 9/11	42	16
Friday 9/18	53	25
Friday 9/25	46	37

5

FIGURE 13.7 *continued*

APPENDIX B

SURVEY TABULATION

CUSTOMER INQUIRY DEPARTMENT

SCHEDULE SURVEY

This survey is being conducted to determine if you would be willing to work on a new rotating schedule. We would also like your comments on the benefits to the department which you foresee if such a schedule is adopted.

1.   Would you favor the following new hours? 10 yes   0 no

              Schedule A   Monday to Thursday   8:00–4:30

                               Friday   8:00–4:00

              Schedule B   Monday to Thursday   8:00–4:00

                               Friday   8:00–6:00

    (Alternate every other week)

2.   Would you like to suggest any other possible new schedule? (Please indicate in the white space.)

                              NONE

3.   If you answered "Yes" to Number 1, please comment on the foreseeable benefits.

    "Better customer service Monday–Thursday"   8 mentions
    "Free up unnecessary personnel on Fridays"   3 mentions
    "Improve morale"   6 mentions
    "Time to catch up on paperwork from 4:00 to 4:30"   5 mentions
    "More flexibility for employees"   1 mention
    "Should stimulate closer attention to detail"   3 mentions

6

FIGURE 13.7   *continued*

A PROPOSAL

THE DEVELOPMENT OF A

REIMBURSABLE EXPENSE POLICY

AND

MONITORING AND CONTROL SYSTEM

Prepared for

Mr. Duane P. Morton

President

Ace Electronics Company, Inc.

Fresno, California

By

Charles E. Smith

Senior Partner

Smith Business Consultants

San Francisco, California

24 February 2000

FIGURE 13.8 *Sample traditional proposal with formal report features*
(Courtesy of Charles E. Smith; adaptation)

SMITH BUSINESS CONSULTANTS
1000 Plaza Court
San Francisco, California 94536
213-659-1222

24 February 2000

Mr. Duane F. Morton, President
Ace Electronics Company, Inc.
2120 West Broward Boulevard
Fresno, California 93710

Dear Mr. Morton:

We have completed the study that you authorized in November 199X concerning the control of reimbursable business expenses at Ace Electronics Company and are pleased to submit the enclosed proposal to make expense practices more equitable.

We appreciate the cooperation of your officers and other employees in conducting our study. All transcribed interviews and records of research findings will be kept in confidence for six months should you need to review this material.

We thank you for this opportunity to be of assistance.

Very truly yours,

Charles E. Smith

Charles E. Smith
Senior Partner

DES:jsv

Enclosure

FIGURE 13.8 *continued*

TABLE OF CONTENTS

Page

FIGURE 13.8 *continued*

SUMMARY

This report for Ace Electronics Company, Inc. proposes the development of a reimbursable expense policy along with a monitoring and control system to make expense practices more equitable, to allow for abuse detection, and to reduce costs.

Extensive interviews and expense records research reveal that expenses are excessive, existing policy is inadequate, and methods of approval and control are deficient. By implementing this proposal the approximately yearly $3.3 million reimbursable expenses may be reduced by as much as $1 million.

We propose the development of a written reimbursable expense policy that incorporates a matrix of all allowable expenses, categorizes employees, sets conditions and restrictions, and clearly states approval individuals. Further, we propose procedural regulations that establish periodic report periods and deadlines, entail new report forms, and institute a definite system of review and analysis. A work plan to carry out these proposals includes the appointment of a director, the establishment of committees, and a listing of chronological duties.

By implementing this proposal reimbursable expenses shall be reduced, and adoption may well lead to cost reduction sensitivity in other expense areas.

FIGURE 13.8 *continued*

PROPOSAL TO DEVELOP A REIMBURSABLE EXPENSE POLICY AND MONITORING AND CONTROL SYSTEM

INTRODUCTION

Purpose

This proposal to develop a reimbursable expense policy and monitoring and control system is designed to make expense practices more equitable, to establish a system which will detect and prevent abuses, and to reduce costs.

Scope

This report presents a description of our study, a review of our findings, a dual proposal, the advantages of implementation, and the conclusions.

BACKGROUND

Procedure

To determine the facts relating to reimbursable expenses we used a sample basis. With the assistance of four senior executives, a sample of 25 of the approximately 100 executives at Ace was chosen. Personal interviews were held with twenty of them, and extensive records research was conducted for fifteen covering the second quarter of 199X, the period chosen for the analysis.

Our first step was to interview these officers, two financial managers, and the personnel manager and to examine a sampling of expense reports. From that work we prepared two documents: a list of discussion questions for the interviews, and an outline of the records research requirement. Appendixes A and B exhibit these documents and indicate the extensive nature of the fact finding.

The Smith consultants conducted the interviews, and an Ace team directed by the assistant corporate controller conducted the records research.

1

FIGURE 13.8 *continued*

2

Findings

A.   Scope of Expenses

Reimbursable business expenses represent a significant controllable cost at Ace. During the first nine months of 199X total
expenses for officers and others was $2.5 million, an annual
expense of over $3.3 million a year. For the officer group alone
the figure was $1 million a year.

B.   Policies and Practices

Numerous examples of excessive expense practices and failure
to control expenses are evident upon examination of the records
and by admission of the officers in confidential interviews. We
believe that reimbursable expenses at Ace could be reduced by
$1 million a year without adversely affecting the business.

As a result of unclear or inadequate policy or poor communica-
tion, most interviewees displayed uncertainty about what was
expected of them. During the interview officers were asked:
"Can you identify the written expense policy/procedure as it
relates to you?" Following are representative answers:
   "One might exist, but I'm not familiar with it."
   "There's only a memo on transportation."
   "There is no written policy."
   "Policies are not spelled out. I'm not familiar with any written policy. I do what I did for my former employer."
Further, existing written expense policies lack clarity.
• A limited number of items that may be authorized are listed,
  but the conditions under which they are authorized are not
  always clear. For example, statements, such as "when time is
  a factor," "where necessary for company business," and "only
  to employees who hold the types of jobs that require such
  meetings," appear in the policies.
• Some expenses require prior approval, but the approving individual is not always identified.
• Some items of allowable expenses, such as home entertainment and telephone answering services, are not treated.

FIGURE 13.8   *continued*

3

- The frequency of reporting is not covered.
- A few items, such as social club memberships and access to the executive dining room, are inadequately treated. In some cases these are considered executive perquisites and in others reimbursable expenses.

C.   Approval and Control

Methods of approval and control are inadequate.

- Officers report they are not comfortable when questioning the expense practices of close associates, and rarely do so.
- Expense documentation and explanation are inadequate. Reporting forms fail to determine such things as class of air travel, cost of overnight accommodations per night, purposes of business meetings, the number of people entertained, and so forth.
- Budgetary control is difficult. Monthly reports do not show expenses by individuals, but by divisions.
- Expense reporting is often tardy. For the sample offices timeliness ran from prompt (2 to 3 days after the reported period) to late (6 months). Many officers reported weekly, others monthly, and some on a less periodic schedule.
- The expense report is not a complete record of expenses. Some officers are reimbursed for expenses directly from petty cash, or the invoice for an expense item is paid directly by Ace and not shown on an expense report.

D.   Variations

The interviews and the records research reveal a wide range of variable practices.

- In the sample of the 15 officers whose expense reports were analyzed, only 7 charged telephone expenses, 11 had social club expenses, 10 were reimbursed for gifts, and 5 incurred expenses for personal entertainment.
- Air travel practices further illustrate the variety that exists. During the analyzed quarter 3 of the officers made at least one flight, but 2 of them did not submit ticket receipts. Of the 11 other employees whose flights could be analyzed, 6 flew first class and only 2 paid the difference between first class and coach fare. Seven wives flew with their husbands at company expense, 2 of them first class.

FIGURE 13.8   *continued*

4

- There is considerable variation in the cost of overnight room accomodations. The range is from $60.00 to $250.00 a night.
- The average per meal cost of business meetings range from $50 to $148. Business meetings were usually conducted at lunch.

The great variation in expense practice is the basis for our conclusion that substantial opportunities for cost reduction of at least $1 million a year exist.

PROPOSAL

To correct these deficiencies in the reimbursable expense system, we propose:

1. The development of a written reimbursable expense policy
2. The establishment of firm procedural regulations

Expense Policy

Ace should prepare a written reimbursable expense policy that is clear with regard to each type of expense, that accomodates differences among employees, and that covers certain procedural requirements. To accomplish this:

1. All personnel should be divided into categories. All employees should not be treated uniformly. Their expense spending requirements and privileges should be recognized in the policy. We recommend that the two categories be designated as follows:

   Members of the Board   All Other
   Vice-presidents   Employees

2. A policy matrix should be devised which lists all reimbursable expenses, employee category differences, restrictions and conditions, and approval authority. Appendix C illustrates a sample policy matrix.

FIGURE 13.8 *continued*

5

3. A clarifying policy should be written for wide distribution and inclusion in the Administrative Policy Manual. The policy should include guidelines and standards, procedures for advances, procedures for preparation, processing, and approval of the expense report, and specific guidelines and standards for each type of reimbursable expense in the matrix. Appendix D shows a sample administrative policy page.

Monitoring and Control

The following procedural regulations should be established:

1. The report period should be monthly; deadline for submission should be one week thereafter.
2. Expense report forms should be revised to require documentation and explanation.
3. The Expense Report must be established as the sole vehicle for recording and reimbursing expenses.
4. Responsibilities of the reporting individual, the controller's department, and the approving individual should be developed into a definite system.
5. The role of the controller's department should be expanded. An editing function there should first evaluate the adequacy of the documentation and explanation, and the Expense Report should be returned to the reporting individuals for correction if there are deficiencies. Next the report should be examined for conformance to policy, and any exception should be noted on a Buck Slip which, along with the Expense Report, should be forwarded to the approving individual.

   The controller's department should prepare a summary analysis of each individual's expenses on a quarterly basis. Quarterly and cumulative expenses should be compared to budget.

FIGURE 13.8   *continued*

6

Work Plan

   I.   Staffing
       A. Appoint the Vice-president of Finance as Project Director with responsibility for implementing all recommendations.
       B. Appoint a Corporate Expense Policy Committee to work with the Project Director.
       C. Assign a small staff to work with the Committee without interruption.
  II.   Expense Policy
       A. Establish the basis for assigning all personnel into the recommended two categories.
       B. Prepare an expense matrix and policy for each type of reimbursable expense for each category of personnel.
       C. Submit the policy to officers for discussion and modification.
       D. Approve the policy.
       E. Disseminate the policy and conduct familiarization training.
 III.   Monitoring and Control
       A. Revise Expense Report format and create Daily Expense Diary, Buck Slip, and Quarterly Analysis.
       B. Establish the edit and analysis functions in each division.
       C. Train expense report editors and analysts.
       D. Test the system and make adjustments.
       E. Commence live operation of the system.

ADVANTAGES

By developing a written reimbursable expense policy and establishing firm procedural regulations, the following advantages should accrue:

1. Reimbursable expense practices should become more equitable over the broad organization due to clarification and control.
2. A system for detection and prevention of abuses will be established.
3. Superiors shall exercise more control of expenses.

FIGURE 13.8 *continued*

7

4. Quarterly analysis should encourage discipline and facilitate budget preparation.
5. Implementation should reduce costs.

CONCLUSION

An attractive opportunity for cost reduction exists and provides the basis for achieving economies through innovations in policy and control. In our opinion, the recommendations in this report are well worth implementing and, in addition, will lead to sensitivity towards cost reduction in other areas.

Charles E. Smith   *2/24/9X*

Charles E. Smith   2/24/9X
Senior Partner

CES:jsv

FIGURE 13.8 *continued*

REIMBURSABLE EXPENSE INTERVIEW TOPICS

1. Position held during second quarter, 199X
   a. Title
   b. Superior, subordinates
   c. Nature of position
   d. Approval authority for expense reports
2. Whose expense reports do you approve?
   a. Are summaries prepared?
   b. How do you control subordinates' expenses?
   c. Do you ever question their expenses? Details.
   d. Have you ever had a drive to reduce expenses? Explain.
3. Who approves your expense reports?
   a. Are summaries prepared?
   b. Are your expenses ever questioned? Details.
   c. Do you ever seek prior approval for expenses? Details.
   d. Do you have any prior understanding with regard to your expenses?
4. What is the budgetary control over your expenses?
   a. In what account in what cost center are your expenses accumulated?
   b. How do your expenses compare to budget?
5. Have you ever been told what the expense policy is as it relates to you? Details.
6. Can you identify the written expense policy/procedure as it relates to you? Details. What are the salient points?
7. Identify the forms that are used: expense reports, petty cash disbursements, report of outstanding advances, summaries.
8. What is the frequency of reporting and approval and what is your timeliness? Any difficulties?
9. What are the controlling policies/procedures with regard to advance accounts, and what are your practices?
10. List all of the categories of your reimbursable expenses.

FIGURE 13.8   *continued*

419

9

11. Do you incur reimbursable expenses that are not shown on your expense report? If so, where are they recorded?
12. What are the significant restrictions on your expense spending practices?
13. Do you ever misrepresent expense items? e.g., combining minor items to recover excessive charges?
14. What is the influence of your status on your expense practices? Is there discrimination by rank? Amplify?
15. How do you view the company attitude concerning expenses? Have there been any changes? How does it compare with other companies in which you have worked?

FIGURE 13.8 *continued*

RECORDS RESEARCH OUTLINE

I. Sources of information
   A. Expense report
   B. Summary business expense report
   C. Petty cash disbursement form
   D. Report of outstanding advances
   E. Approval authority
   F. Quarterly Budget Report for appropriate cost center
II. Information required
   A. Controlling policy/procedure
   B. Approval authority
   C. Summary
   D. Type of form
   E. Completeness and deficiencies
   F. Date and total expenses for period
   G. Documentation check
   H. Assumptions
III. Quarterly Summary
   A. Transportation
   B. Hotel rooms
   C. Meals
   D. Business meetings
   E. Other
IV. Analysis
   A. Personnel category differences
   B. Direct payments
   C. Nature and amount of petty cash disbursements
   D. Timeliness
   E. Discrepancies
   F. Significance

10

FIGURE 13.8 *continued*

SAMPLE REIMBURSABLE EXPENSE POLICY MATRIX

Type of Expense	EMPLOYEE CATEGORY President Vice-pres and above	All Other Employees	Restrictions & Conditions	Approval Required
Airline and Railroad	First Class Reimbursable	a) Coach reimbursable b) First Class Restricted	Restrictions: a) Traveling in company of Pres., V-P, or above b) Duration of trip exceeds 7 hr c) Traveling with customer, etc.	President Vice-Pres or above
Buses	Reimbursable	Reimbursable		
Rented Cars	Reimbursable	Reimbursable	Condition: a) Other forms of public transport cannot meet business requirements b) Standard size low-priced car	
Personal Cars	Reimbursable	Reimbursable	Condition: Must not be used on long trips when other transport is less expensive and obtainable	
Taxi Cabs	Reimbursable	Reimbursable	Condition: Other forms of public transport cannot meet business requirements	
Limousines and Company Cars	Restrictively Reimbursable	Restrictively Reimbursable	Restriction: a) Essential b) Mileage log maintained	President Vice-Pres or above
Air Charter	Reimbursable	Reimbursable	Restriction: Extreme emergency situation only	President Vice-Pres or above
Lodgings Single Rooms First Class Accommodations	Reimbursable	Reimbursable		
Suites	Restrictively Reimbursable	Restrictively Reimbursable	Restriction: Essential for the conduct of business	President Vice-Pres or above

11

FIGURE 13.8 *continued*

SAMPLE ADMINISTRATIVE POLICY PAGE

ACE Administrative Policy Manual

Section	Dist. List	Date Issued	Policy No.
REIMBURSABLE BUSINESS EXPENSES 3		12/30/9X	RB-1

Subject

Ace Reimbursable Business Expense Policy   Page 3
                                                                of 39

III.   TYPES OF REIMBURSABLE BUSINESS EXPENSES
        For ease of reference, a matrix summarizing reimbursable
        expenses has been included in Section XI of this Policy.

   A.   **Transportation Expenses**
        All requests for transportation should be submitted
        through the Transportation Department.

        Where use of the Transportation Department is
        impractical, an employee may arrange his own
        transportation.

        The restrictions and documentation required for
        transportation expenses are as follows:

        1.   **Airlines and Railroads**—Coach type accommodations
             are to be used. All expenses in this category must be
             documented by a ticket stub when submitting the Ace
             Expense Report. When a ticket purchased by Ace is
             not used, it should be returned to the Controller's
             Department with the expense report for a refund. The
             amount of the unused ticket should be included on the
             expense report.

             When traveling to and from the airports and Ace,
             public transportation should be used wherever
             practical.

        2.   **Buses**—The use of long-distance buses is permitted
             when they can reasonably meet the needs of the
             Company.

                                    12

FIGURE 13.8   *continued*

3. **Rented Cars**—Use of rented cars should be limited to those situations where public transportation is not available or cannot meet business requirements. When the situation requires a rented car, short-term arrangements should be made by the individual. On rentals exceeding one month, arrangements should be made through the Ace Purchasing Department. On all rentals, a standard size, low-priced car should be used. In all cases <u>optional insurance should not be purchased</u> and where national car rental services are used, the discount should be obtained. When submitting this expense for reimbursement, a copy of the itemized invoice is required.

13

FIGURE 13.8   *continued*

C H E C✓K L I S T

Longer Reports and Proposals

ALL LONGER REPORTS

- ❏ **1.** Have I identified the type of longer report I am writing?
    - ❏ Quarterly or annual report?
    - ❏ Long-range planning document?
    - ❏ Systems evaluation?
    - ❏ Proposal?
    - ❏ Other?

- ❏ **2.** Have I designed an appropriate title page? Does it include
    - ❏ Name, position, and affiliation of the receiver?
    - ❏ My name, position, and affiliation?
    - ❏ Date?

- ❏ **3.** Have I written an appropriate (letter or memorandum) transmittal correspondence?

- ❏ **4.** Does the transmittal correspondence contain all of its required sections?
    - ❏ Title and purpose of the report?
    - ❏ Identity of the commissioner of the report and when it was requested, or why it is being submitted?
    - ❏ Comments on encountered problems?
    - ❏ Acknowledgment of others who assisted in the report preparation?
    - ❏ Elicitation of feedback?

- ❏ **5.** Have I written an appropriate abstract or summary?

- ❏ **6.** Have I included a table of contents and list of illustrations?

❑ **7.** Have I attached appropriate back matter (glossaries, survey instruments, survey tabulations, financial projections, job descriptions, personnel resumes, equipment brochures, list of references, or other material)?

❑ **8.** Have I designed an attractive document?

PROPOSALS

❑ **1.** Have I developed items 1 to 8 above, which are relevant to all longer reports, in my proposal?

❑ **2.** Have I determined the nature of my proposal?

  ❑ Internal?

  ❑ External?

  ❑ Solicited?

  ❑ Unsolicited?

❑ **3.** Have I included a descriptive or informative abstract for a scientific or technical proposal?

❑ **4.** Have I included an executive summary for management that is persuasive, causally organized, and of appropriate length?

❑ **5.** Have I used a thorough and readable format? Does it include

  ❑ A statement of the objective and purpose?

  ❑ Explanation of the background and problem?

  ❑ Discussion of advantages and disadvantages?

  ❑ Conclusions, recommendations, and/or an action schedule?

❑ **6.** Does the proposal proper cover all necessary aspects?

  ❑ Equipment/procedure?

  ❑ Capabilities?

  ❑ Costs?

  ❑ Personnel?

❑ Timetable?

❑ Other?

❑ **7.** Have I signed and dated the proposal?

EXERCISES

1. **Descriptive Abstract.** Locate and photocopy a scientific or technical article from a professional journal. Write a short *descriptive abstract* that presents an overview of the purpose, scope, and methodology of the article. Do not confuse the abstract with an executive summary.

2. **Informative Abstract.** Using the same article, write a brief *informative abstract* for the same material. Include the major conclusions.

3. **Executive Summary.** Write an executive summary for the following brief proposal:

   Subject: Modification of the North Campus perimeter road and parking lot access road speed bumps.

   Objective: To alleviate complaints and to prevent further damage to student and visitors' automobiles.

   Problem: Since the installation of 80 speed bumps on the campus perimeter road and parking lot access roads in March 199X, more than 100 students and other campus visitors have registered written complaints to the Security Department regarding damage to their automobiles. The height of the bumps is 8 in., compact cars have only an 8-in. clearance. Further, 47 percent of the complaints are by drivers of large and intermediate automobiles. Reported damages include front end misalignments, rear end leakage, and muffler dents.

   Inspection reveals that the bumps are excessively gouged and scraped. Although some automobile damage may be a result of excessive speed, a State Road Department inspector agreed, following his 15 June 199X visit, that the bumps are too high and too abrupt.

   Appendix A exhibits a copy of his letter.

   Proposal: Therefore, we propose that the 80 speed bumps be modified by grinding the peaks to a 6-in. height and sloping the sides with asphalt to a 35-degree slant. These modifications will reduce the shock that automobiles are experiencing yet still deter speeding. Figure 1 shows a cross-section of the speed bumps before and after modification:

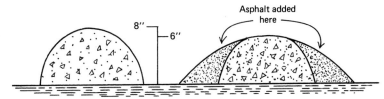

Cross-section of a typical speed bump before and after modification

Equipment: Because the college does not own the equipment necessary for the modifications, the equipment must be rented. Necessary rental equipment includes

- One hand-held, gas-operated, heavy duty grinder with emery stone disc
- One gas-operated hot asphalt mixer
- One water-filled, hand-operated 500-pound roller

Labor: No special skills are needed to operate the equipment. Three Maintenance Department personnel can complete the modifications in an estimated 120 hours over two weeks.

Procedure: One man can grind the bumps at the rate of $1\frac{1}{2}$ hours per bump. Two men can mix and apply the asphalt in five 6-hour shifts.

Cost: Modifications can be completed at an estimated $380.00 as shown in the following table:

Speed Bump Modification Cost Estimate

Item	Cost ($)	Time	Quantity	Total ($)
Grinder	55.00/wk	2 wk	—	110.00
Mixer	140.00/wk	1 wk	—	140.00
Roller	10.00/wk	1 wk	—	10.00
Asphalt Mix	2.00/55 lb bag	—	25	50.00
Pebbles	70.00/load	—	1	70.00
			TOTAL	380.00

Advantages: 1. Speed bumps will still deter speeding.

2. Damage to automobiles should decrease.

3. Project can be accomplished by existing staff.

4. Project can be accomplished prior to Term I traffic.

Action: 1. Approve proposal by 15 July.

2. Arrange for rental equipment and supplies by 1 August.

3. Modify speed bumps 1–15 August while the campus is closed to regular classes.

4. **Title Page.** Title the proposal in Exercise 3 and compose a title page. You are the Supervisor of the Maintenance Department of your college presenting the proposal to the campus provost or dean.

5. **Memo of Transmittal.** Write a memo of transmittal. Include a brief summary of the proposal. Acknowledge an individual or company for assistance in obtaining equipment and supply prices.

WRITING PROJECTS

1. **Group or Individual.** Write a long proposal to solve a specific problem on your campus or at your workplace. Select a subject of sufficient complexity to warrant a proposal in longer report format, but do not tackle something beyond your scope. Chapter 14 discusses information access including library research, surveys, and subsidiary sources. Your proposal should require some research (a survey, interviews with authorities, journal articles, equipment brochure perusal, and related files). To clarify the problem, explain what procedures are being used now and how they are faulty. Document excessive costs, inefficient time use, damages, accidents, robberies, security problems, and so forth. Use survey results, interviews, and other research results to establish the severity of the problem. Present your actual proposal along with capabilities, costs, features, personnel, and a time schedule. Relate the points in your Consequence section to the points you made in the Background or Problem section, emphasizing how each problem will be solved or alleviated. After you complete your proposal draft, prepare all of the appropriate longer report features: title page, transmittal correspondence, table of contents and list of illustrations, summaries, and supplements. Submit your proposal in an attractive folder. Suggested subjects are

On the Campus	*At Work*
new equipment	new equipment
new system of registration	new uniforms

improved parking facilities

picnic or stone tables

snack bars by classrooms

additional computer lab time

additional study carels

new special interest club

Saturday or 5:00 P.M. classes

a women's center

a campus jogging path

student entertainment project

free movie program

improved bomb scare procedures

other?

improved systems of duty roster

improved method of displaying
wares

index for wage increases

improved lounge facilities

improved working conditions

new in-service training program

sports team program

fire evacuation procedures

an advertising program

new staff orientation program

grievance procedures

dental insurance program

other?

NOTES

PART

four

The Research Strategies

CHAPTER 14

Accessing Information

CATHY   by Cathy Guisewite

9:00 PM : CONNECTED TO IRS WEB SITE TO CHECK QUESTION ABOUT TAX FORM.

9:06 PM : EXPLORED JUPITER ON NASA WEB SITE.
9:20 PM : READ LETTERMAN'S "TOP 10" LISTS FOR JANUARY.
9:33 PM : CHECKED WEATHER IN FIJI.
9:46 PM : WANDERED AROUND LOUVRE.
10:15 PM : JOINED CHAT ROOM DISCUSSING BEET FARMING.
10:35 PM : PLAYED ONLINE SCRABBLE WITH COUPLE IN TOLEDO.
11:35 PM : VISITED LIBRARY OF CONGRESS.
12:19 AM : DOWNLOADED RECIPE FOR LOW-FAT GUACAMOLE.
12:30 AM : VISITED HOME PAGE OF GIRL SCOUT TROUP.
12:45 AM : ORDERED SIX CD's AND AN EVENING GOWN.
1:10 AM : HAD TAROT CARDS READ.

1:30 AM : LOGGED OFF THE PROCRASTINATION SUPER HIGHWAY.

S K I L L S

After studying this chapter, you should be able to

1. Access library materials relevant to developing a professional paper.

2. Use a library computer retrieval system to access books, reference works, periodical abstracts, and full-text articles plus other sources.

3. Gain a working knowledge of the Dewey Decimal System and the Library of Congress System for organizing books on library shelves.

4. Name some special reference works (dictionaries, handbooks, manuals, almanacs, and encyclopedias) in your field of study.

5. Access essays within books and abstracts/indexes not included in the computer system.

6. Access and read microform materials.

7. Access information from CD-ROMs.

8. Understand DIALOG databases, USENET "bulletin boards," and mailing lists.

9. Access material in vertical files.

10. Access on-line research materials in the library or on a personal computer.

11. Prepare bibliography cards or a computer file of sources.

12. Understand what constitutes plagiarism and the importance of avoiding it.

13. Record a variety of types of notes to support the thesis of a professional paper.

14. Devise, implement, and utilize results of a survey or questionnaire on a given subject.

15. Use other research sources, such as interviews, letters, lectures, public forums, and direct observation.

INTRODUCTION

Research is the process of investigating and discovering facts. Writers in technology and science are often called upon to write professional papers involving such research. These papers may be designed for information distribution within an organization or for publication in journals of special interest in a particular field: electronics, engineering, all of the sciences, health professions, psychology, teaching, and the like.

As a student, you have already had some experience in writing term or research papers and will continue to sharpen these skills throughout your academic career. The success of your papers will depend, largely, on your ability to research and access the information you need to support your central idea (the thesis). You will find this information in print materials and electronic sources (secondary sources) and through testing, polls, questionnaires, interviews, letters of inquiry, observation, and the like (primary sources). You will eliminate a lot of busywork if you narrow your subject first even if you have not quite formulated your thesis. For instance, if you are interested in cloning, narrow your subject to the aspect that most interests you, such as the medical advantages of cloning certain animal body parts or legislation to limit cloning. After some initial research, you can draft your thesis, which may favor or disfavor such cloning or take a position on legislation. With your narrowed subject and thesis focus, you will direct your research more effectively.

This chapter on strategies for accessing information should be used in conjunction with Chapter 15, which covers writing strategies for professional papers (selecting and limiting the subject, formulating and polishing the thesis, brainstorming the methods of development, writing introductions and closings, and writing drafts and final papers), and Chapter 16, which covers various methods of styling and documenting research papers.

Accessing/researching the source material involves four strategies:

1. Researching print and electronic sources (secondary sources)
2. Preparing surveys, questionnaires, and other primary research (primary sources)
3. Preparing the bibliography cards or computer file
4. Note taking on cards or keyboard

Some instructors may require you to photocopy all materials and print out online sources to submit with a final paper.

RESEARCHING MATERIALS

Objectives

In researching the source materials through library resources and/or by personal computers, you must accomplish five objectives:

1. **Skimming the material.**   Obtain the sources or their abstracts and skim read the material as you locate it to become familiar with the data on your subject. Look over the abstract or the table of

contents and the index. Read the preface and the introduction. Check the appendixes and glossaries for supplemental materials and definitions. Review the source's own bibliography listing, which may direct you to other related sources.

2. **Preparing preliminary bibliography cards or a computer file.** The purpose of the preliminary bibliography cards or bibliography computer file is to compile a collection of possible sources of data for the research project after your skim reading has determined if they address the subject. The information to be recorded for each source is covered in this chapter. Alphabetize the cards or your computer file list for easy reference.

3. **Eliminating redundant and useless sources.** Select those sources that best suit your specific, narrowed topic. Eliminate those that present redundant data, are outdated, or are of questionable value.

4. **Preparing a working bibliography.** Alphabetize the sources (cards or computer file) remaining in the preliminary bibliography set and prepare a working bibliography page (see later and Chapter 16).

5. **Refining your thesis.** Refine and amplify your thesis as you learn more about your subject.

Libraries

Familiarize yourself with your school, company, or public library. Locate the computer retrieval system workstations, the photocopy machines, the special indexes, the book stacks, the vertical files, and the microform files and readers. The computer system is usually the Public Access Catalog (PAC). Some libraries may still rely on the traditional card catalog for books and printed indexes for other materials (periodicals, essays within books, special subject abstracts and indexes, and government publications). Even those libraries that now catalog their holdings on a computer system may still maintain small card catalogs on special subjects and still subscribe to, and hold back issues of, indexes and periodicals.

Computer cataloging systems are still being installed around the world and are constantly improving. Therefore, it is impossible for any textbook to be up-to-date on all the ins and outs of computer retrieval, but a general overview follows.

COMPUTER RETRIEVAL SYSTEMS

Most college, university, high school, large public, and business libraries now catalog their stock by computer retrieval systems that are replacing the card catalog drawers and periodical indexes. The systems list books, journals, magazines, newspapers, and special reference books, as well as reports, government documents, manuscripts, typescripts, prints, photographs, maps, posters, animated illustrations, CD-ROM listings, and more. Libraries subscribe to any number of the databases that have been developed around the world. Databases are special-subject listings on the sciences, technologies, business, health, government documents and statistics, literature, education, psychology, and so on, and most provide abstracts and full-text articles. Many computer retrieval systems provide access to other library holdings, such as multicampus libraries, city or county public systems, and state systems. Most systems now allow free searches on the Internet using keywords, World Wide Web addresses, and hyperlinks to access and download research materials. Consult the librarians in any library for instruction.

Determine if your library catalogs its collection of books by the Dewey Decimal System or the Library of Congress System. In order to actually retrieve books on specific subjects from the shelves, it is helpful to have a good working knowledge of these two systems.

The **Dewey Decimal System** arranges all books and journals on the shelves by the following general subject categories:

000–099 General works (bibliographies, library sciences, general encyclopedias, publishing, manuscripts, and rare books)

100–199 Philosophy and psychology (metaphysics, paranormal phenomena, logic, ethics, ancient and Oriental philosophies)

200–299 Religion (theory of religion, Bible, Christian studies, comparative religion and other religions)

300–399 Social sciences (political science, economics, law, public administration and military science, education, commerce, and customs)

400–499 Language (linguistics and all languages)

500–599 Natural sciences and mathematics (mathematics, astronomy, physics, chemistry, earth sciences, paleontology, biology, plants, and animals)

600–699 Technology/Applied sciences (medicine, engineering, agriculture, and home economics)

700–799 The Arts/fine and decorative arts (civic and landscape art, architecture, sculpture, drawing, painting, graphic arts, photography, and music)

800–899 Literature and rhetoric (American, English, Germanic, Romance, Latin, Greek, and other culture literature)

900–999 Geography and history (travel, genealogy, geography, and Ancient World, European, Asian, African, and other continental general histories)

Each of these general categories is, in turn, divided into ten smaller categories. For example, the Technology/Applied sciences classification (600–699) is divided into the following:

600–609 Technology (Applied Sciences)

610–619 Medical sciences/Medicine

620–629 Engineering and allied operations

630–639 Agriculture and related technologies

640–649 Home economics and family living

650–659 Management and auxiliary services

660–669 Chemical engineering

670–679 Manufacturing

680–689 Manufacture for specific use

690–699 Buildings

Under each of these categories even more precise classifications apply. Familiarize yourself with the primary, secondary, and even third-level classifications within your field.

The **Library of Congress System** classifies books on the shelves by 21 letters to designate the following major subject categories:

A	**General Works**
B	**Philosophy and Religion**
C	**History and Auxiliary Sciences**
D	**Universal History and Topography**
E/F	**American History**
G	**Geography, Anthropology, Folklore**
H	**Social Sciences**
J	**Political Science**
K	**Law**
L	**Education**
M	**Music**
N	**Fine Arts**
P	**Language and Literature**
Q	**Science**
R	**Medicine**
S	**Agriculture**
T	**Technology**

    U **Military Service**
    V **Naval Services**
    Z **Library Science and Bibliography**

Each of these classes is subclassed by a combination of letters for sub-topics. For example, T (Technology) contains 16 subclasses:

    TA **General engineering, including general civil engineering**
    TC **Hydraulic engineering**
    TD **Sanitary and municipal engineering**
    TE **Highway engineering**
    TF **Railroad engineering**
    TG **Bridge engineering**
    TH **Building construction**
    TJ **Mechanical engineering**
    TK **Electrical engineering, Nuclear engineering**
    TL **Motor vehicles, Aeronautics, Astronautics**
    TN **Mining engineering, Mineral industries, Metallurgy**
    TP **Chemical technology**
    TR **Photography**
    TS **Manufactures**
    TT **Handicrafts, Arts and crafts**
    TX **Home economics**

A number notation is also assigned to each book.

    At your computer workstation begin your search for materials on your selected subject seeking books, bibliography listings, abstracts, periodical abstracts, and full-text articles to skim read. Print out selected lists and articles that support your narrowed subject and tentative thesis. Use your printed lists and notes to prepare bibliography cards or a computer file.

    The computer will first display a screen that lists the categories of inclusion, such as the overall catalog, magazine and newspaper indexes, abstracts, and other cataloging classifications. This is usually called the **Home Screen** as shown in Figure 14.1:

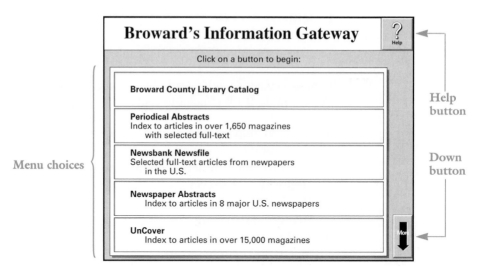

FIGURE 14.1   *A typical computer retrieval system*
Home Screen

When you click your mouse or trackball (a mouse embedded to one side of the keyboard) on a selected classification, the next page displayed is a **Dialog Box,** which allows you to enter an author, title, or subject for a specific search. Most libraries use the Library of Congress Subject Headings (LCSH) for classifying book, video, film, and other tangible holdings by subject, but you may, of course, use common sense to list these keywords. Your system will probably allow you to make numerous combinations using the bullean operators "and," "or," and "not." That is, if your subject is "cloning," you might type just "cloning" for a complete listing of library holdings on the subject. You might type "cloning *and* human body parts" to narrow the search to just those works dealing with cloning of animals to produce body parts for human transplants. Similarly, you may type "cloning *or* biological reproduction," which may locate works on cloning and other DNA reproduction studies. "Cloning *not* human" will locate just those works on animal cloning. The Dialog Box page will include icon buttons for Enter, Search Tips, Erase, Help, and so on. Figure 14.2 shows a typical Dialog Box page.

If you have typed a subject into the Dialog Box page, your next displayed page will be a **Title List Page,** which shows authors and titles, publication dates, and the type of holding (book, video, films, CD-ROMs, or audiocassettes) for each listing. Figure 14.3 shows a typical Title List page of general holdings.

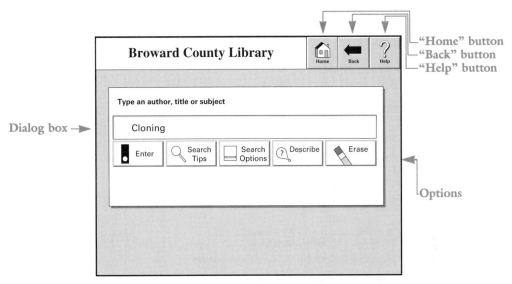

FIGURE 14.2 *A typical computer retrieval system Dialog Box with typed subject word*

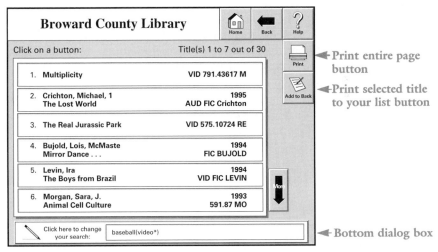

FIGURE 14.3 *A typical computer retrieval system Title List Page*

If this is a long list, the computer will invite you to scroll down for additional readings. You may print out selected titles or the entire list for future reference by clicking on the appropriate icons.

By clicking on a specific title on the Title List Page, the computer will display an **Individual Entry Page** with additional information for that one title, such as the location (which library in the system actually holds the material), publisher, number of pages, content headings, and/or subjects covered. Figure 14.4 shows a typical Individual Entry Page:

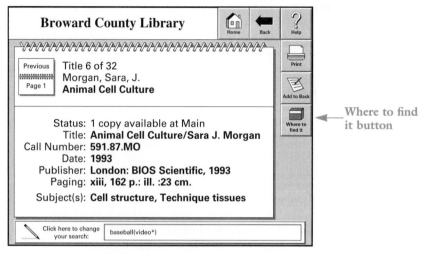

FIGURE 14.4   *A typical computer retrieval system Individual Entry Page*

If you actually print one of the Title List Pages, the printout will look somewhat different and will include even more information. Figure 14.5 shows a full, second-page printout of a Title List Page:

cloning			Broward County Library
8.	Author: Title: Date: Call #:	Berg, Paul 1926– Dealing with genes the language of heredity 1992 MN 575.1 BE	
9.	Author: Title: Date: Call #:	Drlica, Karl. Understanding DNA and gene cloning a guide for t 1992 MN 574.87328 DR	
10.	Title: Date: Call #:	Evolution man takes a hand 1990 VIDEO R 575.10724 EV	
11.	Author: Title: Date:	Carmen, Ira H. Cloning and the Constitution an inquiry into gov 1985 MN	
12.	Author: Title: Date: Call #:	Singer, Peter. Making babies the new science and ethics of conc 1985 MN 613.94 SI	

"MN" is location Key=Main

FIGURE 14.5   *A typical printout of a Title List Page; the printout includes full information*

If you have selected a different category on the Home Screen, such as Newspaper Abstracts, and then typed in a subject "cloning and medicine" on the Dialog Page, and finally clicked on "Enter," the monitor will display a list of periodical articles by author, title, date, and the name of the periodical. Figure 14.6 shows a Newspaper Abstract printout:

---

cloning and medicine   **Periodical Abstracts**

1. **Author:** Winston Robert
   Title: The promise of cloning for human medicine
   Date: 03/29/97 Text
   Call #: *British Medical Journal (Internatio

2. **Author:** Roberts Melinda A
   Title: Human cloning: A case of no harm done?
   Date: 10/96
   Call #: *Journal of Medicine & Philosophy

3. **Author:** Schick Barbara P
   Title: Hope for treatment of thrombocytopenia
   Date: 09/24/94
   Call #: *New England Journal of Medicine

4. **Author:** Brenner Sydney Lerne
   Title: Encoded Combinatorial Chemistry
   Date: 06/15/92
   Call #: *Proceedings of the National Academy

FIGURE 14.6   *A typical newspaper periodical abstract listing*

Figure 14.7 shows an abstract page for one of the selected articles:

The promise of cloning for human medicine   **Periodical Abstracts**

Title: The promise of cloning for human medicine

Magazine: British Medical Journal (International)
Mar 29, 1997

Page: 913–914

Volume: V314n7085

Abstract: The production of a sheep clone, Dolly, from an adult somatic cell is a stunning achievement for British science and holds great promise for human medicine as well. The media's treatment of this experiment is discussed.

Subject(s): 2
Editorials Media coverage Cloning Sheep Medical research Editorial

Graphics: References

FIGURE 14.7   *A typical actual abstract for a selected periodical title*

Finally, Figure 14.8 shows a section of a full-text article:

The promise of cloning for human medicine	Periodical Abstracts

Article Title: The promise of cloning for human medicine

Copyright British Medical Journal 1997 Words: 0000976 Headnote: Not a moral threat but an exciting challenge. The production of a sheep clone, Dolly, from an adult somatic cell is a stunning achievement of British science. It also holds great promise for human medicine. Sadly, the media have sensationalised the implications, ignoring the huge potential of this experiment. Accusations that scientists have been working secretively and without the chance for public debate are invalid. Successful cloning was publicised in 1975, 2 and it is over eight years since Prather et al published details of the first piglet clone after nuclear transfer. Missing from much of the debate about Dolly is recognition that she is not an identical clone. Part of our genetic material comes from the mitochondria in the cytoplasm of the egg. In Dolly's case only the nuclear DNA was transferred. Moreover, we are a product of our nurture as much as our genetic nature. Monovular twins are genetically closer than are artificially produced clones, and no one could deny that such twins have quite separate identities. Dolly's birth provokes fascinating questions. How old is she? Her nuclear DNA gives her potentially adult status, but her mitochondria are those of a newborn. Mitochrondria are important in the aging process because aging is related to acquired mutations in mitochondrial DNA, possibly caused by oxygen damage during an individual's life. Experimental nuclear

FIGURE 14.8   *A typical portion of a full-text article*

The process is similar for all of the Home Screen listings.

REFERENCE BOOKS

Reference books—general encyclopedias, technical encyclopedias, almanacs, handbooks, dictionaries, histories, and bibliographies—are listed in both computerized and traditional catalogs. Many systems allow you to read encyclopedia text at your workstation or by CD-ROM access. There are reference books for every discipline. Following is a partial listing of technical reference books:

Dictionaries

Oxford English Dictionary

Random House Dictionary of the English Language

Webster's Third New International Dictionary of the English Language

Webster's New World Dictionary, Third College Edition

American Heritage Dictionary

Dictionary of Engineering and Technology

Dictionary of Architecture and Construction

Dictionary of Telecommunications
Dictionary of Scientific Biography
Dictionary of Nutrition and Food Technology
Dictionary of Practical Law
Duncan's Dictionary for Nurses
Funk & Wagnall's Dictionary of Data Processing Terms
McGraw-Hill Dictionary of Scientific and Technical Terms
Paramedical Dictionary
Stedman's Medical Dictionary

Handbooks, Manuals, and Almanacs

Book of Facts
Almanac for Computers
CRC Handbook of Marine Sciences
Fire Protection Handbook
Nurse's Almanac
U.S. Government Manual
The World Almanac
Civil Engineering Handbook
The McGraw-Hill Computer Handbook

Encyclopedias

Encyclopedia American
Encyclopedia Britannica
Collier's Encyclopedia
Encyclopedia of Careers and Vocational Guidance
Encyclopedia of Computer Science
Encyclopedia of Food Technology and Food Service Series
Encyclopedia of Management
Encyclopedia of Marine Resources
Encyclopedia of Materials Handling
Encyclopedia of Modern Architecture
Encyclopedia and Dictionary of Medicine, Nursing, and Allied Health
Encyclopedia of Photography
Encyclopedia of Textiles
Encyclopedia of Urban Planning
Grolier's Multimedia Encyclopedia
McGraw-Hill Encyclopedia of Energy
McGraw-Hill Encyclopedia of Environmental Science
McGraw-Hill Encyclopedia of Food, Agriculture, and Nutrition

INDEXES TO ESSAYS WITHIN BOOKS

Following your search of general books, reference books, and periodicals, you may want to refer to one or more indexes of essays on your subject that are published within books of collected essays. Your computer search may not identify these essays unless the entire collection is on one subject. To check that you have not overlooked a collected essay on your subject, refer to the printed *Essay and General Literature Index,* which catalogs essays by author and subject. The index directs you to essays within books and collections of a bibliographical and/or critical nature. Figure 14.9 shows a section of a page from the *Essay and General Literature Index:*

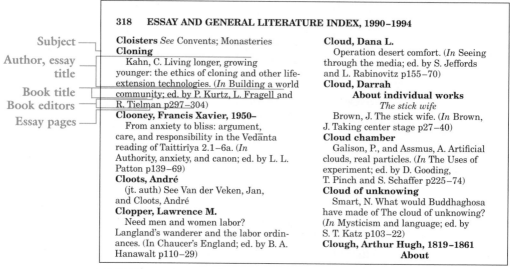

FIGURE 14.9   ***Section of a page from an Essay and General Literature Index***

ABSTRACTS

Abstracts are newspaper and technical indexes that, in addition to listing authors' names, publishers, titles, and publication data, include a brief summary of the content and scope of books, articles, or pamphlets in particular fields. These abstracts are available both on the computer and in printed form, and sometimes both, in a given library. Newspaper indexes include abstracts on commentaries, editorials, editorial cartoons, features, obituaries, reviews, and more and are updated monthly. Among the newspapers included are the "big eight":

The Atlanta Constitution and Journal   *Boston Globe*

Chicago Tribune   *New York Times*

Christian Science Monitor   *Wall Street Journal*

Los Angeles Times   *Washington Post*

Some technical abstracts, which are typically on the computer system, may be available only in print format. These include

Abstracts on Criminology and Penology   *Environment Abstracts*

Abstracts of Health Care   *General Science Index*

Air Pollution Abstracts   *Index to Publications of the United States Congress*

Biological Abstracts   *International Aerospace Abstracts*

Computer Abstracts

Criminal Justice Abstracts   *Nursing Research*

Current Index to Journals in Education   *Oceanic Abstracts*

Engineering Abstracts   *Solar Energy Update*

Figure 14.10 shows a partial page from the *General Science Index* with very brief abstracts:

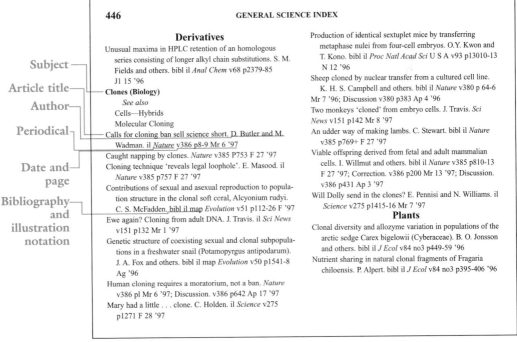

446   **GENERAL SCIENCE INDEX**

Derivatives

Unusual maxima in HPLC retention of an homologous series consisting of longer alkyl chain substitutions. S. M. Fields and others. bibl il *Anal Chem* v68 p2379-85 Jl 15 '96

Clones (Biology)
See also
Cells—Hybrids
Molecular Cloning
Calls for cloning ban sell science short. D. Butler and M. Wadman. il *Nature* v386 p8-9 Mr 6 '97
Caught napping by clones. *Nature* v385 P753 F 27 '97
Cloning technique 'reveals legal loophole'. E. Masood. il *Nature* v385 p757 F 27 '97
Contributions of sexual and asexual reproduction to population structure in the clonal soft coral, Alcyonium rudyi. C. S. McFadden. bibl il map *Evolution* v51 p112-26 F '97
Ewe again? Cloning from adult DNA. J. Travis. il *Sci News* v151 p132 Mr 1 '97
Genetic structure of coexisting sexual and clonal subpopulations in a freshwater snail (Potamopyrgus antipodarum). J. A. Fox and others. bibl il map *Evolution* v50 p1541-8 Ag '96
Human cloning requires a moratorium, not a ban. *Nature* v386 pl Mr 6 '97; Discussion. v386 p642 Ap 17 '97
Mary had a little . . . clone. C. Holden. il *Science* v275 p1271 F 28 '97

Production of identical sextuplet mice by transferring metaphase nuclei from four-cell embryos. O.Y. Kwon and T. Kono. bibl il *Proc Natl Acad Sci* U S A v93 p13010-13 N 12 '96
Sheep cloned by nuclear transfer from a cultured cell line. K. H. S. Campbell and others. bibl il *Nature* v380 p 64-6 Mr 7 '96; Discussion v380 p383 Ap 4 '96
Two monkeys 'cloned' from embryo cells. J. Travis. *Sci News* v151 p142 Mr 8 '97
An udder way of making lambs. C. Stewart. bibl il *Nature* v385 p769+ F 27 '97
Viable offspring derived from fetal and adult mammalian cells. I. Willmut and others. bibl il *Nature* v385 p810-13 F 27 '97; Correction. v386 p200 Mr 13 '97; Discussion. v386 p431 Ap 3 '97
Will Dolly send in the clones? E. Pennisi and N. Williams. il *Science* v275 p1415-16 Mr 7 '97

Plants

Clonal diversity and allozyme variation in populations of the arctic sedge Carex bigelowii (Cyberaceae). B. O. Jonsson and others. bibl il *J Ecol* v84 no3 p449-59 '96
Nutrient sharing in natural clonal fragments of Fragaria chiloensis. P. Alpert. bibl il *J Ecol* v84 no3 p395-406 '96

Subject
Article title
Author
Periodical
Date and page
Bibliography and illustration notation

FIGURE 14.10   *Section of a page from the General Science Index with very brief abstracts*

PERIODICAL INDEXES

Although periodical indexes for magazines, journals, and newspapers are now included in the computer catalog, a library may still hold selected periodicals or purchase inexpensive microform versions of the same material. The computer database indexes may include more than 15,000 magazines with abstracts and selected full-text articles. If the periodical or specific articles you seek are not on the computer, the material may be on a CD-ROM or on microforms (microfilm and microfiche).

Microforms are either reels (microfilm) or flat sheets (microfiche) of filmed information. You will need to locate the microform files and use either a microfilm or microfiche reader. You may make copies of microform materials or articles from actual periodicals that cannot be checked out of the library.

Some periodical indexes for general and technical subjects which are on computer databases or available in print or microform include:

Magazine and Journal Indexes

The Reader's Guide to Periodical Literature

Social Sciences Index

Biological and Agricultural Index

Cumulative Index to Nursing and Allied Health Literature

Energy Index

Engineering Index

Environmental Index

Hospital Literature Index

Index to U.S. Government Periodicals

Index Medicus

International Nursing Index

Microcomputer Index

Newspaper Indexes

The New York Times Index

Newspaper Index (Washington Post, The Chicago Tribune, New Orleans Times-Picayune, and The Los Angeles Times)

Wall Street Journal Index

Boston Globe Index

These databases or printed indexes are updated every few weeks. *The New York Times Index* is the most comprehensive of the large periodical indexes and is most likely to be included in a computer system. It indexes stories and articles that have appeared in the *Times* since 1851 by subject and author. The index also summarizes articles and occasionally reprints photographs, maps, and other illustrations that accompanied the original article. Back issues will be more likely accessed by referring to the printed index included on CD-ROMs and microfiche. Figure 14.11 shows a section of a page from the printed *New York Times Index:*

- 64 -

REPAIR SERVICES. See also Automobiles, F 23

REPLICAS. Use Models and Replicas

REPORTERS AND REPORTING. Use News and News Media

Subject——**REPRODUCTION (BIOLOGICAL). See also** Birds.

See also note——┌ F 18.23.24

Birth Control and Family Planning

Pregnancy and Obstetrics

Abstract——**British researchers report cloning adult mammal for first time, genetic engineering feat anticipated and dreaded more than any other;** group led by Dr Ian Wilmut embryologist at Roslin Institute in Edinburgh, created lamb using deoxyribonucleic acid (DNA) from adult sheep: achievement shocks researchers, who said it could not be done on assumption that DNA of adult cells would not act like DNA formed when sperm's genes mingle with those of egg; researchers say technique could, theoretically, be used to create genetically identical human, or time-delayed twin; Wilmut's experiment described: he comments; foresees benefits for medicine, notably aiding in study of genetic diseases; other researchers comment: thorny ethical and philosophical questions raised by experiment

Date, Page,——discussed; diagram (M), F 23, I, 1:6

Column

Dr Andrew Y. Silverman, fertility specialist, describes sperm separating technique he uses to help couples improve odds of predetermining sex of baby; questions have been raised about efficacy and ethics of procedure; photo (M), F 23.XIII-WC,1:1

Cloning of an adult sheep raises thorny ethical questions, especially on prospect of making carbon copies of humans; another issue is genetic diversity of livestock, if breeders start to clone animals; Dr Stanley Hauerwas, divinity professor, says those who want to clone will sell it with wonderful benefits for medicine and animal husbandry; Dr Kevin FitzGerald, Jesuit priest and geneticist, says clone of human being would have different environment than person whose DNA it carried, and so would be different person; other researchers have produced genetically identical animals by dividing embryos, but Dr Ian Wilmut is believed to be first to create clone using DNA from adult animal; Wilmut says he used udder, or mammary, cells from 6-year-old adult sheep; he is publishing his results in British journal Nature; photo; drawing (M), F 24, A, 1:2

First products to emerge from remarkable cloning of an adult sheep by British researchers will probably be animals that can serve as drug factories (M), F 24, B, 8:2

Man in the News profile of Dr Ian Wilmut (M), F 24, B, 8:5

Pres Clinton, reacting to what White House calls 'very starting news' that scientists in Scotland have successfully cloned sheep, asks Federal bioethics advisory commission to review research implications for human beings and report to him in 90 days (M), F 25, A, 1.5:7.

Editorial contends that startling news that scientists have cloned adult sheep, producing younger, genetically identical 'twin,' is reminder that reproductive technologies are advancing far faster than our understanding of their ethical and social implications; maintains that immediate goal is to produce better animals for agriculture, but most troubling issues involve potential for cloning adult humans; asserts that it is hard to escape thought that most such experiments in humans could backfire, given crucial role that environment plays in human development (M), F 25, A, 26:1

Experts in United States say Dr Ian Wilmut of Scotland would never have won Federal grant for sheep-cloning project because goal of making sheep that are drug factories was too practical and because he comes from world of animal science, not high-technology world of molecular biology; success of effort astounds biologists, who believed cloning adult animals was impossible and pay little attention to animal science issues (M), F 25, C, 1:1

Shares in PPL Therapeutics PLC soar on London Stock Exchange following news that scientists in the tiny Scottish biotechnology company successfully cloned a sheep, in collaboration with researchers at Roslin Institute in Edinburgh; despite investors' enthusiasm, there is often vast gulf between scientific breakthrough and marketable product; analysts note that biotechnology industry's short history is littered with remains of companies that succeeded in laboratory, but came up short on route to sales success; analysts warn that investment in PPL is subject to technology risk, because use of genetically altered animals may not be best way to make drugs (M), F 25, D, 1:2.

Daniel J Kevles Op-Ed article says cloning should be regulated but not prohibited; drawing (M), F 26, A, 23:2

Dr Daniel Callahan Op-Ed article opposes cloning; says human species does not need it and will not benefit from it (M), F 26, A, 23:2

New York State Sen John J Marchi, jumping into debate started by recent cloning of sheep, says that he wants New York State to outlaw cloning of human

FIGURE 14.11   *Section of* **The New York Times Index**

In addition, every newspaper indexes its own publication. Your major area newspapers will probably be included on your computer catalog. Each newspaper maintains an office section called "the morgue," which includes a computer index of its editions plus past editions in print. Companies also index their newsletters. Both newspaper offices and companies will assist you with their indexes and supply you with copies or printouts of the articles you need.

CD-ROM INDEXES

Index CD-ROMs are electronic bibliographies of thousands of article titles that can be accessed by subject words or keywords. There are two types of CD-ROM indexes: index only and full-text. The computer general catalog listings will list these CD-ROMs. There are hundreds of files now available. Some CD-ROM diskettes include

> *Applied Science and Technology*
>
> *Government Publications Index*
>
> *Infotrac* (monthly listing of articles from 900 business, technical, and general magazines and journals)
>
> *Investext* (index of company, industry, and market analysis with more than 1 million research reports from more than 500 investment firms, brokerage houses, market research organizations, trade associations, and worldwide companies)
>
> *LegalTrac* (monthly listing of entries from 750 legal publications)
>
> *Population Statistics*

DIALOG is an on-line database using telephone modems to connect a library with a national vendor of databases. Available files include

> SCISEARCH (articles on scientific topics)
>
> GEOBASE (articles on geology and earth sciences)
>
> ERIC (articles on education and related topics)

There are hundreds more on biological sciences, business, chemistry and chemical engineering, computer science, ecology, economics, ethnic studies, geography, geology, law, medical studies, physics, and all of the humanities, social sciences, and philosophy. They may, however, be expensive to use and are not necessarily recommended for undergraduates if other sources can be used. Some libraries offer limited free searches whereas others charge.

VERTICAL FILE

Many libraries maintain a vertical file on timely subjects covered in pamphlets, booklets, bulletins, and clippings that would not otherwise be cataloged. Ask the reference librarian to assist you in locating the file and the material contained within it.

ONLINE SEARCHES

Most library computer systems will allow you to search for materials online. These searches may be conducted on personal computers. (See the later section on "Personal Computer Searches")

TRADITIONAL CARD CATALOGS

In those libraries that still use traditional card catalogs (an alphabetized compilation of cards) instead of the PAC computerized system for reference and general books, you may locate sources on cards under three separate headings: the author's name, the title of the work, and the subject heading. Thus, a book by Donn B. Parker titled *Fighting Computer Crime* will be cataloged under *P* for Parker, Donn B., under *F* for *Fighting Computer Crime,* and under *C* for Computer Crimes—Prevention. Figure 14.12 shows a typical author card with explanations of the data and Figure 14.13 is the title card for the same book.

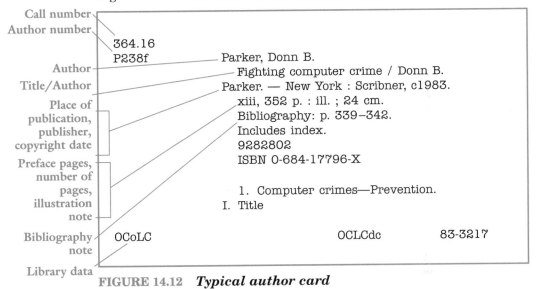

Call number

Author number

364.16
P238f

Author

Title/Author

Place of publication, publisher, copyright date

Preface pages, number of pages, illustration note

Bibliography note

Library data

Parker, Donn B.
Fighting computer crime / Donn B.
Parker. — New York : Scribner, c1983.
xiii, 352 p. : ill. ; 24 cm.
Bibliography: p. 339–342.
Includes index.
9282802
ISBN 0-684-17796-X

1. Computer crimes—Prevention.
I. Title

OCoLC   OCLCdc   83-3217

FIGURE 14.12 *Typical author card*

364.16
P238f

Parker, Donn B.
Fighting computer crime / Donn B.
Parker. — New York : Scribner, c1983.
xiii, 352 p. : ill. : 24 cm.
Bibliography: p. 339–342.
Includes index.
9282802
ISBN 0-684–17796–X

1. Computer crimes—Prevention.
I. Title

OCoLC   OCLCdc   83–3217

FIGURE 14.13 *Typical title card*

If you do not know the authors or titles of books on your subject, you may locate books by looking up *subject headings.* The subject heading "Computers" will be further divided into subcategories, such as "Computer Crimes—Prevention" and "Computers—Access Control." Figure 14.14 shows the subject card for the same book:

```
                      COMPUTER CRIMES—PREVENTION.
     364.16
     P238f              Parker, Donn B.
                          Fighting computer crime / Donn B.
                        Parker. — New York : Scribner, c1983.
                          xiii, 352 p. :ill. ; 24 cm.
                          Bibliography: p. 339–342.
                          Includes index.
                          9282802
                          ISBN 0-684-17796-X


                          1. Computer crimes—Prevention.
                        I. Title
     OCoLC                       OCLCdc                      83–3217
```

FIGURE 14.14   *Typical subject card*

PERSONAL COMPUTER SEARCHES

You may conduct online research on the library computer system or on your personal computer. Library systems include the database indexes to which you may subscribe individually, but they tend to be expensive. Nevertheless, you may locate valuable research materials on the Internet. You may access subjects by keywords, specific Internet addresses (usually World Wide Web addresses), or hyperlinks (text and pictures that link to other pages in the same web site or even to other web sites). Materials that you locate online may be downloaded directly to your computer's memory, disk drive, or printer. One caution applies, however. Establish that the sources of articles and information are reputable before you include them as professional paper resources. Anyone can post almost anything on the Internet.

Several encyclopedias are online, such as *The Concise Columbia Encyclopedia, The New Grolier's Multimedia Encyclopedia,* and *Compton's Living Encyclopedia.* These are good general references to get your research started, and the articles also include bibliographies for further research.

Articles from your own local newspaper plus those of other nearby large cities (and other major U.S. newspapers) are accessible. Some other full-text periodicals (newspapers and magazines) not previously mentioned are

Air Force Times

Army Times

Business Week

Congressional Quarterly

Health Magazine

Nature Conservatory

Navy Times

Newsweek

Scientific American

Smithsonian Magazine

Others are being added constantly. You may also access newspapers from Africa, Asia, Australia, Canada, Central America, South America, Europe, and elsewhere.

You may purchase your own diskettes or CD-ROMs for ongoing reference availability. For example, all 43 volumes of the *Encyclopedia Britannica* are on one CD-ROM, which scans a 44 million-word database; indexes more than 400,000 references on more than 65,000 subjects with hypertext links to the actual text articles; provides more than 10,000 links to images and tables; and includes the *Merriam-Webster's Collegiate Dictionary, Tenth Edition.* Other CD-ROMs for your personal library include *The New York Public Library Desktop Reference, The Grolier Interactive Multimedia Encyclopedia,* and *Microsoft Encarta.*

You can subscribe to electronic "bulletin boards" (interactive specific-subject news groups) on the USENET. Using special news reader software, you may post questions on a specific subject and retrieve responses from all over the world. Free specific-subject mailing lists are also available for interactive discussions. Using the World Wide Web, you may look for Listservers that provide a list of the topics they service and subscription instructions. There are thousands of such listed subjects on which you may post questions and then access responses.

Using the Web affords the broadest range of access and the greatest possibility of finding information relating to your subject. You may accomplish your search utilizing any number of Web browsers, such as Netscape, Microsoft Internet Explorer, or Mosaic. Although Web addresses are always subject to change or deletion, and others are added constantly, some currently helpful addresses for student researchers in the sciences and technology follow:

www.refdesk.com

My Virtual Reference Desk site is the bible for hyperlink access to text and graphics on weekly U.S. and worldwide wire services, news services (print, radio, and television), special reference sources (astronomy, finance, politics, weather, etc.), facts reference sources, and weekly updates on encyclopedia entries plus access to *Encyclopedia Britannica, World Fact Book, Roget's Thesaurus, Bartlett's Quotations,* and so much more.

www.elibrary.com

The Electric Library site is a moderately priced subscription database that is undoubtedly on your library computer system. However, you may subscribe as an individual. Designed for student research, it includes more than 900 full-text periodicals, 9 international newswires, 2,000 classic books, hundreds of maps, thousands of photographs, as well as major works of literature, art, and reference from publishers such as Archive Photos, Reuters, Simon & Schuster, Gannett, and *Times Mirror, U.S. News & World Report, National Review Publisher, Inc., New Republic, Psychology Today,* and hundreds more.

www.answers.com

A question/answer service site that is run by retired librarians, college professors, and other professionals who monitor your posted questions and respond with several paragraphs or more including hotlinks to Web sites if they are included in the research. Fees are charged according to the extent of research required and range from under $2.00 to $15.00 or more.

www.loc.com

The Library of Congress site with resources for researchers and information professionals including the catalogs of the Library of Congress and other libraries plus databases on special topics.

www.access.GPO.gov

The United States Government Printing Office site with links to official documents covering all areas of government.

www.whitehouse.gov

The White House site with information on the executive branch, cabinet posts, and independent federal agencies including the president's speeches, publicly released documents, executive orders, copies of the budget, and other major publications.

www.NASA.gov

The National Aeronautic and Space Administration site with text, audio, and visual access to its projects.

www.nsf.gov

The National Science Foundation site with research information and press releases on a wide variety of program areas including biology, computer/information sciences, education, engineering, geosciences, math/physical sciences, polar research, and social and behavioral sciences.

www.snre.umich.edu/nppc/

The National Pollution Prevention Center (NPPC) site established at the University of Michigan, representing a collaborative effort between the Environmental Protection Agency, business, industry, nonprofit organizations, and academia to educate faculty, students, and professionals about pollution prevention.

www.noaa.gov

The National Oceanic and Atmospheric Administration site with information on seasonal and annual climate forecasts, environmental information, information on sustaining healthy coasts and fisheries, protected species information, navigation studies, and more.

http://quake.wr.usgs.gov/

The U.S. Geology Survey site with information about earthquakes, earth crustal studies, seismology studies, and more.

www.geo.mtu.edu/weather/aurora/

The Aurora Borealis information site including solar and geophysical reports, Hubble Space Telescope images and information, and space physics information and links to comet information and other astronomical phenomena, Rice University's Department of Space Physics and Astronomy, the U.S. Naval Research Laboratory research activities, and more.

http://user88.lbl.gov/NSD_docs/abc/home.html

A nuclear science site with information on radioactivity, fission and fusion, atomic structure, half-life studies, cosmic rays, and experiments, and a link to an interactive question/answer site.

www.srl.caltech.edu/

The California Institute of Technology Space Radiation Laboratory site with information on its projects, cosmic ray physics and astrophysics experiments, X-ray and Gamma-ray astronomy, computational astronomy research studies on energetic particles and cosmic rays, a bibliography, and links to other physics departments around the country.

www.nejm.org/JHome.htm

A weekly *New England Journal of Medicine* site with results of important medical research and abstracts on articles from current and past issues, complete text of many articles and partial text of others, plus information on ordering entire articles by fax or e-mail.

www.medsite.com/

A comprehensive site allowing keyword searches for medical topics and diseases with a rating system for the linked sites.

www.CNN.com

The CNN Television Network site containing world, U.S., and local news and links to current scientific/technical full-text stories and pictures plus links to research sources.

www.sec.gov/edgarhp.htm

The Security and Exchange Commission's site with annual reports, 10K or 10Q filings, and other information.

You may also connect to most colleges and universities in the United States. They will provide directories for their special scientific and technical departments and schools listing research activities and including library links. Usually you need only type http://www.college name.edu. Try Oberlin College, Harvard, Yale, Cornell, Rutgers, Stanford, Amherst, Massachusetts Institute of Technology, and so on. William and Mary's address is **www.wm.edu.** Cornell's is **www.cornell.edu/uao/intro.html,** and Johns Hopkins University is **www.jhu.edu.** You may even connect to the great universities of Great Britain; Cambridge University is **www.cam.ac.uk/,** Oxford is **www.ox.ac.uk/,** and the University of Edinburgh is **www.ed.ac.uk.**

A few more valuable sites for up-to-the-minute periodical articles are **www.altavista.digital.com,** a search site to the Web index of 31 million pages on 670,000 servers and 4 million articles from 14,000 news groups; **www.newsworks.com,** a compendium of news articles from the newspapers of nine major media companies with hyperlinks to special subjects (Sci/Tech, News, Business, and more), and **www.hotwired.com** which reaches out to more than 100 of the Web's best news sites and locates the articles by keywords. Newsworks includes the *Philadelphia Enquirer, Houston Chronicle, Detroit Free Press, Boston Globe, Miami Herald,* and the *Los Angeles Times.* Hotwired includes the *New York Times, The Washington Post, Advertising Age, The Chicago Tribune,* and *Wired* magazine. If you are on the Internet, you will discover many other sites with information within your research field.

BIBLIOGRAPHY CARDS OR FILES

Each time you locate relevant source material, create a separate bibliography card or add to your computer file of sources. Each entry should contain

1. Author's name
2. Title of the work
3. Publication information
4. Library call number, database or Internet identification code, or other identifier
5. An annotation about the contents of the source (optional, but strongly recommended)

Later you will transform this information into the documentation style you will be using (see Chapter 16 for APA, MLA, CBE, and other styles). The various styles include the proper formats for citing books, periodicals, government documents, pamphlets, and so on plus the formats for electronic sources, such as online Internet sources, CD-ROMs, e-mail, and databases. Each style also has separate forms for advertisements, art works, broadcast interviews, bulletins, computer software, conference proceedings, dissertations, film or video recordings, personal interviews, letters, microfilm, public addresses or lectures, and recordings on record/tape/disk.

Figures 14.15 and 14.16 (see p. 458) show two typical hand-written bibliography cards. You may, of course, key this same information into a bibliography file.

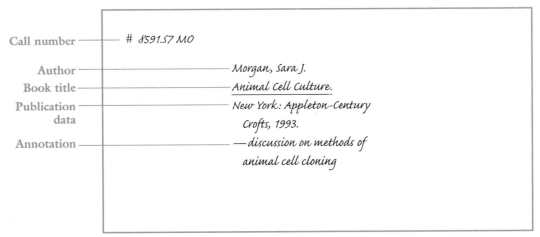

FIGURE 14.15   *Annotated bibliography card for a book*

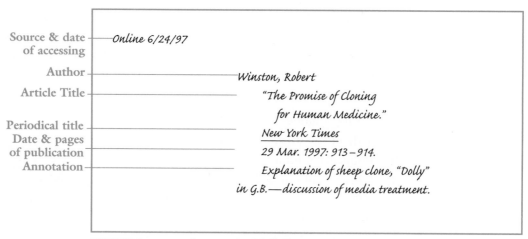

Source & date — Online 6/24/97
of accessing

Author —

Article Title —   Winston, Robert

                                                         "The Promise of Cloning
                                                              for Human Medicine."
Periodical title —   New York Times
Date & pages
of publication   29 Mar. 1997: 913–914.
Annotation —   Explanation of sheep clone, "Dolly"
                                              in G.B.—discussion of media treatment.

FIGURE 14.16   *Annotated bibliography card for a periodical
located on the Internet*

You will use these cards or file entries to relocate material for note taking and for compiling into the appropriate format for a complete bibliography, works cited, or reference listing at the end of your paper.

PLAGIARISM

Notes must be recorded carefully to avoid plagiarism. The word *plagiarism* derives from the Latin word *plagiarus,* which means "kidnapper." In effect, plagiarism in writing is kidnapping the words (apt phrases or entire sentences) and ideas (the organization of material, another's argument, or a line of thinking) and then presenting them as if they were your own original thinking, organization, or phrasing. The published words, ideas, and conclusions of an author are protected by law; therefore, the use of the material of others must be acknowledged whenever it appears in your work. Failure to provide acknowledgment may result in a failed paper, failure in the course, expulsion from school, and lasting impairment of your scholarly reputation. To avoid plagiarism, follow five rules:

1. Introduce all borrowed material by stating in the text the name of the authority from whom it was taken.
2. Enclose any exact wording within quotation marks.
3. If you summarize or paraphrase material, be sure that the information is written in your own style and language.
4. Provide a parenthetical reference note for each borrowed item.
5. Provide in your bibliography, works cited, or reference list an entry for every source that is referenced in your text (see Chapter 16).

Following is an original excerpt followed by three note cards. The first two (Figures 14.17 and 14.18) are unacceptable plagiarisms, while the last (Figure 14.19, see p. 460) is an acceptable note.

Original

The explosive development of the computer has created change both inside and outside an organization. Inside the organization the computer has altered work assignments and provided management with new tools for decision making. Outside the organization the computer is greatly changing our way of life and will continue to exert a force for change. Some

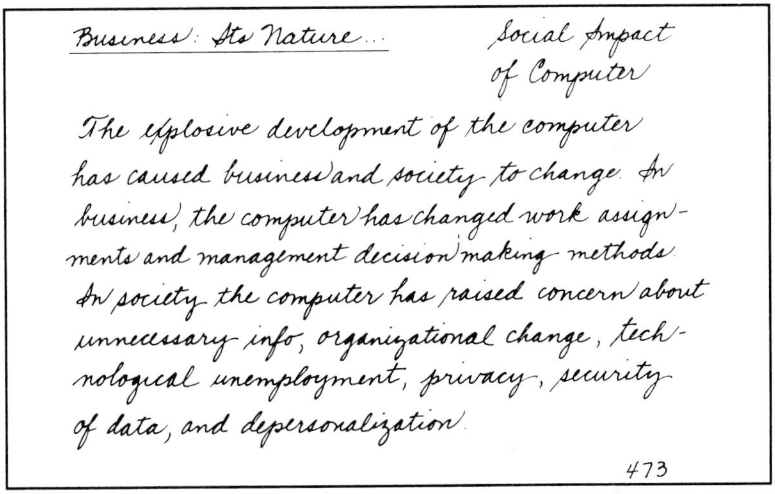

FIGURE 14.17   *Plagiarized note*

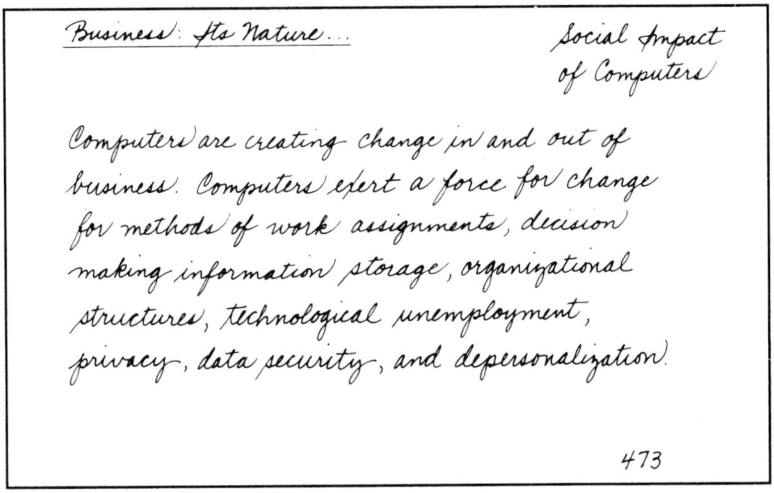

FIGURE 14.18   *Less obvious, but plagiarized note*

Business : Its Nature...   *Social Impact*
                                                                      of Computers

Computers are revolutionizing society. Professors
Glos, Steade, and Lowry point out that " the
explosive development" of computers affects not
only the businessman but also all of us. They
list seven major concerns including changes in
management methods, excessive information, em-
ployment layoffs, and privacy and security
worries.

                                                                              473

FIGURE 14.19   *Acceptable note*

of the social concerns related to the computer are: generation of
unnecessary information, organizational changes, fear of technological
unemployment, invasion of privacy, security of data, and depersonalization.

In the acceptable note: (1) the information is summarized, (2) the language is the note taker's, (3) the author's exact phrases or sentences are
in quotation marks, and (4) the source is acknowledged in the text. In
addition, a parenthetical reference to the source and a complete "Works
Cited" entry will appear in the final draft of the paper.

TAKING NOTES

With your thesis firmly in mind at all times and with a working outline
as a guide (see Chapter 15), begin now to record the notes from each of
your sources. Notes are written on 3 × 5 in. or 4 × 6 in. index cards or
entered into computer files. If you keyboard your notes into a computer,
write each note as a separate temporary file under a common directory
so that you may move it later as appropriate. Alternatively write all your
notes into a single file using code words or phrases that can be located by
a SEARCH command as you write the actual paper. Save all of your note
cards or printed copies of your computer files to submit with your final
research paper should your professor require them.

Note-Taking Guidelines

Use a consistent system to record your information:

1.   Enter only one item of information on each card or under each
     computer file notation so that you may continuously organize

and rearrange the cards or computer notes during all stages of the research.

2. If you use cards, write on only one side of each. A note continued on the back may be overlooked. If an item of information is too lengthy for one card, use two or more and staple them together.

3. Record the topic notation or code words or phrases at the top left or top right of the card or computer note. This topic notation or code may correspond with an entry in your outline.

4. Record the source at the top right or left of the card or file. This notation may be an abbreviated form of the bibliography card, such as the author's last name, the book or article title, or both.

5. Write or type the note in the center of the card or computer list. Refer to the types of notes and their explanations in the next section of this textbook.

6. Record the exact page number(s) or online computer material identifier(s) from which you took the note.

Types of Notes

There are specific strategies for taking notes:

1. **Quotation notes.**   Copy the exact words and syntax of an unusually well-expressed opinion from an authority to support your own thesis or topic sentences.

2. **Precis notes.**   Copy a review note, abstract, or bibliographic annotations for later development.

3. **Summary notes.**   Record factual data or opinions to support your own ideas.

4. **Paraphrase notes.**   Use your own words entirely to interpret and express the opinions and explanations of an authority. It is likely that most of your notes will be paraphrase notes.

5. **Personal notes.**   Record your own ideas for further development or inclusion as you research your materials.

Sample Note Cards

Most of your notes will consist of summaries or paraphrases, that is, condensations or restatements of the source material in your own words. Following is an excerpt as it appears in the original source material. Notice that the note (Figure 14.20, see p. 462) both summarizes the information and restates the message in the note taker's own language.

Original

For workers one of the most fearful aspects of the computer is its capacity for controlling machinery to perform tasks that were formerly done by human labor. The unemployment that results is called *technological unemployment*. Employees who are eliminated by automation are generally unskilled and semiskilled workers who find difficulty in seeking new employment. Because of the growing use of computerized checkouts in supermarkets, the Retail Clerks Union fears a loss of 25 to 30% of supermarket jobs. Perhaps many of these fears are unjustified. The telephone companies now employ more persons than they did before the use of computerized switching services.

Source ┃ Glos and others 1995 ┃ Technological Unemployment ┃ Topic notation

Note ┃ Because computers can perform tasks previously done by people, technological unemployment results. Unskilled and semiskilled workers lose their jobs and cannot find other employment. Despite the fact that telephone companies hire more people due to computerized switching services, workers' unions fear that employment levels will fall due to computerized services ┃ 309 ┃ Page #

FIGURE 14.20   *Summary and paraphrased note*

Occasionally, you will want to quote the exact words of an author. Usually no more than 10 percent of the text should consist of direct quotation. Take notes of direct quotation only if the material conveys a highly original idea, opinion, or conclusion of the author that, if paraphrased, would lose its striking effect or distort the meaning.

Following is an original excerpt plus two sample notecards (Figures 14.21 and 14.22), one employing full sentence quotation and the other employing partial sentence quotation. Note the use of the ellipsis (. . .) to indicate omitted words and the brackets ([   ]) to indicate a slight addition by the note taker.

Original

Because of the sea of dangers, past practice has been to isolate sensitive computers. That strategy proved itself this summer at the Los Alamos National Laboratory in New Mexico when young raiders from Milwaukee succeeded only in cracking an unclassified computer. There is no evidence that they actually obtained classified information or gained access to the top-secret computers, which remain unconnected to networks.

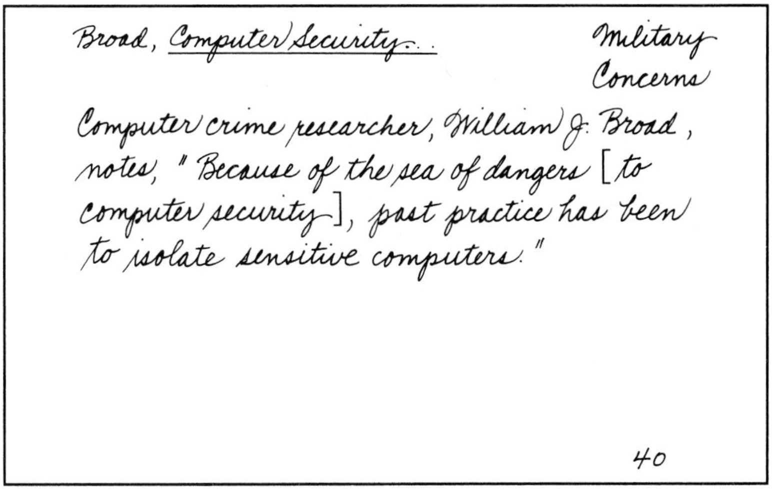

Broad, Computer Security... Military Concerns

Computer crime researcher, William J. Broad, notes, "Because of the sea of dangers [to computer security], past practice has been to isolate sensitive computers."

40

FIGURE 14.21 ***Full sentence quotation note***

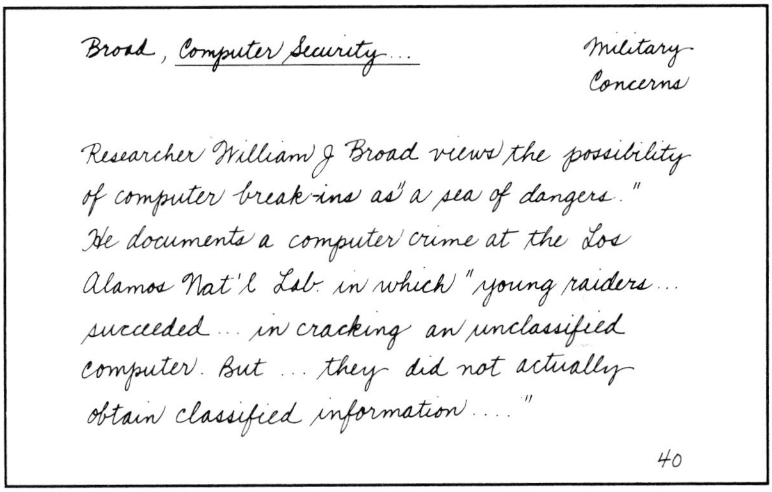

Broad, Computer Security... Military Concerns

Researcher William J Broad views the possibility of computer break-ins as "a sea of dangers." He documents a computer crime at the Los Alamos Nat'l Lab. in which "young raiders... succeeded... in cracking an unclassified computer. But ... they did not actually obtain classified information...."

40

FIGURE 14.22 ***Partial quotation note***

SURVEYS AND QUESTIONNAIRES

Surveys and questionnaires are often used to gather data to guide the writer and to back up conclusions. Essentially these primary sources serve three purposes:

- Gathering opinions of many people
- Determining solutions to a problem
- Substantiating predetermined positions.

Care must be taken in selecting the test group, motivating a response, constructing the questions, and compiling the results. The survey sheet itself is referred to as the **survey instrument** or the **questionnaire.** Figure 14.23 shows a sample survey instrument:

AAA's Traffic Safety Survey

Since its founding in 1902, traffic safety has been one of the building blocks of AAA's foundation. Proactive AAA traffic safety programs start in elementary school and continue through mature vehicle operator education.

The School Safety Patrol program supports over 48,000 children in the Florida, Louisiana, and Mississippi public and private school systems. AAA also has proactive alcohol programs that start in kindergarten, along with supplemental printed materials that can be used for vehicle operators of all ages.

Today traffic safety programs include pedestrian, bicycle, school bus, and a variety of driver safety materials. AAA investigates member concerns on local enforcement and traffic engineering issues as well as serves community organizations dealing with transportation and traffic safety issues.

This survey will help us to understand better the traffic safety concerns of AAA members, both now and for the future. Please answer these questions as accurately as possible with the information provided and mail or fax us your answers. Please choose or write in only one answer per question unless otherwise instructed.

1. Should AAA offer a driver improvement refresher program for those over 55?
_____yes
_____no
_____undecided

2. If yes, what would you be willing to pay for this type of course?
_____$10
_____$15
_____$20
_____Other (please specify in $5 increments) $_____

3. Should AAA become involved in a vehicle theft deterrent program, such as Combat Auto Theft?
_____yes
_____no
_____undecided

4. If yes, would you like to see AAA (check all that apply):
_____Provide a brochure on vehicle theft deterrence?
_____Provide vehicle identification number (VIN) etching?
_____Provide decals for a theft deterrent program?
_____Other (please specify)
_____

5. Please indicate the programs you would like to see AAA continue (check all that apply):
_____ Pedestrian safety
_____ Bicycle safety
_____ Alcohol education

_____ Safety Patrol
_____ School bus safety
_____ Safe driving for mature operators
_____ Enforcement investigations
_____ Traffic engineering investigations

6. Should AAA observe major highway construction projects to make sure they adhere to federally established guidelines for motorist safety?
_____yes
_____no
_____undecided

7. Would you like AAA to provide effective reflective materials that can be used by pedestrians, bicyclists, and others to increase their visibility to motor vehicle operators?
_____yes
_____no
_____undecided

8. Would you like local-issue traffic safety articles and/or a column in each issue of *Car & Travel*?
_____yes
_____no
_____undecided

9. Would you like traffic safety information offered on the Member Services page of *Car & Travel* (see page 34)?
_____yes
_____no
_____undecided

10. Sex
_____Male
_____Female

11. Age
_____18–34
_____35–54
_____55–64
_____65 or above

12. Marital status
_____Married
_____Single

13. Annual household income
_____$25,000 or below
_____$25,001–$40,000
_____$40,001–$55,000
_____$55,002–$75,000
_____$75,001 or above

Thank you for your participation in this survey. Please return your completed survey by **Jan. 31** to *Car & Travel*, 1000 AAA Drive, MS73, Heathrow, FL 32746-5080; or fax it to us at 407/444/4140. As always, the information provided in this survey is confidential, but if you would like to be entered in a drawing to win a *Car & Travel* golf shirt, please provide your name, address, and size.

Name_____

Address_____

City/state/Zip_____

Golf shirt size (circle one)

Medium   Large   XLarge

FIGURE 14.23 **Sample survey instrument** (By permission AAA Florida *Car & Travel* magazine [January/February 1997]: 17. Adaptation.)

Advantages

Written surveys have several advantages:

1. Carefully formatted questions require only a check-mark in a box to speed up the survey response process.
2. A large sample of data may be collected in a short time.
3. Even distribution (employee mailboxes, targeted mailings) ensures that all relevant people are actually contacted within a given time frame.
4. Printing and distribution are less costly and more efficient than person-to-person inquiries and interviews.
5. Written responses are usually less biased than those garnered by personal interviews.
6. The data can easily be tabulated by counting or computer keypunching.
7. The data are on hand to point out other problems and avenues for future study.

Disadvantages

A survey is not always advantageous. Certain disadvantages or draw-backs may result from the use of written surveys or questionnaires:

1. A *low* response rate (less than 10 percent of those both affected by the study and actually surveyed) will result in unreliable data.
2. A *slow* response rate slows up your findings.
3. Certain members of your group may not respond to written surveys because they are uninformed, disinterested, careless, lazy, prejudiced, or offended.
4. Others may not respond accurately due to inability to read or to understand the precise meaning of your terms.
5. Some people may be reluctant to respond to questions regarding personal data or to offer opinions for fear of repercussions.

Test Group

Selecting the size of the test group and determining the number of responses required to assure validity are problems for the true statistician. For the purposes of accessing information for your reports, try to survey all of the people who will probably be affected by your proposal for change, such as all of the employees in an office or department, all of the users of an inadequate parking lot, all of the residents of your apartment or condominium complex, or all of the neighbors touched by a city project. Remember, you need a 10 percent response rate of those affected and surveyed for validity. Consider whether you wish to collect numbers or percentages, or both, as well as comments, suggestions, and the like.

Familiarize yourself with your survey group. Are they knowledgeable about your subject? Are you looking for new data, substantiation of a position, or opinions and suggestions to determine direction? What prejudices might influence answers (age, sex, race, organization rank, religion, politics, length of involvement, etc.)? Should your tone be formal or folksy?

Motivating Responses

You need a clearly defined goal first. What are you trying to determine? Do your own research of the problems and develop possible solutions before developing your survey. Consider the following motivating factors.

Time.   Do not begin to write your proposal or other type of report until you have your survey results. Consider a schedule for devising, typing, printing, distributing, collecting, and tabulating your data. Distribute forms at an opportune time, not when people are on vacation or occupied with IRS returns, Christmas, and the like. Sometimes distribution at a large meeting of the affected people will guarantee a quick, high return on the spot. Preference and opinion questions do not require research on the part of the respondents. Urge your group to return the form(s) to you quickly by hand, by mailing a preaddressed envelope, or by collecting the forms at a designated spot at a set date and time.

Sponsorship.   Make it clear that you are seeking the information under the auspices of an institution, organization, professor, manager, or the like. Gathering data for a college professional and technical writing course project should lend prestige to your inquiry.

Cover Sheet.   A cover letter, memo, information sheet, or simply an explanatory paragraph above the survey questions may accompany the survey instrument stating the purpose, nature of the report, and how the data will be analyzed. This is a logical place to state the sponsorship of the survey. Emphasize the importance of the survey and offer to share results of the study if that is feasible. Above all, stress that you need a return as soon as possible.

Anonymity.   Allow for anonymity by guaranteeing confidentiality. Do not ask for names on the forms, particularly if the material covers touchy subjects such as opinions about supervisors or administrators, work schedules, and raises. Ensure anonymity also for questionnaires about highly personal feelings on such issues as abortion, capital punishment, personal habits, and so on.

Format

The overall appearance and arrangement of the instrument should motivate responses. Follow these suggestions:

1. Select a good quality, possibly colored, paper, which suggests care in preparation.
2. Title the survey carefully to clarify the content.
3. Keep your questions as brief as possible.
4. Provide tick boxes for responses and printed lines for comments.
5. Do not crowd your questions. Allow for plenty of white space.
6. Use language carefully. Such words as *should, might*, and *could* often "lead" answers. Do not ask offensive or very personal questions if possible. (Avoid: *Should employees be blood-tested for AIDS?* [ ] yes [ ] no)
7. Ask for only one piece of information per question. (Avoid: *Do you favor a personnel reorganization of your department or would you be interested in a new position?* [ ] yes [ ] no)
8. Ask for comments, opinions, or other alternatives when it is appropriate to do so. Type in lines rather than just leaving white space for the answers.

Types of Questions

There are six conventional types of questions. Each has its own advantages and disadvantages in tabulation.

1. **Dual alternatives.** This is the easiest type of data to tabulate in that there are only two choices: yes or no, positive or negative, true or false, and so on. Take care to ensure that there really are only two possible answers to your question.

   Sample: Have you eaten purchased cookies in the past month:

   [ ] yes   [ ] no

2. **Multiple choice questions.** This type also produces easy-to-tabulate data in that you provide a number of alternatives to indicate a fact, preference, or opinion. Your survey may ask for a single tick or multiple ticks.

Sample: Check the **one** cookie brand you most prefer.

[   ] Hydrox   [   ] Fig Newtons
[   ] Oreo   [   ] Pepperidge Farm
[   ] Duncan Hines   [   ] Almost Home

or

Check **all** of the cookie brands you purchase on a regular basis.

[   ] Hydrox   [   ] Fig Newtons
[   ] Oreo   [   ] Pepperidge Farm
[   ] Duncan Hines   [   ] Almost Home

3. **Rank ordering.**   This format provides respondents with a series of items to rank according to preference, frequency of use, or other criteria. Tabulating this data is easy. However, distortion may occur if the mean (the average) differs from the mode (the most frequently occurring response). The items must be significantly different to make rank choices.

> **Sample:**   Rank (1 highest to 6 lowest) the following cookie brands in order of your preference.
>
> _____ Hydrox
>
> _____ Oreo
>
> _____ Duncan Hines
>
> _____ Fig Newtons
>
> _____ Pepperidge Farm
>
> _____ Almost Home

4. **Continuum scales.**   This format is similar to rank ordering in that it provides a method for respondents to express opinions by rank ordering numerically or verbally on a continuum. Scales are most valid if they have an *even* number of choices (usually 4–6). If a scale has a middle choice, respondents choose it a disproportionately large percentage of the time.

> **Sample:**   Mark the response which best indicates the frequency of your cookie purchases per grocery trip.
>
> [   ] always   [   ] often   [   ] seldom   [   ] never

5. **Completions.**   This format asks respondents to provide facts and/or opinions in either fill-in or open-ended responses. Questions on age, frequency, or amount are easier to tabulate than are questions which allow for open-ended responses.

> **Samples:**   (Fill-in type) I usually buy _____ cookie products a month.
>   number

or

(Completion type) The quality which most determines my cookie product choice is

_____

6. **Essays.** This format asks respondents to express fully opinions or facts. The results may be difficult to tabulate and arrange into groupings. Such questions are more effective if they urge a focus on certain criteria.

Sample: Suggest criteria which would motivate you to purchase Fig Newtons more frequently. Consider butter content, sugar content, thickness, consistency, and amount of filling.

_____

_____

_____

_____

Data compiled from a survey may be presented within a paper in tables, bar charts, circle charts, line graphs, and the like. Figure 14.24 shows a sample survey instrument to determine the popularity of a particular cookie, who buys it, and how the product should be improved. It asks for respondent classification data and employs a variety of types of questions.

PRODUCT PREFERENCE SURVEY

———————— Classification ————————

(multiple
choice)

1. Please indicate your age bracket.
    [ ] 16–20   [ ] 21–30   [ ] 31–40   [ ] 41 or over

(dual
alternative)

2. Are you [ ] male   [ ] female?

(multiple
choice)

3. What is your primary occupation?
    [ ] student   [ ] housewife
    [ ] professional manager   [ ] farmer
    [ ] operative/laborer   [ ] retired
    [ ] foreman/craftsman   [ ] not employed
    [ ] clerical/sales

(completion)

4. Fill in the blank with a number. I generally shop
    for groceries_____times a month.
                    number

———————— Cookie Questions ————————

(dual
alternative)

1. I am the primary cookie shopper in my household.
    [ ] yes   [ ] no

(continuum)

2. In the span of a year I purchase cookie products
    [ ] often   [ ] frequently [ ] seldom   [ ] never

(multiple
choice)

3. Check all of the brands of cookies which you
    purchase on a regular (once or more a month) basis.
    [ ] Hydrox   [ ] Fig Newtons
    [ ] Oreo   [ ] Pepperidge Farm
    [ ] Duncan Hines   [ ] Almost Home

FIGURE 14.24   *Sample survey/questionnaire*

(rank order)

4. Rank your order of preference (1–6) for each brand.
   [ ] Hydrox   [ ] Fig Newtons
   [ ] Oreo   [ ] Pepperidge Farm
   [ ] Duncan Hines   [ ] Almost Home

(completion)

5. Fill in the number. I usually buy _____ cookie products a month.   number

(multiple choice)

6. Which factor most determines your cookie brand choice?
   [ ] packaging   [ ] overall taste
   [ ] calorie content   [ ] cost
   [ ] additives (nuts,   [ ] consistency (chewy,
       raisins, fillings)   crisp, crunchy)

(essay)

7. Please comment on improvements which would motivate you to purchase Fig Newtons more often.

   _____

   _____

   _____

   _____

FIGURE 14.24   *continued*

Figure 14.25 suggests ways that such data may be incorporated into the text of your report:

Female respondents indicate a lower incidence of cookie purchases than do males. Figure 1 shows the frequency of cookie purchases by sex:

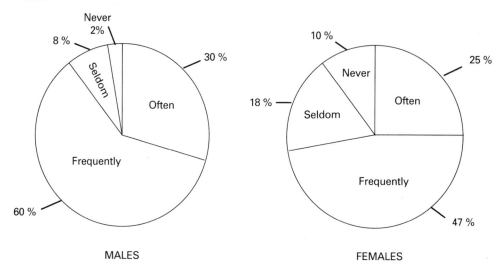

Figure 1   Frequency of cookie purchases per month by sex

Seventy-two percent of female respondents purchase cookies often or frequently; whereas, a full 90 percent of male respondents purchase cookies often or frequently. The mean number of cookie sales per month is two for all occupational categories except for students who indicate a mean number of six purchases per month. Table 1 shows the number of purchases a month by occupational categories:

Table 1   Mean number of purchases per month—600 respondents

Category	Number of purchases
Professional/manager	4
Operative/laborer	2
Foreman/craftsman	2
Clerical/sales	2

FIGURE 14.25 *Samples of data incorporation into text using graphics*

Housewives	3
Farmers	1
Retirees	2
Not employed	0
Students	6

The 600 respondents indicate that taste is the major factor in determining cookie brand selection. Figure 2 shows the percent of respondents who indicate other determining factors:

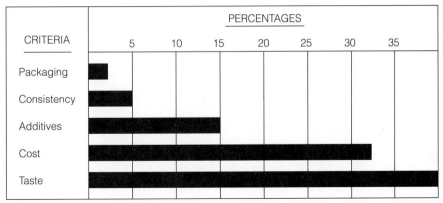

Figure 2   Determining purchase factors—600 respondents

As shown, a full third of the respondents indicated that cost is the major factor. Packaging has the least influence on cookie purchasing.

FIGURE 14.25   *continued*

OTHER PRIMARY SOURCES

Personal testing, interviews, letters and their responses, lectures, public forums, and direct observation may also provide material for a carefully researched paper. Again, you should include an explanation of the test and its results, the interview questions and answers, the letter of inquiry and its response, a thorough explanation of public forum material, and your direct observation.

CHECKLIST

Accessing Information

THE SOURCES

❑ **1.** Have I accessed all of the possible sources that pertain to my limited subject and that will define the subject and support my thesis?

    ❑ General books?

    ❑ Special reference books (dictionaries, handbooks, manuals, almanacs, and encyclopedias)?

    ❑ Periodical literature?

    ❑ Essays within books?

    ❑ Abstracts/indexes?

    ❑ Vertical files?

❑ **2.** Have I examined all of the possible media that might contain sources for my paper?

    ❑ Specific subject databases

    ❑ CD-ROMs

    ❑ Audiocassettes?

    ❑ Videocassettes?

    ❑ Film?

    ❑ Other?

❑ **3.** Have I accessed the Internet for additional sources and text?

    ❑ Specific addresses?

    ❑ Keyword searches?

    ❑ Bulletin boards and mailing lists?

    ❑ Subscription databases?

    ❑ Personal CD-ROMs?

❑ On-line periodicals?

❑ Other?

❑ **4.** Have I skimmed the material to eliminate redundant and useless sources?

BIBLIOGRAPHY CARDS OR FILES

❑ **1.** Have I prepared a separate bibliography card or computer file entry for each source that I will use?

❑ **2.** Have I included all of the necessary information on each card or file listing?

❑ Author's name?

❑ Title of the work?

❑ Library call number, database, Internet address, and source type?

❑ An optional personal note for future use?

❑ **3.** Have I alphabetized these cards or files?

❑ **4.** Am I prepared to assemble a final bibliography or works cited page(s)?

NOTE TAKING

❑ **1.** Have I recorded all of my notes with an eye to avoiding plagiarism?

❑ **2.** Have I recorded the information on each card (or computer file) so that I may access it when I write my paper? Have I

❑ Recorded only one item per card?

❑ Used just one side of each card?

❑ Stapled together multiple cards dealing with one item?

❑ Recorded a topic notation at the top left of each card?

❏ Recorded the source in abbreviated form at the top right of each card?

❏ Recorded the exact page number or other similar notation of the item?

❏ 3. Have I included a variety of types of note cards?

❏ Precis notations of summaries, abstracts, and listed bibliographies?

❏ Summary notes?

❏ Direct quotations that are unusually well expressed?

❏ Paraphrased information?

❏ Personal notes and ideas?

❏ 4. In taking my notes have I employed ellipses to indicate omitted words, phrases, or sentences in a direct quotation?

❏ 5. Have I employed brackets to add clarifying words within a direct quotation?

❏ 6. Have I incorporated display quotations correctly?

SURVEYS, QUESTIONNAIRES, AND OTHER PRIMARY SOURCES

❏ 1. If I am devising a survey instrument or questionnaire, have I given careful consideration to

❏ Advantages?

❏ Disadvantages?

❏ Test group selection?

❏ 2. Have I motivated a response to my instrument?

❏ Date deadline?

❏ Sponsorship authority?

❏ Cover sheet explanation?

❏ Anonymity guarantee?

❏ Other?

❏ **3.** Have I formatted my instrument carefully?

    ❏ Paper quality concerns?

    ❏ Clarifying title?

    ❏ Brevity?

    ❏ Tick boxes and printed lines for responses?

    ❏ White space?

    ❏ Avoidance of leading, offensive, and too personal questions?

    ❏ One query per question?

    ❏ Room for comments, opinions, or other alternatives?

❏ **4.** Have I used a variety of types of questions?

    ❏ Dual alternatives?

    ❏ Multiple choice?

    ❏ Rank ordering?

    ❏ Continuum scale?

    ❏ Completions?

    ❏ Essays?

❏ **5.** Have I considered any other primary source research for the paper?

    ❏ Personal testing?

    ❏ Interviews (see Chapter 11)?

    ❏ Letters of inquiry (see Chapter 10)?

    ❏ Public forums?

    ❏ Direct observation?

❏ **6.** Am I prepared to incorporate the data by a number of appropriate graphics and a commentary?

❏ **7.** Am I prepared to include the survey instrument or questionnaire and the results or other primary source material in my final paper?

EXERCISES

1. **Subclassifications of a Broad Subject.** Locate by using computer system databases or printed indexes the periodicals in your library that are most appropriate to your field of study *(Applied Science and Technology Index, Business Periodicals Index, Microcomputer Index, Education Index, General Science Index,* etc.). Look up a general subject, such as nutrition, medicine, fire science, radiology, capital punishment, electronics, dentistry, push technology, fiber optics, solar heat, cloning, space exploration, landscape design, tourism, cardiology, dental hygiene, pollution, aviation, and so on. List five subclassifications that could be researched for the development of a 1,000-word paper. Consider if the material is recent, not too technical, and available in sufficient amount.

2. **Limiting Your Subject Search.** Using your library computerized catalog, look up the same broad subject as in exercise 1. Limit your search in three different ways using "and," "or," and "not." For instance, if you select cloning, search "cloning not human," "cloning and medicine," and "cloning or animal reproduction." Use the bullean operators for your broad subject. Print out the bibliographies located under each. Submit these along with the headings you used to derive the bibliographies.

3. **Internet Search.** If you have access to the Internet, locate five sources through the Web or keywords that contain information on your subject. List the addresses plus a brief summary of the material available.

4. **Bibliography Cards.** Prepare four separate bibliography cards on your subject, one each for a book, a magazine article, a newspaper article, and an encyclopedia entry. You may either use common sense in deciding what information (author and editor names, title, publisher, place of publication, year) to include on each card, or preview Chapter 16 to select the AMA, APA, Number System, or other style suitable for your subject.

5. **Note taking.** Read the following excerpt from an article on the environment:

UTILITIES

Antidote for A Smokestack

Question: What do 52 million trees in Guatemala have to do with one coal-burning power plant in Uncasville, Conn.? Answer: they form a healthy environmental equation. That is the hope of Virginia-based Applied Energy Services, a builder and operator of power plants in Texas, Pennsylvania and California. Like any other coal-fired generator, the 180-megawatt plant now under construction in Uncasville will spew carbon dioxide, the chief culprit in the globe-warming greenhouse effect. But acting on a recommendation from the World Resources Institute, a Washington environmental-policy research center,

AES has voluntarily donated $2 million in seed money to a CARE project in Guatemala designed to stave off the climatic crisis by replanting depleted forests. The AES donation, along with help from the Peace Corps and the Guatemalan forestry service, will help an estimated 40,000 local farmers plant some 52 million seedlings that eventually will absorb a quantity of CO_2 roughly equal to the amount generated at Uncasville over the 40-year lifespan of the facility. Says AES chief Roger Sant, "Given the scientific consensus on the seriousness of the greenhouse problem, we decided it was time to stop talking and act."

a. Write a summary note of the entire excerpt.

b. Write a paraphrase note of the sentence: "Like any other coal-fired generator, the 180-megawatt plant now under construction in Uncasville will spew carbon dioxide, the chief culprit in the globe-warming greenhouse effect."

c. Write a note containing a logical *partial* quotation from the sentence: "But acting on a recommendation from the World Resources Institute, a Washington environmental-policy research center, AES has voluntarily donated $2 million in seed money to a CARE project in Guatemala designed to stave off the climatic crisis by replanting depleted forests."

6. Note taking.   Obtain a copy of an article from your bibliography searches on your subject and write three types of note cards: a partial direct quotation, a summary note, and a paraphrase note. Record the six necessary pieces of information on each card. If the note requires more than one card, staple the cards together. If you enter your notes in a computer file, submit a printed copy.

WRITING PROJECTS

1. **Paraphrasing.**   In your own words paraphrase the section on plagiarism in this chapter.

2. **Group Project: Survey/Questionnaire.**   In groups of three devise a survey or questionnaire to determine facts about on-campus or at-work parking facilities. Use a variety of question types. Determine the number of spaces and the hours per day lot(s) are used, and the problems encountered (too few spaces, puddles, distance, accidents with car doors, vandalism, muggings, etc.). List options to improve the facilities and determine a "rating of the options" procedure for the respondents. Seek additional comments. Record statistics, such as whether the respondent is a student, faculty, or staff member. Consider listing compact, medium, luxury, and recreation vehicle ownership to determine if responses differ for each type of vehicle. What other control data might you include? Title your survey or questionnaire and write a brief purpose statement that will also motivate serious and timely responses.

3. **Group Project: Survey Analysis Paper.**   In the groups of three established for writing project 2, administer the survey to three classes other than this one by obtaining another professor's permission to do so. Use the data that you compile and write a two- or three-page paper that analyzes the data. Include at least two graphics and a commentary that interprets your findings.

4. **Group Project: Mini Research Paper.**   In groups of three select a subject, limit the subject, and devise a thesis to give direction to your paper. Locate at least three sources that provide the best information to support your thesis and compile three separate bibliography cards. Make copies of the research material. Prepare note cards on these three sources, and using the note cards *only,* write a brief (five-paragraph) report. Make use of the information in Chapter 15 for structural guidance (subject selection, thesis statement, writing strategies, outline, introductions, and closing). If your professor asks you to document the paper, use one of the methods discussed in Chapter 16.

CHAPTER 15

Writing Professional Papers

SHOE   by Jeff MacNelly

S K I L L S

After studying this chapter, you should be able to

1. Understand the nature of professional scientific and technical papers.
2. Understand the difference between scientific writing and science writing.
3. Locate and be familiar with journals in your field.
4. Discover the editorial policy of three or more technical journals in your field.
5. Understand how to submit an article for publication.
6. Critique the overall organization and content of a scientific/technical paper.
7. Select and limit scientific/technical and academic subjects.
8. Formulate a suitable thesis for a professional paper.
9. Prepare a working and polished outline.
10. Brainstorm a report by considering specific expository development methods.
11. Develop a suitable introduction and closing to a professional paper.
12. Write a brief, documented professional paper.
13. Begin to recognize different styles of documentation by examination of the sample papers.

INTRODUCTION

Professional papers include documented scientific and technical articles intended for publication in special interest journals and documented academic papers. The latter include undergraduate term papers, graduate theses, and dissertations, intended to hone the writer's research skills and to present experimental results and original insights into existing knowledge in a given field. Both types, scientific/technical articles and academic papers, are characterized by documentation, accuracy, and clarity.

Science writing and scientific writing are not quite the same thing in that they address different audiences. *Science writing* explains scientific matters, such as new discoveries, theories, and research, to laypeople in magazines and journals aimed at a general audience. Excellent examples of science writing can be found, for example, in the magazine *Discover, The New Yorker* magazine section "Annals of Science," *The New York Times,* and other major newspapers. Courses in science writing are taught in journalism schools. *Scientific writing* is done by scientists writing about science (those same discoveries, theories, and research) for other scientists in specialized journals aimed at specific audiences. Technical writing involves not only all of the strategies covered in previous chapters but also, as in scientific writing, preparing articles for publication. Moreover, these technical articles are designed for peers, rather than laypeople. Generally science writing is not documented and uses more expressive language than scientific/technical and academic papers. Compare the differences in language in the following two examples of writing about photographs from the Hubble Space Telescope:

Example 1—Scientific Writing

The photograph taken inside the Eagle Nebula depicts three columns of dark dust and gas rising into a blue sky, which reveals new-forming stars. The columns are illuminated from the heat of the new stars; a red turns to gold at the edges. Spirals of interstellar material form into steep peaks and disperse into weblike forms.

Example 2—Science Writing

The photograph [taken inside the Eagle Nebula] must be the Hubble Space Telescope's most emotion-laden image yet—an icon to rival the Apollo photograph of the Earth taken from the surface of the moon. It depicts three eerie pillars of dark dust and gas thrusting up into a blue iridescence sprinkled with newborn stars. The pillars glow from the heat of star birth: a deep red fades to halos of gold at their margins. Whorls of interstellar dust alternately billow into cliff-top peaks and trail off in wispy webs of stellar Spanish moss.[1]

Scientific/Technical Articles

Professional papers written for publication are prepared by researchers to inform other specialists and those in similar areas of current developments in the field. These papers report on new research, discoveries, new methodology, and inventions to keep colleagues abreast of the latest advances. Because the articles have been reviewed by an editorial board of specialists, readers may rely on the authority of the material. Students should seek out the journals that print papers in their field and read at least one or two of them regularly. The most recent findings in any field will be in the journals, not in textbooks.

Academic Papers

As a student, you probably think that you will not be required to write scientific and technical papers until much later in your career. However,

---

[1] © 1996 *The New Yorker* magazine. Davis Sobel, "A Reporter at Large: Among Planets," December 9, p. 90.

it is likely (especially if you are majoring in one of the sciences or technologies) that you will be required to write professional-level papers while still in college. The papers you may write as an undergraduate student are most likely to be confined to either class-directed laboratory reports or papers reporting on secondary research you have done for a particular class. On the other hand, papers written at the graduate level are often the result of original, primary research and experimentation. These papers are often submitted for publication by the student, the professor, or both and must be in perfect form, as your reputation (and future) may be greatly hindered by a poorly written paper. Conversely, an excellent paper will enhance your employment and advancement potential.

THE SCIENTIFIC/TECHNICAL PAPER

The professional writing a scientific/technical paper must consider four elements for successful submission and/or publication:

- Designing, researching, and completing the project
- Selecting the intended journal for publication
- Writing the paper
- Submitting the paper for publication

Designing, Researching, and Completing the Project

The scientific/technical paper usually requires primary and secondary research (see Chapter 14). The interpretation or thesis of a scientific/technical paper often comes from reading conflicting technical reports and articles by other authors or from one's own intuition. Unlike secondary research in which the writer proves a premise with facts determined by other scholars' research, primary research involves using original sources, such as experimentation, surveys, or other means, to obtain first-hand data.

Some projects may involve pre-approval and funding for primary research. One may seek approval for the project from an institution, a targeted journal, or a variety of funding sources in the field. If the research is primary and is funded by a university or other outside source, accurate records must be maintained along with any combination of laboratory reports, progress reports, and even EPA (Environmental Protection Agency) and OSHA (Occupational Safety and Hazard Act) reports. Once the research has been completed and verified, the writer needs to compose the paper for submission to the university or the appropriate journal.

Selecting the Journal

The first step in reviewing the journals in a field in order to select one for article submission is to determine which are the most prestigious, which are most likely to publish your type of material, and which are aimed at the desired audience. By selecting a journal first, the writer may tailor the article to its requirements. Most journals publish statements that define the intended audiences, comment on their editorial policy for selecting unsolicited articles, and specify style and documentation guidelines. Figures 15.1 and 15.2 (see pp. 486 and 487) are examples of author guidelines for the *Journal of Dental Research* and the *Technical Communication Quarterly,* respectively.

As is evident, guidelines may specify general length, margins, pagination, spacing, number of copies to submit, methods for handling graphics, and other details. In addition, a clear policy will suggest the content, organization of materials, the acceptable level of technical language, and expected prose style (paragraph length, passive or active voice, British or American English, etc.). A close review of published articles will also help you to ascertain the acceptable style.

The journal's published guidelines will also clarify which documentation style to use, how to title the paper, and whether or not to use topical headings. The published statement may direct you to include an abstract (see Chapter 13), reference lists, your credentials in a formal or informal resume, institutional affiliation, or other support materials.

If you are seeking to publish in a journal, you may wish to approach those journals that seem most promising. In a brief letter, you should ask if the journal is interested in your intended paper by clarifying the specific subject, emphasizing the importance of your article, explaining the approach you have used to develop the paper (for example, lab findings and research), and summarizing the particulars (length, graphics). You should also include your credentials, organization affiliation, address, phone number, fax number, and even e-mail address.

Writing the Report

The professional scientific/technical paper usually includes the following elements:

- A precise title
- Definition of the problem
- Review of the literature
- Methods and materials
- Results
- Discussion
- Conclusions and implications
- Documentation

INSTRUCTIONS TO AUTHORS

Manuscripts: Reports of original research and special review articles that have been submitted solely to the Journal are considered for publication. Manuscripts devoted to the description of a procedure or an apparatus, to improvement of a procedure or apparatus, to nonexperimental research, or to case reports are not acceptable. Manuscripts should be written in clear, concise, and grammatically correct English. Those not adequately prepared will be returned to authors, since it is not possible for the editors to revise extensively or to rewrite manuscripts. Contributors who are unfamiliar with English usage are encouraged to seek the help of colleagues in preparing manuscripts for submission. Only those papers that meet high standards of scientific quality will be accepted for publication.

Submit manuscripts in triplicate (one original and two copies), double-spaced on $8^{1}/_{2} \times 11$-inch bond paper.

After September 1, 1979, the submission fees of $45.00 for a standard paper and $25.00 for an Annotation will be discontinued. There will be no charge for papers that do not exceed five Journal pages (approximately 13 typewritten pages, including charts and photographs). A page charge of $100.00 will be assessed for each page over five.

All manuscripts should be submitted to Barnet M. Levy, Editor, The University of Texas Dental Branch, P.O. Box 20068, Houston, Texas 77025, U.S.A.

The editor reserves the right to edit manuscripts to ensure conciseness, clarity, and stylistic consistency. The corresponding author will receive page proofs for checking and reprint order information. Changes in page proofs should be minimal. Authors will be charged for excessive changes.

Length of Manuscripts: Manuscripts should not exceed five Journal pages or about 13 typewritten pages, *including* tables, charts, illustrations and references. Manuscripts requiring more than five Journal pages will incur a charge of $100.00 for each extra page.

Title Page: List a concise title, author(s), their professional address(es), and a short title for running heads. Include footnotes acknowledging sources of support on this page.

Synopsis: The second page should be a synopsis of 75 words or less.

Key Terms: A list of five or six terms descriptive of the subject matter should be supplied on the synopsis sheet for use in indexing and abstract services.

Text: Follow this standard form: Sections should be labeled Introduction, Materials and methods, Results, Discussion, and Conclusions. Proprietary names and sources of all commercial products must be given in footnotes; the text should contain only generic names. Follow the Council of Biology Editors Style Manual, 4th ed., 1978.

References: Type double-spaced on a separate sheet. Follow the Journal style, keeping the references to the text in numerical (not alphabetical) order. Limit references to those specifically referred to in the text. Lengthy reviews of the literature are not appropriate. Examples of the Journal form for bibliography are:

1. JONES, B. D.; BANNA, C. S.; and CISCO, F. O.: Experimental Induction of Bone, *J Dent Res* 58:22-24, 1959.
2. BINNER, B. A. and JONES, R. M.: **Book on Dental Materials,** 5th ed. Philadelphia: W.D. Leabiger Co., 1944, p. 359.

Illustrations: Submit in triplicate. Photographs should be unmounted black and white glossy prints of good quality. On the back of each, lightly pencil the author's name, the figure number, and indication of the top edge. Unmounted glossy black and white prints of diagrams, charts, graphs, and drawings are preferred. Lettering should be large enough to be read after the drawings are reduced to column or page width. All legends should be grouped on a separate sheet in the numerical order referred to in the text.

Tables: Type each table on a separate sheet. Tables should be titled and numbered consecutively. Do not use vertical rules. Footnotes should be brief. If the above format is not followed, the manuscript will be returned to the author for correction prior to review by the referees.

Brief Communications and Annotations: Brief communications reporting experimental data of immediate importance to scientists will be considered for accelerated publication after the usual review and acceptance. Standards for acceptance of brief communications are rigorous. The paper, including tables, illustrations and references, must not exceed three printed pages (about seven and a half typewritten sheets).

Annotations are limited to one printed page of 650 words. If illustrations or tables are included, the text must be correspondingly shorter. If the annotation exceeds the required length, processing will be delayed while it is cut by the author. Insert references within the text following the style used in current issues of the Journal.

Letters to the Editor: The editor invites concise letters commenting on papers published in recent issues. Those letters selected for publication will first be referred to the author, whose response may also be published. The editor reserves the right to reject any letter or to make appropriate editorial changes.

Rejected manuscripts should not be resubmitted to the Journal. All manuscripts will be quality-rated by at least two referees. Manuscripts with low ratings will not be accepted.

FIGURE 15.1 *Typical editorial policy statement* (From *Journal of Dental Research*)

Journal Guidelines

Manuscript Submissions

Send submissions directly to

Mary M. Lay, TCQ
Department of Rhetoric
201 Haecker Hall
1364 Eckles Avenue
University of Minnesota
St. Paul, MN 55108-6122

All submissions are assumed to be original, verified, unpublished, and not under present consideration by any other publisher.

Follow these guidelines when preparing manuscripts for submission:

- Send an original and three copies of manuscript. Prepare a floppy disk for submission when article is accepted. Place entire manuscript in one document using ASCII, Word, WordStar, WordPerfect, or Macintosh text files.
- Prepare cover sheet with title of the manuscript, author's name, place of affiliation, mailing address, phone number, and a 35-word biographical sketch.
- Do not make references to the author in the text or on any page besides the cover sheet.
- Center title on top of first page of text.
- Do not exceed 6000 words in the text.
- Provide a 50–75 word informative abstract.
- Tables and figures should appear on separate pages at the end of the text.
- Provide camera-ready quality copies of all figures.
- Include no footnotes.
- Indicate italics by underscoring or italicizing.
- Indicate a first-level, or primary, header by boldface or all capital letters.
- Indicate a second-level, or secondary, header by indenting and boldface or all capital letters.
- Use the parenthetical method for citing a reference in the text, according to The MLA Style Manual (1985).
- Provide a list of works cited at the end in a section called "Works Cited."
- Follow The MLA Style Manual in the "Works Cited" section.
- Include first names of all authors in "Works Cited" and first mention in text.

FIGURE 15.2 *Another editorial policy statement* (From *Technical Communication Quarterly.* By permission.)

TITLE

Often a journal will require both a short title, such as "Improving Testing Components," and a longer title, such as "Environmental Stress Testing: Improving the Quality and Reliability of Testing Components." Essentially, devise a precise title that clearly indicates the focus and the content to follow.

DEFINITION OF THE PROBLEM

Defining the problem is the major task of the introductory material. What weaknesses or shortcomings in previous techniques and methods have prompted the investigation that has led to the preparation of the scientific/technical paper? State the purpose of the research that focuses the resultant paper. State your thesis clearly.

REVIEW OF THE LITERATURE

Review the results found in other literature on the subject. Summarize the findings to provide a context for your new discussion; the summary lends authority by indicating that you have a command of the subject.

METHODS AND MATERIALS

Next, explain the methods and materials that you have used in your research. Decide whether to discuss your methods in the passive voice ("Next, the leg bone was connected to the thigh bone"), in the active voice ("The leg bone connects to the thigh bone"), or in the instructional mode ("Connect the leg bone to the thigh bone"). Carefully explain the steps taken in the investigation.

RESULTS

Discuss the exact results of your investigation. The Methods section and Results section are the major components of your article.

DISCUSSION

Explain any steps, results, or information about the product or procedure that have not yet been clarified.

CONCLUSIONS

Finally, draw valid conclusions and suggest implications for future use of the product or procedure. These closing strategies are similar to the conclusions and recommendations of other reports.

DOCUMENTATION

Include the internal references and a final bibliography, which may be titled *Endnotes, Bibliography, References,* or *Works Cited,* depending on the documentation style you are using (see Chapter 16).

Figure 15.3 (see pp. 490–494) shows a professional technical article written for the *Journal of Dental Research* and includes the four headings: Introduction (the problem and review of the literature), Materials and methods, Results, and Discussion. It is documented in the CBE System style with variations in the reference listings requested by the publishing journal.

SUBMITTING THE REPORT
FOR PUBLICATION

Once you have written the paper and determined the submission guidelines of the selected journal, you should mail two copies of the article with a cover letter and either a self-addressed, stamped envelope (SASE) or loose postage, to ensure that the editors will acknowledge receipt of your manuscript. Usually the manuscript is not bound, is in proper manuscript form, and has the author's name at the top, right-hand corner of each page.

Increasingly journals request a computer disk of the document, either along with the hard copy or in place of it.

Pack the cover letter and manuscript (or disk) between cardboard covers and mail it in a padded envelope (jiffy bag) to protect the contents. Do not despair if your article is not accepted; fewer than 5 percent are. Instead, use the suggested revision notes (if included) as free advice and keep trying.

ACADEMIC PAPERS

As an undergraduate student, you are probably not involved in developing new technical products and materials on which you can report. Nevertheless, you can research the latest findings within your field and develop a research paper of publishable value. As a graduate student, you will be required to write a number of documented papers throughout your studies. The academic paper process requires a review of general expository writing skills along with the documentation techniques that will help you to write clearly. The skills include

1. Selecting and limiting the subject
2. Formulating the thesis
3. Preparing the preliminary bibliography
4. Brainstorming the methods of development

MATERIALS SCIENCE

Microscopic Study of Smooth Silver-plated Retention Pins in Amalgam

Y. GALINDO,* K. McLACHLAN,** and Z. KASLOFF***

*Université Laval, Ecole de Médecine Dentaire, Ste-Foy, Québec, Canada G1K 7P4; **University of Manitoba, Faculty of Civil Engineering, Winnipeg, Manitoba, Canada R3T 2N2; and ***University of Colorado, School of Dentistry, Denver, Colorado 80262

A silver-plating technique was developed in an effort to produce good mechanical bonding characteristics between stainless steel pins and amalgam. Metallographic microscope and scanning electron microscope (SEM) studies were made to assess the presence, or otherwise, of such a bond between (a) the silver layer plating and the surface of the stainless steel pins, and (b) and silver plating and the amalgam. Unplated stainless steel and sterling silver pins were used as a control and as a comparison, respectively. A "rubbing" technique of condensation was devised to closely adapt amalgam to the pins. It is concluded that there is strong evidence for the existence of a good bond between the plated pins and amalgam. The mechanical performance of the bond is discussed elsewhere.[1]

J Dent Res 59(2):124-128, February 1980

Introduction.

Metallic pins are used in dentistry to retain large amalgam restorations and to oppose tilting forces when the restoration is subjected to occlusal loads. The inability of these pins to bond with amalgam contributes to the formation of high stress concentrations around the top of the pin.[2,3] Fracture of the amalgam is likely to ensue.

Previous investigations[2] have shown that stress concentrations around smooth pins possessing bonding properties with amalgam are significantly less than is the case with non-bonded pins. Likewise, such pins increase the retention of the restoration[3] and markedly reduce the occurrence of fracture and crack propagation.[4,5] The reduction of strength, due to the inclusion of pins in amalgam specimens, has been shown to be much less when bonded, rather than unbonded pins are used.[2,4,6,7,8]

Received for publication November 29, 1978
Accepted for publication February 29, 1979
This study was supported in part by the Medical Research Council of Canada, Grant MA-4065.
*Present address: Canadore College, Health Sciences, P.O. Box 5001, North Bay, Ontario, Canada P1A3X9

124

This evidence led to the investigation and development of smooth pins likely to provide a metallurgical bond with amalgam, and be clinically acceptable in all other respects. A microscopic evaluation was selected for establishing the goodness of the pin-amalgam metallurgical adaptation which, in turn, would indicate the likely quality of any bond.

Materials and methods.

Smooth, silver-plated, stainless steel pins were the primary subject of this study. They were compared with sterling silver and unplated stainless steel pins. Sterling silver pins were used only as a mode for comparison, since it has been shown that they adapt well and form a metallurgical bond with amalgam.[1,4,5,7,8] Unplated stainless steel smooth pins were also included as a control, because it is well known that they do not form a bond with amalgam.

Silver-plated stainless steel pins were made with a core of 18-8 stainless steel orthodontic wire. In order to achieve the plating, the passive layer which forms on stainless steel was first eliminated by pickling the previously cleaned wire in an acid chloride solution. To prevent the passive layer from forming again, a very thin, microscopically non-detectable flash of nickel was applied by a plating process. This process consisted of immediately transferring the wire to an acid nickel-chloride solution at room temperature, and using a nickel anode for the electroplating procedure. Once the passive layer had thus been eliminated, and its formation avoided by the nickel flash, successful electroplating of silver onto the surface was readily performed. In this case, an electrolytic, low concentration, silver-cyanide plating solution was used with a stainless steel strip anode. This was followed by the final plating process, involving a highly con-

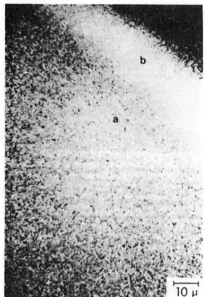

Fig. 1 – Nickel-plated stainless steel wire (X-ray microscope, scanning for nickel, 1250 X). (a) nickel contained in stainless steel wire (white dots), (b) nickel contained in layer of nickel plate.

centrated silver plating solution and a pure silver anode. In this manner, the good bonding characteristics expected from pure silver and amalgam could be combined with the desirable physical properties of stainless steel as a pin material.[3] This plating method is the one used for certain airplane sections, and was adapted for pins in dentistry.*

Because the thickness of the nickel flash was too minute to be observed directly, three separate stainless steel specimens were made. Each of these specimens was subjected to the nickel plating procedure for a period ten times as long as that used in preparing pins for the silver plating process. Under these conditions, the thickness of the nickel layer could be measured. Thus, an indirect measure of the thickness of nickel under the silver plating of the experimental pins could be obtained.

The method of making sterling silver pins is now described. Pin patterns were first made from the blue inlay wax and then cast

*Bristol Aerospace, Winnipeg, Manitoba, Canada

in sterling silver, using standard methods. Sterling silver pins, as cast, yielded properties which were poor in strength and in modulus of elasticity. These properties were greatly improved by heat-treating the pins at 650°F for one hour.[3,9]

Amalgam specimens containing each type of pin were prepared and all specimens were constructed in the same way. The mold utilized was similar to that used for dental amalgam specimens, described in the American Dental Association Specification No. 1. During the course of the investigation, a technique for achieving good adaptation of the amalgam to the pin was developed. It consisted of thoroughly rubbing the first portion of the freshly mixed amalgam around the pin with a plastic instrument. This procedure assured a reaction between the whole surface of the pin and the amalgam, thus eliminating voids around the pin. All specimens were between 24 hours and five days old when viewed microscopically.

The three special nickel-plated stainless steel pin specimens were mounted to be analyzed under X-ray microprobe. Four specimens of each type of pin used in this research were mounted on bakelite bases for metallographic microscope examination, and similar ones were mounted on stubs with a conductive silver com-

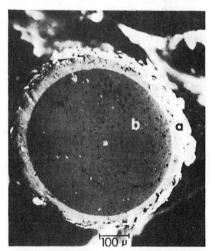

Fig. 2 – Silver-plated stainless steel wire (SEM, 170 X) (a) silver plate, (b) stainless steel wire. Transversal section.

FIGURE 15.3 *continued*

126   GALINDO ET AL.   J Dent Res February 1980

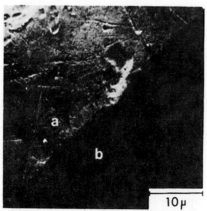

10μ

Fig. 3 – Silver-plated stainless steel wire (SEM 3000 X), (a) silver-plate layer, (b) stainless steel wire. Transversal section.

Fig. 4 – Silver-plated stainless steel pin in amalgam. Conventional method of amalgam condensation (metallographic microscope, 350 X). (a) stainless steel pin, (b) silver-plate layer, (c) void, (d) amalgam. Transversal section.

pound** for scanning electron microscope (SEM) study. All specimens within each group presented similar, reproducible characteristics. Therefore, only microphotographs of typical specimens of each type were made.

Results.

Results obtained from the pin-amalgam specimens used as controls confirmed two expected and well known facts: a total lack of bond between stainless steel and amalgam, and a very good adaptation of amalgam to sterling silver. The latter strongly implies the presence of a bond, and is consistent with findings of other researchers mentioned earlier. It was, therefore, judged unnecessary to present, in this paper, micrographs of these specimens.

An illustration of the nickel coating on a stainless steel pin is shown in Figure 1. This picture was obtained with an electron probe scan for nickel on one of the stainless steel specimens that was subjected to ten times the normal nickel-plating period. From this it can be seen that complete continuity between the nickel in the pin and that of the plated layer exists.

Silver-plated stainless steel wire specimens as seen under SEM are presented in Figures 2 and 3. The results of this examination were consistent and confirmed that the

**Silver Dag - Acheson Colloids Canada Ltd., Brantford, Ontario, Canada.

plated layer of silver adapts very well to the stainless steel.

The effectiveness of the "rubbing" technique prior to condensation is confirmed in Figures 4 and 5, which are views under the metallographic microscope. A continuous, intimate adaptation of the amalgam to the pin and an absence of voids at the interface is evident. Similar results were obtained with the sterling-silver pin and were consistent with all specimens of both types used in this experiment.

Pin-amalgam specimens were likewise viewed under SEM. Figures 6 and 7 illustrate the close adaptation of the amalgam to the surface of the silver plating, so that a definite boundary between them is not readily discerned. In order to confirm the above results, a longitudinal pin-amalgam section, seen under a metallographic microscope, is shown in Figure 8.

Discussion.

In establishing the quality of any bond for the stated purposes, mechanical tests will be the final arbiter. However, microscopic study of the interface regions was necessary to confirm that metallurgical conditions existed, and that satisfactory adaptation had been achieved.

From the indirect evidence of the X-ray microscope (Fig. 1), the thickness of the nickel flash upon which the silver plating was performed was about one micrometer. The same evidence also shows that the nickel contained in the stainless steel and that from the nickel-plated flash are

FIGURE 15.3 *continued*

Fig. 5 – Silver-plated stainless steel pin in amalgam. Rubbing method of amalgam condensation (metallographic microscope, 305 X). (a) stainless steel pin, (b) silver-plate layer, (c) amalgam. Transversal section.

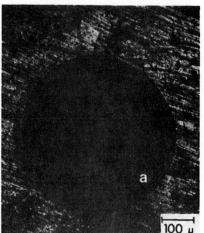

Fig. 6 – Silver-plated stainless steel pin in amalgam (SEM, 175 X). (a) stainless steel pin, (b) silver-plate layer, (c) amalgam. Transversal section.

mass can be detected at the surface of the pin.

Conclusions.

Specimens utilized in this study confirmed the bonding potential between smooth pins and amalgam under the conditions of this research. The plating technique

continuous and, in all probability, chemically bonded.

Microscopic examinations (including SEM) of the layer of silver-plating on stainless steel (Figs. 2 and 3, typical) showed extremely good adaptation with no voids, therefore, strongly suggesting the presence of a good mechanical bond. Because silver is known to bond with amalgam, it was expected that the silver-plating would act as a "soldering" agent between stainless steel and amalgam. Micrographic evidence (Figs. 4 to 8) confirmed this expectation, particularly when compared with that from sterling silver pins in amalgam. It is, then, reasonable to suggest that good mechanical bonding between stainless steel and amalgam can be achieved through the plating method presented here.

When amalgam is condensed without using the "rubbing" technique, large voids are readily seen at both the silver-plated and sterling silver pin and amalgam interface. On the other hand, when rubbing is used, close adaptation is obtained, and only the small and expected voids of the amalgam

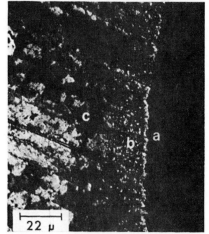

Fig. 7 – Silver-plated stainless steel pin in amalgam (SEM, 1050 X). (a) stainless steel pin, (b) silver-plate layer, (c) amalgam. Transversal section.

FIGURE 15.3 *continued*

128   GALINDO ET AL.   *J Dent Res February 1980*

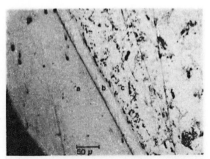

Fig. 8 – Silver-plated stainless steel pin in amalgam (metallographic microscope, 350 X). (a) stainless steel pin, (b) silver-plate layer, (c) amalgam. Longitudinal section.

used produced evidence of the strong possibility of a bond being present between silver and stainless steel. The technique of rubbing amalgam on the pin during condensation proved beneficial in achieving an excellent adaptation between the surface of the pin and amalgam. This technique is also useful in keeping stress concentration influences around the pin to a minimum, when retaining an amalgam restoration. The results of mechanical tests are needed to confirm the clinical usefulness of these conclusions.

Acknowledgments.

The authors are indebted to Mr. G. Freedman and Mr. Gordon Richardson from Bristol Aerospace, Winnipeg, Manitoba, for their guidance regarding the plating technique. Special thanks to Mr. B. Bergman for his help in constructing the special devices used in this research. We wish to acknowledge Dr. Peter Williams from the School of Dentistry and Dr. J. R. Cahoon from the Faculty of Engineering of the University of Manitoba for their advice.

REFERENCES

1. GALINDO, Y.; McLACHLAN, K.: and KASLOFF, Z.: Mechanical Tests of Smooth Silver-plated Retention Pins in Amalgam, *J Dent Res,* (in press).
2. DHURU, V.: A Photoelastic Study of Stress Concentration Produced by Retention Pins in an Amalgam Restoration, MSc Thesis, University of Manitoba, 1972.
3. GALINDO, Y.: The Development and Testing of Retention Pins which Metallurgically Bond with Dental Amalgam, MSc Thesis, University of Manitoba, 1973.
4. MOFFA, J. P.; GOING, R. R.; and GETTLEMAN, L.: Silver Pins: Their Influence on the Strength and Adaptation of Amalgam, *J Prosth Dent* 28:491-499, 1972.
5. LUGASSY, A. A.; LAUTENSCHLAGER, E.P.; and HARCOURT, J. K.: Crack Propagation in Dental Amalgam, *Aust Dent J* 16:302-306, 1971.
6. CECCONI, B. T. and ASGAR, K.: Pins in Amalgam: A Study of Reinforcement, *J Prosth Dent* 26:159-169, 1971.
7. DUPERON, D. F.: The Effect of Selected Pin Retention Materials on Certain Properties of Dental Silver Amalgam, MSc Thesis, University of Manitoba, 1970.
8. PETERSON, E. A. and FREEDMAN, G.: Laminate Reinforced Dental Amalgam, *J Dent Res* 51:70-87, 1972.
9. METALS HANDBOOK COMMITTEE: **Metals Handbook,** American Society for Metals, 6th ed., 1960.

FIGURE 15.3 *continued*

5. Consulting with professors
6. Preparing the working outline
7. Conducting the research and taking notes
8. Writing the rough draft
9. Writing suitable introductions and closings
10. Providing the documentation

Selecting and Limiting the Subject

In the professional world the selection of a subject is usually yours, but as a student a supervisor or professor may assign you a research report topic, such as "New Findings of the Influence of Temperature on Viruses," "The Practicality of Laser Surgery on Heart Patients," or "Secure Financial Electronic Transactions on the Internet." If the subject selection is entirely yours, consider the following three principles:

1. Select a subject that can be thoroughly investigated within the confines of your length limitations. "Computers" is too vast a subject and fails to suggest a particular focus. But subjects such as "Methods to Combat Computer Viruses" or "The Advantages of Vocal Software" are both more narrow and focused.

2. Select a subject on which you either are an authority or are able to access a wide variety of published research material, interviews, or survey responses. "Cures for the Ebola Virus" will not be productive because research is inconclusive. "Procedure for a Magnetic Circuit Test on Transformer Performance" may be too specialized for you and your audience. "Methods to Reduce Pollution in the Aral Sea (Uzbekistan/Kazakhstan)" is too far removed for you to obtain published, up-to-date material.

3. Select a subject that will allow you to formulate a judgment. A study on "The Smartcard Debit System versus Traditional Credit Cards" will yield information on which you can formulate an informed comparative judgment.

Formulating the Thesis

The next step is to develop a preliminary thesis, a statement that focuses the purpose of your paper. This statement of the central idea to be developed in your paper will help you to organize facts, limit your note taking, and eliminate needless research.

Consider what you know about your limited subject. You probably already have some opinions or new information or you would not have selected the subject in the first place. State in one sentence (or two if necessary) an opinion, conclusion, generalization, or prediction about your subject. For instance, if you have narrowed your subject to "The Effectiveness of Computer Access Control Systems," you may tentatively state, "Computer access control systems are improving." If you are an engineer who has been working on an improved system, your thesis may state, "A fail-safe method is now available to eliminate computer viruses."

Often the thesis of a technical paper refutes the benefits of a present system and points out new findings. Such theses frequently contain the words *however, instead, nevertheless,* or *consequently.*

Examples

For decades Medic Alert neck tags or bracelets have provided instant medical history; however, laser optic technology can now produce revolutionary medical data-memory cards containing up to 800 pages of information.

The consumer no longer has to withdraw money from a wallet or suffer through a time-consuming check approval process; instead, she can pay a retailer directly out of her checking account via an electronic fund transfer with a debit card.

While you are taking notes and extending your research, you should revise or refine your thesis. After you have assembled all of your data, reduce the thesis to one sentence to unite your findings. You may wish to add an organizational clue to the structure of your paper. For instance, our final thesis may read, "Despite advances, computer access control systems demand constant upgrading due to the increasing amounts of sensitive computerized data, the expansion of networking systems, and the growing number of sophisticated, computer-competent criminals." This thesis not only contains an overall inference on the need for constant improvement, but also provides a clue to the sections of the report: the problems of data proliferation, networking, and unethical programmers.

Preparing the Preliminary Bibliography

Once you have formulated a thesis for your report, the next logical step is to locate possible sources of information. Determine if you can locate an adequate number of timely resources and provide a working list of sources from which you can choose. Chapter 14 discusses in detail how to access your sources both in the library and through a personal computer and indicates methods of primary research.

Brainstorming the Methods of Development

There are several major strategies for approaching the organization of material. These include *narration* (telling of events), *description* (recounting in precise detail), and *exposition* (explaining certain points).

The professional paper usually requires expository strategies, although it may contain elements of narration and description.

There are eight methods of developing expository material, which may be used singularly but are likely to be used in combination:

- Definition
- Description
- Classification and division
- Exemplification
- Comparison and/or Contrast
- Cause and/or Effect
- Instructions
- Process analysis

Chapters 6 through 9 examine in detail strategies for definition, description, instructions, and process analysis. The student writer should employ a number of these expository strategies in producing effective academic papers. Ask yourself these questions:

1. Which terms will I need to define?

2. What descriptions are necessary?

3. How will I classify (place into major sections) and divide (separate into subsections) my material?

4. What specific examples will support my thesis?

5. What comparisons (likenesses) and/or contrasts (differences) will help to explain my material?

6. Are cause and/or effect factors an essential part of my evidence?

7. Are instructions for operation of any procedure necessary?

8. Will I need to analyze a process in my paper?

Search for material (see Chapter 14) that will support the pertinent strategies in the preceding lists. Then organize the strategies into a logical presentation.

Consulting with Professors

Once a preliminary bibliography of sources has been developed and the methods of development decided upon, it is wise to consult with one or more professors prior to continuing with the research. Professors are experts on the subjects they teach and can advise you on whether you are on the right track and offer help with sources and development. At the graduate level, a written proposal and preliminary bibliography should be submitted for the professor's approval.

Preparing the Working Outline

Develop a preliminary or working outline of your materials. Consider the writing strategies that you have brainstormed. Your first draft may be sketchy, but it will serve as a guideline to your note taking. Without an outline you are likely to take notes on materials irrelevant to your purpose or overlook an area that should be explored.

You should not be totally bound by your working outline. As you think over your subject and take notes, you will want to expand, rearrange, discard, and subordinate your outline topics. Your final research report should include a final outline.

FORMAT

Select a traditional or decimal outline format:

I. A.		
1.	**1.0**	
2.	**1.1**	
B.		**1.1.1.**
1.		**1.1.2.**
2.		
	1.2	
a.		**1.2.1.**
b.		**1.2.2.**
(1)	**2.0**	
(2)	**2.1**	
(a)		**2.1.1.**
(b)		**2.1.2.**

Although the Roman-numeraled, traditional outline allows for five levels of subordination, you will seldom find it necessary to use subheads beyond two subordinates. The decimal outline format is popular for technical presentations.

TOPIC OUTLINE

Although each entry in an outline may be a sentence, a topic outline is not only easier to develop in the preliminary stage but is also more common in the final draft. Major topics name the divisions of your paper. Next, subtopics are subordinated and indented under each major topic, and subsequent developmental topics are subordinated beneath each subtopic.

Partial Preliminary Outline

1.0 Introduction (thesis)
    1.1 Computer definition

         1.2 Access control definition
           1.2.1 Access control examples
           1.2.2 Access control processes
     2.0 Causes of computer security violations
         2.1 Cause 1
         2.2 Cause 2
         2.3 Cause 3
     3.0 Effects of computer security violations
         3.1 Effect 1
         3.2 Effect 2
         3.3 Effect 3
     4.0 Examples of computer crimes
         4.1 Industrial crimes
           4.1.1 Company A
           4.1.2 Company B . . . (etc.)

Keep working on your outline as you take notes.

Partial Final Outline

     2.0 Causes of computer security violations
         2.1 Faulty access control devices
           2.1.1 Telephone networking
           2.1.2 Passwords
         2.2 Excessive computerized information
           2.2.1 Pentagon's 8,000 computers
           2.2.2 Rand Corporation statistics
         2.3 Number of trained computer specialists
           2.3.1 Number of military contractor clearances
           2.3.2 Number of U.S.-trained computer operators
         2.4 Amateur experimenters
           2.4.1 Milwaukee amateur club incident
           2.4.2 Russians in Applied Systems Analysis group

Your professor may require you to include your polished outline with the final paper. Figure 15.4 shows a complete outline for a student academic paper:

OUTLINE

Thesis: Tomography, the innovative creation of a combination of scientific minds of our generation, is used to help cardiologists analyze the heart, to aid brain surgeons in the detection of cerebral disorders, to assist geologists in the study of earth's composition, and to permit aerospace technicians to inspect the MX missiles for malfunctions.

   I. Introduction

  II. Background

      A. Definition

      B. Origin

 III. Cardiological applications

      A. Heart

          1. Disorders

          2. Normalities

      B. Cardiovascular system

          1. Arteries

          2. Veins

 IV. Cerebral applications

      A. Brain scans

          1. Normal children

          2. Impaired children

      B. Oxygen/glucose loss

          1. Causes

          2. Effects

  V. Geological applications

      A. Earth's core

      B. Earth's surface

 VI. Aerospace applications

      A. Study of MX missiles

          1. Disorders

          2. Remedies

      B. Approval of MX missiles

FIGURE 15.4   *Sample student review article outline*

Conducting Research and Taking Notes

With your thesis in mind, your outline to guide you, and your preliminary bibliography, you now proceed with the research and note-taking tasks. Chapter 14 covers this material thoroughly.

Writing the Rough Draft

Using your outline as a structural guide and your organized note cards (Chapter 14) as the content guide, write a rough draft of your research paper. Do not forget to use graphics and other visuals when they clarify your data. Follow these guidelines for your rough draft:

1. **Double space your draft** in order to permit expansion or revision of your draft. Determine if your professor wants you to single or double space your final draft.

2. **Write your introduction.** Be sure your limited subject and polished thesis are clearly stated. A clue to the overall organization of the paper is needed if it is not indicated in your thesis. Clearly define your key terms before beginning to present the evidence that supports your thesis. (The next section of this chapter contains additional information on writing introductions.)

3. **Write the body paragraphs.** Each body paragraph or grouping of paragraphs should begin with a topic sentence (a sentence that clarifies the topic of the paragraph and states your opinion or generalization about it) based on the organizational parts as included in your thesis or revealed in your introduction. Each paragraph must have unity (address itself to the same subtopics of the overall subject), coherence (logical progression of evidence), and transitions (words and phrases that connect the evidence, such as *for instance, in the first place, furthermore, finally*). Review Chapter 2. Include your graphics and visuals, but do not use them to just "decorate" your paper; use them only where appropriate.

4. **Include the author's name within the text** when paraphrasing opinions and conclusions or directly quoting any source material. All of your researched data will require internal and end documentation. Chapter 16 must be studied before writing the draft. Check the five safeguards in Chapter 14 on p. 458 in order to avoid a plagiarism charge and to ensure that all of your evidence is documented.

5. **Insert your internal reference notes in parentheses** as you write, and circle each in order not to overlook one when you prepare your final draft and bibliography listing (see Chapter 16).

6. **Write your conclusion.**   This may be one or several paragraphs that summarize your findings, present solutions, or recommend a particular action. See the next section in this chapter for some other tips on closings.

7. **Edit all mechanics** (spelling and punctuation) and style considerations (sentence construction and variation, grammar, usage, and other language concerns).

Writing Introductions and Closings

Consider the introduction and closing of your paper. The introduction of the truly professional paper is serious and forthright. If you have a particular broad-interest journal written for laypeople in mind, you may want to write a catchier introduction. The major purpose of the introduction is to state your thesis, but you may consider including

- A reader-attention device
- A context in which to consider your ideas
- Definition of key terms
- The thesis itself
- Organizational clues

Engaging introductions might include a direct quotation of a well-expressed statement, a startling fact or statement about your subject, a rhetorical question, or a related anecdote. Here are some sample introductions for articles on point-of-sale transaction systems:

1. One of life's great excuses—"The check is in the mail"—may be heading for extinction. Financial institutions and retailers want to stop handling cash and checks, so they are switching to cheaper, faster electronic point-of-sale (POS) transactions. Point-of-sale is the electronic transfer of money from a consumer's bank account to a merchant's bank account with the use of a plastic card and merchant computer terminals. The new system is creating a gradual revolution in the way people buy items and deal with their banks.

2. "In our opinion it is impossible to exist in the record industry without some sort of computerized point-of-sale system," says Irving Heisler, president of Montreal-based Discus Music World. Discus operates more than 100 units across Canada. The major advantages of the point-of-sale system lie in being able to serve customers better while at the same time lowering operating costs by making both inventory management and labor scheduling more efficient.

3.   A shopper in a Dublin, California, Lucky Store wants to pay cash for her purchases. She no longer has to suffer a time-consuming check approval process. Instead, by using a debit card, in just seven seconds she can pay the retailer directly out of her checking account with an electronic funds transaction. This point-of-sale system effectively speeds the sale, reduces paperwork, and costs less than traditional credit card transactions.

Your closing will usually restate your thesis in its original or in paraphrased terms and reach beyond it to conclusions and suggested further logical implications of your findings. Your final few sentences may

- Draw a stronger conclusion than already expressed
- Artfully summarize your material or emphasize just one strong aspect of it
- State a climactic fact
- Urge your reader to action

A few brief examples follow:

1.   As well as benefiting individual consumers, point-of-sale systems speed up corporate, and even international, financial transactions.

2.   The POS system enables management to reduce losses by thousands of dollars. Ace Business Supplies increased its sales by 20 percent in the first year of use. You can't ask for more than that.

3.   Quicker consumer services and lower operating costs have persuaded merchants to use the POS system. Shouldn't you install the system at your place of business?

Providing the Documentation

There are several styles and guides to documentation. Chapter 16 covers the styles for internal and end documentation of all of your research sources. It will help you to determine in advance the required documentation style in order to label your note cards, write your notes, write your draft, and polish the paper.

Figure 15.5 (see pp. 504–510) reproduces a student research paper on a new medicine. The article is documented in this case in the APA (American Psychological Association) format for a working draft, explained in Chapter 16.

MEXILETINE HYDROCHLORIDE:

AN INVESTIGATION INTO

ANTIARRHYTHMIC TREATMENT

Submitted to

Professor Julia K. Nelson

Technical Writing 2210

Submitted by

W. Jeanne Early

28 October 199X

FIGURE 15.5   *Sample research paper in APA Working Draft documentation style writing* (By permission)

Mexiletine 2

Outline

Thesis: Mexitil is a promising new oral antiarrhythmic that is proving effective with few adverse effects and minimal drug interaction problems.

I. Introduction
II. Mechanism of action
   A. Lidocaine comparison
   B. Action potential
   C. Atrial, nodal, and ventricular fibers
III. Pharmacokinetics
   A. Absorption
   B. Half-life
   C. Excretion
IV. Dosage
   A. Campbell
   B. Heger and others
   C. Woolsey
V. Adverse effects
   A. Cardiological
   B. Neurological
   C. Gastrointestinal
VI. Drug interactions
   A. Antacids, cimetidine
   B. Analgesic narcotics
   C. Phenobarbital, dilantin, rifampin
   D. Cigarette smoking

FIGURE 15.5 *continued* (outline is optional)

Mexiletine   3

Abstract

Mexiletine hydrochloride (Mexitil-Boehringer Ingelheim), recently
approved by the U.S. Food and Drug Administration (USFDA) for
treatment of ventricular arrhythmias as an oral agent in the practice of cardiology, is a promising new oral antiarrhythmic effective
with few adverse effects and minimal drug interaction problems.
Similar to lidocaine, it allows greater movement in and out of sodium
channels, decreases the rate of depolarization, and decreases the
refractory period of cardiac electrical cells; therefore, it is faster acting. Its absorption rate is near 100 percent with a peak action in 1.5
hours and a half-life of 8–15 hours. Dosage recommendations vary
from 400–600 mg to 450–1200 mg to 10–14 mg depending on the
division of dosages. Adverse effects have proven minimal. Various
other drugs—antacids, cimetidine, narcotics, phenobarbital, dilantin,
and rifampin plus cigarette smoking—may delay absorption.

FIGURE 15.5   *continued*

Mexiletine   4

MEXILETINE HYDROCHLORIDE: AN INVESTIGATION INTO ANTIARRHYTHMIC TREATMENT

Mexiletine hydrochloride (Mexitil-Boehringer Ingelheim) was recently approved by the U.S. Food and Drug administration (USFDA) for treatment of ventricular arrhythmias as an oral agent in the practice of cardiology. Mexitil has been approved for use in countries other than the United States since 1969 and was first used as an anticonvulsant (Campbell, 1987). But after other anticonvulsants were found to have antiarrhythmic effects, Mexitil was also tested in laboratory animals and was found to have an effect against ventricular arrhythmias (Woolsey, 1984). Mexitil is a promising new oral antiarrhythmic that is proving effective with few adverse effects and minimal drug interaction problems.

Mexitil is similar to another common antiarrhythmic. Mexiletine and lidocaine have close structural similarities, which are shown in Figure 1:

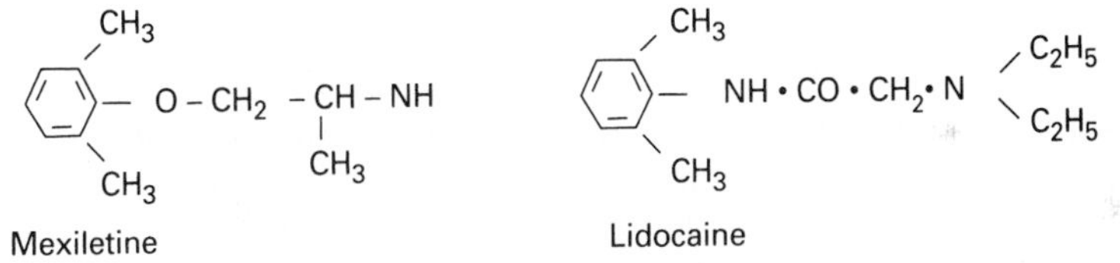

Mexiletine   Lidocaine

Figure 1   Structures of Mexiletine and Lidocaine (Campbell, 1987)

Campbell (1987) has stated that Mexitil is a primary amine, and its chemical structure much resembles lidocaine, but Mexitil is significantly smaller (Woolsey, 1984). Woolsey (1984) stated, "This may allow greater movement in and out of sodium channels which is essential in effecting the action potential" (p. 1058). Mexitil decreases the rate of depolarization and decreases the refractory period of Purkinje fibers (cardiac electrical cells). Thus the action potential duration is shortened. Campbell (1987; cf. Woolsey, 1984, esp. p. 1061) has also stated that the electrophysiologic effect is to slow the maximal rate of the action potential. Campbell (1987) also determined that in human studies there was no consistent effect on the sinus node or atrium unless there was preexisting disease. Woolsey (1984) also noted that Mexitil's main area of activity is on

FIGURE 15.5   *continued*

Mexiletine   5

ventricular arrhythmias. Mexitil's action is rate dependent indicating better action for fast ventricular tachycardias (Johansson, Stavenow, & Hanson, 1984).

Mexilitene absorbs well into the system. Campbell (1987) noted that Mexitil is absorbed well in the gastrointestinal tract, but most of the absorption is in the upper small intestine. It has a great absorption nearing 100 percent (Johansson et al., 1984). Campbell's (1987) study showed that the peak plasma level is usually in 1.6 hours, and the elimination half-life can be 8–15 hours. Woolsey (1984) has stated, "Agents that induce drug-metabolizing enzymes in the liver, such as rifampin and phenytoin, decrease the elimination half-life of mexiletin" (p. 1060). The liver is the main organ of metabolism although there is some renal excretion (Heger, Nattel, Rinkenberger, & Zipes, 1980). According to Johansson and his associates (1984), the liver metabolism is mainly by oxidation and reduction and can be influenced by exposure to substances that can stimulate enzyme production in the liver.

The recommended dosage of Mexitil varies little among different sources. Campbell (1987) recommends starting with a single loading dose of 400–600 mg to achieve a therapeutic level quickly. Then Campbell (1987) and Heger et al. (1980) agree on 450–1200 mg in three equivalent divided doses every eight hours. Woolsey (1984) believes therapeutic concentration can be achieved with 10–14 mg/kg/day in divided doses. Studies continue.

Adverse effects to Mexitil that have caused discontinuation of the treatment have been infrequent. But ventricular arrhythmias have occurred in clinical trials requiring the discontinuation of Mexitil (Heger et al., 1980). Campbell (1987) has suggested that adverse cardiac effects on the sinus node, the A-V node, and cardiac contractility are uncommon with Mexitil. Heger et al. (1980) discovered that neurologic and gastrointestinal effects are the most frequent limiting effects of Mexitil. Heger (1980) has reported, "The most common adverse reactions to mexiletine were nausea, vomiting, tremors, and dizziness" (p. 1060). Gastrointestinal side effects were diminished with smaller doses given more frequently and within a few weeks disappeared (Johansson et al., 1984).

Several different drugs affect different actions of Mexitil. Antacids, cimetidine, and narcotics can delay absorption of Mexitil (Campbell, 1987). Woolsey (1984) pointed out that use of narcotic analgesics, especially taken during an acute myocardial infarction, can delay or impair Mexitil absorption and should be closely

FIGURE 15.5   *continued*

Mexiletine   6

monitored. Campbell (1987) discovered that other drugs that decrease the half-life of Mexitil by inducing liver enzymes are phenobarbital, dilantin, and rifampin. Also, cigarette smoking decreases plasma levels of Mexitil (Johansson et al., 1984).

Mexiletine appears to be effective for the treatment of symptomatic ventricular arrhythmias. Lidocaine is usually given intravenously, and procainamide and quinidine have undesirable side effects (Johansson et al., 1984). Therefore, Mexitil is a much needed drug that should be used as a second-line treatment of ventricular arrhythmias.

FIGURE 15.5   *continued*

Mexiletine   7

References

Campbell, R. (1987). Mexiletine. The New England Journal of
Medicine, 316, 29–34.

Heger, J. J., Nattel, S., Rinkenberger, R. L., & Zipes, D. P. (1980).
Mexiletine therapy in 15 patients with drug-resistant ventricular
tachycardia. The American Journal of Cardiology, 45, 627–632.

Johansson, B., Stavenow, L., & Hanson, A. (1984). Long term clinical
experience with mexiletine. American Heart Journal, 107,
1099–1102.

Woolsey, R. L. (1984). Pharmacology, electrophysiology, and pharma
cokinetics of mexiletine. American Heart Journal, 107,
1058–1065.

FIGURE 15.5   *continued*

CHECKLIST

Writing Scientific/Technical Articles and Academic Papers

SCIENTIFIC/TECHNICAL ARTICLES

☐ 1. Have I designed the research article as a study of a product or a procedure?

☐ 2. Have I received approval from superiors (or funding sources)?

☐ 3. Have I completed the research?

☐ Primary sources?

☐ Secondary sources?

☐ 4. Have I used multiple sources and checked the research for accuracy?

☐ 5. Have I selected a journal in the field most likely to publish the paper?

☐ 6. Have I obtained the instructions to authors for the selected journal?

☐ 7. Have I provided a precise title?

☐ 8. Does my introduction clarify the purpose of the paper?

☐ Does it define a problem?

☐ Does it review the existing literature?

☐ 9. Have I thoroughly discussed the methods/materials used in the investigation of the problem?

☐ 10. Have I included a thorough discussion of other aspects of the investigation?

☐ 11. Have I discussed the results of the investigation?

❑ **12.** Does my conclusion make valid judgments and suggest implications for the future use of the product or problem solution?

❑ **13.** Have I provided the appropriate internal and end documentation (see Chapter 16)?

❑ **14.** Have I submitted the article for publication appropriately?

   ❑ Have I provided a cover letter?

   ❑ Have I included two copies and/or a disk of the paper?

   ❑ Have I included a self-addressed, stamped envelope for acknowledgment of receipt of the manuscript?

   ❑ Have I packaged the manuscripts and/or disk carefully?

ACADEMIC PAPERS

❑ **1.** Have I selected a scientific or technical subject and limited it sufficiently for length considerations?

❑ **2.** Have I formulated a preliminary thesis to focus the purpose of the paper?

❑ **3.** Have I prepared a preliminary bibliography and cards or computer file (see Chapter 14)?

❑ **4.** Have I decided on the appropriate method(s) of development?

   ❑ Definition?

   ❑ Description?

   ❑ Classification and division?

   ❑ Exemplification?

   ❑ Comparison and/or contrast?

   ❑ Cause and/or effect?

   ❑ Instructions?

   ❑ Process analysis?

❏ **5.** Have I consulted with my professors on my progress?

❏ **6.** Have I developed a working bibliography that I will polish for inclusion?

❏ **7.** Have I conducted the research and taken notes (see Chapter 14)?

❏ **8.** Have I written the rough draft?

    ❏ Have I included an introduction?

    ❏ Have I written the expository body paragraphs?

    ❏ Have I incorporated partial and sentence direct quotations appropriately?

    ❏ Have I written a closing?

❏ **9.** Have I provided all of the appropriate documentation (see Chapter 16)?

❏ **10.** Have I edited all of the mechanics of the paper?

❏ **11.** Have I assembled the paper correctly (title page, outline, paper, bibliography)?

EXERCISES

1. **Professional Journals.** Compile a list of at least five journals in your field.

2. **Journal Guidelines.** Obtain the instructions to authors from three of the journals in your field.

3. **Thesis Statements.** Rewrite the following broad thesis statements to indicate a more limited focus and purpose for each:

    a. The Internet has revolutionized academic research.

    b. Many diet prescriptions now available have harmful side effects.

    c. Macroengineering will improve continental travel in the future.

    d. The twenty-first century is ushering in changes in the job market.

    e. New medicines to lower cholesterol are emerging.

    f. Desalinization systems will provide additional usable water.

    g. Mars exploration is providing benefits here on Earth.

    h. Computers are changing the nature of international telephone calls.

    i. Our knowledge is increasing in the study of humanity's origins.

    j. Global warming is threatening our coastal cities.

4.   **Outlining.** Revise the following outline from a traditional sentence format to a decimal, topic outline:

Thesis: Changes in automobile design are predicted for the next century.

    I. Automobiles will remain the main mode of transportation.
       A. Cars will be totally computerized.
          1. Voice commands will lock and unlock the car, operate the radio, and operate the climate control system.
          2. All vehicles will come with telephones.
          3. All systems will be electronically monitored.
       B. Design will be aerodynamic.
          1. Future cars will have sharply raked windshields and aerodynamic bodies.
          2. Undercarriages and wheels will be enclosed.
          3. Windows will be flush with the body.
          4. Grills and outside ornaments will become obsolete.
       C. Alternative fuels will be employed.
          1. Hydrogen fuel will replace gasoline.
          2. Electric cars will be developed.
   II. Rapid transit will ease the burden in cities.
       A. Urban rail systems will expand between cities.
       B. Bus systems will expand within cities.
       C. Bullet trains will curb air travel.
          1. Fast trains will replace airplanes for moderate trips.
          2. The high cost of air travel will become inhibitive.
  III. High-speed jetliners will make long-distance travel easier.
       A. Supersonic jets will continue to replace slower jumbo-jets.
       B. The number of flights to various points will become greater.
       C. New airlines will offer long-distance flights.

WRITING PROJECTS

1.   **Scientific/Technical Articles.** Using the information in the Scientific/Technical Articles section of this chapter, write a one-page evaluation of the style, clarity, and organization of a published article of your choice. Include a copy of the article with your evaluation.

2.   **Group Project-Academic Research Paper.** Select two classmates to collaborate on an academic research paper concerning a
breakthrough discovery in your field. Develop a bibliography that
includes a variety of sources (encyclopedia, magazines and journals,
newspapers, and brochures and pamphlets). Take notes. Develop a
thesis, a preliminary outline, and methods of development.
Investigate who made the discovery, how it was made, how it was
tested, what benefits it offers, what problems it poses, and what
future applications are possible. Use the documentation techniques
assigned by your professor (see Chapter 16). At minimum include
an alphabetized list of references. Submit your bibliography and
note cards with the paper.

3.   **Scientific/Technical Paper.** Using the organization of a scien-
tific/technical article, research and write a paper on some problem
of the late twentieth or early twenty-first century. In your paper
delineate the problem by reviewing the existing literature. Discuss
the problem, the methods being explored to alleviate the problem,
and the results obtained by these methods. Draw conclusions on the
findings and point out future implications. Following is a list of
broad subjects that can be narrowed to focused topics:

education	pollution
transportation	fuels
merchandising	the shape of cities
endangered species	new computer applications
the economy	travel modes
satellites	electronic marvels
genetic engineering	weather modification
architecture	home appliances
population	space exploration
new occupations	health care
the American family	medicine
crime	life expectancy
food production	minerals and materials

other?

CHAPTER 16

Documenting Reports

DUFFY   by Bruce Hammond

S K I L L S

After studying this chapter, you should be able to

1. Name five systems for professional paper documentation.
2. Name five documentation strategies that apply to all systems.
3. Prepare both a working draft and a paper for publication documented in the two American Psychological Association (APA) systems.
4. Prepare a paper documented in the Modern Language Association (MLA) system.
5. Prepare papers in both variations of the Council of Biology Editors (CBE) documentation style.
6. Prepare a documented paper in the Number System consistent with the requirements of a specific field.
7. Prepare a documented paper in the Chicago Style.

INTRODUCTION

Your scientific/technical articles and academic papers (Chapter 15) require formal documentation of the sources that have been quoted, paraphrased, or summarized. Documentation involves making parenthetical or footnoted references within your works to the sources cited and preparing a concluding list of references. There are a variety of accepted documentation styles and systems endorsed by different organizations:

- The **American Psychological Association (APA),** an author/date system used for the social sciences, and similar versions for the biological and earth sciences, education, linguistics, and business
- The **Modern Language Association (MLA),** a system used for language and literature
- The **Council of Biology Editors (CBE),** a name/year system used for biology, botany, and zoology, with a second protocol, the Number System (following), for the applied and medical sciences
- The **Number System (and its variations),** used for the applied sciences (chemistry, computer science, engineering, mathematics, and physics) and the medical sciences (bio-medicine, health, medicine, and nursing)
- The **Chicago Manual of Style,** used for some of the humanities (history, philosophy, religion, and theology) and for the fine arts (art, dance, music, and theater)

Comprehensive manuals of these and other discipline-specific systems may be purchased in any bookstore. The MLA and APA systems are online on your computer; use the keywords Modern Language Association or American Psychological Association. As mentioned in Chapter 15, every scholarly journal has its own requirements, and there are slight variations in the procedures of every academic discipline. Although the APA and its variations are required for the majority of papers in most technical and scientific fields, the MLA style is the most popular for undergraduate assignments.

The styles, though varied, all include the name of the author, the title of the work, the date of publication, the place of publication, and the publisher (or Internet address or CD-ROM access).

The documentation strategies for all systems are similar:

1. Compiling an annotated working bibliography on cards or computer files (Chapter 14)

2. Taking notes (Chapter 14)

3. Providing in-text, numbered, or footnoted documentation, depending on the system in your rough and final drafts

4. Preparing a final list of references or works cited as determined by your system

5. Presenting the paper in a formal format

COMPUTER ASSISTANCE

Computer software can assist you in a number of professional paper tasks, for example, in accessing the American Psychological Association, the Modern Language Association, or the Chicago Manual of Style systems for title pages, abstracts, outlining, tables of contents, in-text documentation, footnoting, and a final bibliography list. Some software incorporates formatting for headings, the handling of capitals and lowercase features, quotation styles, outlining, and more. Most word processing software will also handle superscript numbers. It is your responsibility, however, to become totally familiar with one or more systems without reliance on software.

APA SYSTEM

The fourth edition of the *Publication Manual of the American Psychological Association* (1994) provides documentation advice for writers in the social sciences and is written primarily for authors preparing manuscripts for professional publication in journals. It covers manuscript content and organization, writing style, and manuscript preparation. An appendix offers advice for student writers. Even if you are not going to submit a professional paper for publication, it is valuable to gather experience in using the APA system for papers in the social sciences.

General Conventions

There are a number of general conventions in the APA system in addition to the internal referencing and final reference list systems:

1. The final paper must contain a title page, an abstract (limited to 960 characters including spaces—approximately 80 to 120 words), the in-text documented paper, and the final reference list. Include an outline only if your professor requires one.

2. The paper should be double-spaced throughout with the pages numbered consecutively in the upper, right-hand corner (with the exception of the title page). The abstract will be on page 2 following the unnumbered title page, and a page number is included on the reference listing.

3. Use last names only within your text (Kline, not Robert J. Kline).

4. The year of the research is primary, so it is featured in the text immediately after the named source ["Kline (1998) . . ." or "Woolsey and his team (1997) . . ."].

5. Because one purpose of a scientific paper is to review the work of previous researchers, the past tense is required when you cite their findings ("Kline (1998) **showed** . . ." or "Woolsey and his team (1997) **demonstrated** . . .").

6. Title your list of sources "References"; do not use all capital letters nor boldface the word.

7. Refer to the manual for the handling of abbreviations, racial references, capitalization, punctuation, numbers, and other style questions.

Internal References

When you write your paper in the APA style or its variations, several internal referencing practices apply. Essentially you must reference the sources within your text by indicating within parentheses the author and date and sometimes the exact page number. Examples of internal referencing follow:

If the author's name(s) occurs in the text, then all that is needed is the publication date in parentheses:

According to Campbell (1996) the use of melatonin is . . .

If the author's name(s) does not appear in the text, place it and the date in parentheses at the end of the paraphrased material:

Studies proved that melatonin is often taken in unnecessarily large doses (Michaels, 1997; Vann, 1998).

If the text contains a direct quotation, you must include the page numbers with "p." or "pp." The pagination note may occur with the date notation or at the end of the direct quotation:

Campbell (1996, p. 107) stated, "Melatonin, while it may have some efficacy as a sleep or jet lag aid, is totally overrated for its other so-called benefits."

or

Campbell (1996) stated, "Melatonin, while it may have some efficacy as a sleep or jet lag aid, is totally overrated for its other so-called benefits" (p. 107).

Write a quotation of 40 words or more as a separate block, which is indented five spaces from the left margin. Introduce the quotation. Do not indent the first line, but do indent the first line of additional paragraphs five more spaces. Omit the quotation marks; the indentation indicates that the material is a quotation:

Smith (1998) reported the following:

Melatonin, the much touted herb of the 90's, has not been found to slow aging as its manufacturers claim. It does stimulate the thymus to release seratonin, but not in any significant amounts.

Melatonin may be considered a natural hormone, but as such not one test has indicated its efficacy for anything beyond a sleep aid. (pp. 314–315)

Notice that the parenthesized page numbers come *after* the final period in a block quotation. Paraphrases *may* be handled like quotations. Give the author's last name in the text, the date in parentheses, and the appropriate page number(s) in parentheses at the end:

Campbell (1996) suggests that the advantages of melatonin are overrated aside from the fact that melatonin may help one get to sleep or overcome jet lag (p. 107).

Notice that the period comes *after* the page reference for a short direct quotation or paraphrase page reference.

Corporate authors must be spelled out in full in the first reference, followed by the appropriate abbreviation. Use the abbreviation only in subsequent references:

One study has questioned that any benefits at all accrue from melatonin (American Medical Association [AMA], 1997).

The AMA (1997) concluded that . . .

The foregoing is not an exhaustive coverage of all the requirements you may encounter, but rather a sampling of the most commonly used internal referencing methods. Figure 16.1 shows some sample text documented in the APA style.

Modern scientific methods, invented in the 16th century, were not only a stunning technical innovation, but a moral and political one as well, replacing the sacred authority of the Church with science as the ultimate arbiter of truth (Grant, 1987). Unlike medieval inquiry, modern science conceives itself as a search for knowledge free of moral, political, and social values. The application of scientific methods to the study of human behavior distinguished American psychology from philosophy and enabled it to pursue the respect accorded the natural sciences (Sherif, 1979).

The use of "scientific methods" to study human beings rested on three assumptions:

> (1) Since the methodological procedures of natural science are used as a model, human values enter into the study of social phenomena and conduct only as objects; (2) the goal of social scientific investigation is to construct laws or law-like generalizations like those of physics; (3) social science has a technical character, providing knowledge which is solely instrumental. (Sewart, 1979, p. 311)

Critics recently have challenged each of these assumptions. Some charge that social science reflects not only the values of individual scientists but also those of the political and cultural milieux in which science is done, and that there are no theory-neutral "facts"

FIGURE 16.1   *Sample documented text in APA style* (From Stephanie Riger,
"Epistemological Debates, Feminist Voices—Science, Social Values, and the Study
of Women," *American Psychologist, 57* (June 1992): 78–92. By permission.)

(e.g., Cook, 1985; Prilleltensy, 1989; Rabinow & Sullivan, 1979; Sampson, 1985; Shields, 1975). Others claim that there are no universal ahistorical laws of human behavior, but only descriptions of how people act in certain places at certain times in history (e.g., K. J. Gergen, 1973; Manicas & Secord, 1983; Sampson, 1978). Still others contend that knowledge is not neutral; rather, it serves an ideological purpose, justifying power (e.g., Foucault, 1980, 1981). According to this view, versions of reality not only reflect but also legitimate particular forms of social organization and power and asymmetries. The belief that knowledge is merely technical, having no ideological function, is refuted by the ways in which science has played handmaiden to social values, providing an aura of scientific authority to prejudicial beliefs about social groups and giving credibility to certain social policies (Degler, 1991; Shields, 1975; Wittig, 1985).

Within the context of these general criticisms, feminists have argued in particular that social science neglects and distorts the study of women in a systematic bias in favor of men. Some contend that the very processes of positivist science are inherently masculine, reflected even in the sexual metaphors used by the founders of modern science (Keller, 1985; Merchant, 1980). To Francis Bacon, for example, nature was female, and the goal of science was to "bind her to your service and make her your slave" (quoted in Keller, 1985, p. 36). As Sandra Harding (1986) summarized . . .

FIGURE 16.1   *continued*

Final Reference List

The APA style manual distinguishes between a reference listing system for a working draft and that for a manuscript for publication in a professional journal. Determine which form your professor prefers. The differences include the following:

1. The working draft will be written in Courier typeface, found on most typewriters, or in Times Roman, 12 points, a standard for computer word processing. A manuscript for publication permits more freedom in the selection of fonts, margins, borders, hanging indentations, and other document design features, in accordance with the publication guidelines and your own preferences.

2. The working draft requires underscoring of titles, whereas the manuscript for publication requires italics for the titles.

3. The working draft will have an unjustified right margin; the manuscript for publication may use right justification.

4. The working draft will use paragraph indentation for the final reference list whereas the manuscript for publication will use hanging indentation.

Figure 15.5 (p. 510) shows a reference listing for a *working draft* with paragraph indentation and title underscoring. All of the following examples are for a *published paper,* that is, with hanging indentation and italicized major titles. Pay close attention to the capitalization conventions. For a complete explanation of all possible entries, consult the APA style manual.

Book

Burns, M. (1995). *Handbook of psychology* (2nd ed.). Englewood Cliffs, NJ: Prentice Hall.

Two Authors

Fitzpatrick, J., & Swenson, T. (1998). *Studies in sociobiology.* Washington, DC: American Psychological Association.

Book with Editors or Translators

Leonard, R. T., & Coonz, C. R. (Eds.) (1992). *Bilingual education: Teaching English as a second language.* New York: Praeger.

Lura, A. R. (1996). *The mind of a mnemonist* (L. Solotaroff, Trans.). New York: Avon Books. (Original work published in 1965)

Article in a Book

Gurman, A. S., & Kniskern, D. P. (1995). Family therapy outcome
    research: Knowns and unknowns. In A. S. Gurman & D. P. Kniskern
    (Eds.), *Handbook of family therapy* (pp. 742–776). New York:
    Brunner/Mazel.

Unknown Author

Study finds free care is used more. (1992, April). *APA Monitor,* p. 14.

Encyclopedia

Carey, C. W. (1995). Bonding. In *Encyclopedia Britannica* (Vol. 2, p. 42).
    Chicago: Encyclopedia Britannica, Inc.

Journal

Gingro, J. (1997). Television in the classroom. *Social Education, 52,*
    221–225.

Magazine

Arial, S. K. (1996, August). Computers in the social studies classroom.
    Wired, 37–38, 101–102.

Newspaper

Lublin, J. S. (1996, December 5). On idle: The unemployed shun much
    mundane work. *The Wall Street Journal,* pp. B14, B18.

Electronic Sources (the system for electronic source documentation
is not yet standardized; consult your most up-to-date sources, including
online discussions. One excellent site for news of updated formats is on
www.refdesk.com; click on the hyperlink "Citing electronic citation style
for Internet sources").

General Online Source

Whistler, D. K. (1998). Librarians answer questions.
    <http://www.library.com/Whistler/easyref.html> (1998, May 26).

E-Mail

Foster, L. <fosterl@prodigy.com> (1996, April 27). Personal e-mail.
    [Manager of Media Consultants, Motorola, Boynton Beach, FL.]

Bulletin Boards (LISTSERV or USENET)

Kline, P., (1998, March 3). [LOGSERVE%LSULOG.BITNET@MJR.IRA.EDU]
    Testing engineer, Allied Bendix, Fort Lauderdale, FL.

Figures 16.2 and 16.3 show the same reference list in two formats:
one for a working draft and the second for a manuscript for publication.

References

Adams, J. K. (1997). Laboratory studies of behavior without awareness. Psychological Bulletin, 54, 383–406.

Breuer, J., & Freud, S. (1955). Studies on hysteria. In J. Strachey (Ed.), The standard edition of the complete psychological works of Sigmund Freud (Vol. 2, pp. 1–307). London: Hogarth Press. (Original work published 1893–1895)

Bruner, J. (1992). Another look at new look. American Psychologist, 47, 780–783.

Hilgard, E. R. (1986). Divided consciousness: Multiple controls in human thought and action (rev. ed.). New York: Wiley-Interscience.

Jacob, L., Lindsay, D. S., & Toth, J. P. (1992). Unconscious influences revealed: Attention, awareness, and control. American Psychologist, 47, 802–809.

Sanderson, K. L. Priming and human memory systems. <http://www.sci.com/sanderson/mem/lrs.html> (1996, May 2).

Tulving, E. (1976). The evolution of intelligence and access to the cognitive unconscious. In E. Stellar & J. M. Sprague (Eds.), Progress in psychobiology and physiological psychology (Vol. 6, pp. 245–280). San Diego, CA: Academic Press.

FIGURE 16.2 *Sample APA reference list for a working draft. Note the first line indentation and the underscored major titles.*

References

Adams, J. K. (1997). Laboratory studies of behavior without awareness.
Psychological Bulletin, 54, 383–406.

Breuer, J., & Freud, S. (1955). Studies on hysteria. In J. Strachey
(Ed.), *The standard edition of the complete psychological works of
Sigmund Freud* (Vol. 2, pp. 1–307). London: Hogarth Press. (Original
work published 1893–1895)

Bruner, J. (1992). Another look at new look. *American Psychologist, 47,*
780–783.

Hilgard, E. R. (1986). *Divided consciousness: Multiple controls in human
thought and action* (rev. ed.). New York: Wiley-Interscience.

Jacob, L., Lindsay, D. S., & Toth, J. P. (1992). Unconscious influences
revealed: Attention, awareness, and control. *American Psychologist, 47,*
802–809.

Sanderson, K. L. Priming and human memory systems.
<http://www.sci.com/sanderson/mem/lrs.html> (1996, May 2).

Tulving, E. (1976). The evolution of intelligence and access to the cognitive
unconscious. In E. Stellar & J. M. Sprague (Eds.), *Progress in psychobiology and physiological psychology* (Vol. 6, pp. 245–280). San Diego, CA:
Academic Press.

FIGURE 16.3 *Sample APA reference list for a manuscript for publication.
Note the hanging indentation of second and consecutive lines
and the italicized major titles.*

MLA SYSTEM

The fourth edition of the *Modern Language Association Handbook for Writers of Research Papers* by Joseph Gibaldi (1995) provides documentation advice for writers in languages and literature, and in some of the humanities if you decide not to use the *Chicago Manual of Style.* Some journals carry articles about technical communications but are not truly technical journals. An example is the *Technical Communication Quarterly,* which also uses the MLA style. Your instructor may want you to be able to prepare a research paper in the MLA style as well as in your appropriate discipline style. The *MLA Handbook* includes sections on the purposes of research; suggestions for choosing topics; outline guidance; internal reference and bibliography systems; advice on spelling, punctuation, abbreviations, and other stylistic matters; and manuscript preparation. The handbook covers a number of conventions that are particularly important if you are writing about literature.

General Conventions

A number of general conventions apply in the MLA system in addition to the internal referencing and final bibliography listing issues:

1. The paper consists of a title page, the text, content endnotes (optional), and a Works Cited page. Include an outline and an abstract only if your professor requires either of them.

2. The paper should be double-spaced throughout and the pages (with the exception of a separate title page and an outline) numbered in the upper, right-hand corner in Arabic numbers. If you keyboard the manuscript, you may use Times Roman, Helvetica, Arial, or other clear and legible fonts.

3. If a separate title page is not used, number the outline pages and/or a separate abstract in the upper, right-hand corner in lowercase Roman numerals.

4. If you use an abstract, you may (a) place it on a separate page between the title page and the first page of the text, or (b) place it on the first page of the text one double-space below the title and before the first lines of the text. If you place it on the first page of the text, indent the first sentence 10 spaces; then indent the rest 5 spaces as a block. Use quadruple spacing at the end of the abstract to set it off from the text.

5. If you use content endnotes (your own explanatory comments not included within the text), label a separate page "Notes," place it before the Works Cited listing, and provide raised superscript numerals within the text to match your superscript numeral notes. Double-space all entries and double-space between them.

6. Title your compilation list of sources "Works Cited." Do not use all capitals nor boldface. Center the title.

7. Unlike the APA system, which reports on researchers' findings in the past tense, the MLA system uses the present tense to report universal assertions: ("Ernest Hemingway **writes** . . ." or "Patricia O'Malley **proclaims** . . .").

Internal References

When writing your paper in the MLA style, several referencing practices apply. If the author's name or publication title is in the text, then add only the page reference in parentheses after the quotation, summary, or paraphrase:

> McCarroll reports that a fax machine sold for as much as $10,000 and up in 1971 (38).

If the author's name (article or book title if there is no author) is not included in the text, insert both the author and the page number in parentheses directly after the paraphrase, direct quotation, or summary:

> Actually, the origin of the fax machine dates to 1926 when it was invented by Radio Corporation of America (Costigan 2).

> Financial centers in New York, London, and the Far East fax material before and after office hours ("The Fax Revolution" 14).

If a summary refers to an entire work, page numbers may be omitted:

> In his article "The New Fax," Gerald Smith points out that the older fax systems used electromechanical scanning techniques which converted visual tonal variations for transmission to a receiver and were read by scanners, but that modern machines send electronics copies over ordinary telephone lines to the fax machine even at the opposite end of the world.

If quotations are used, the name of the author is included in the text, and the page number is in parentheses at the end with a following period. If the quotation is 40 words or more, indent all of the quotation, omit the quotation marks, and place the parenthetical reference after the period:

> "It's the electronic boom," declares John A. Widlicka, fax marketing manager for Sharp Electronics Corporation (qtd. in Gelfond 59).

> Thomas McCarroll reviews the benefits of the fax machine:

>> It eliminates the need for a typist to forward something, as you do when you telex. There are no anxieties about typing or keyboard errors—especially worrisome if you're sending critical financial numbers. You can maintain some degree of confidentiality if the document goes directly from you to a protected receiver. In contrast to e-mail, you don't have to be computer literate to send or receive. (16).

Figure 16.4 shows a sample page of document text in the MLA system:

Librarian Joan Rutledge helped a student to locate a bibliography in a book in another library of a multi-campus university. In the past, Rutledge would have to telex a request to the other library to send the book to the first campus. But now she can request the other campus to send a copy of the bibliography immediately by use of a facsimile machine and receive it within minutes (Mills 2). This process will save the student hours, if not days, in her research activities. The facsimile machine transmits documents electronically. All one needs are a telephone line and two parties with machines. Faxing popularity is growing to auspicious heights due to evolution of its capabilities, the simplicity of the transmission process, innovative features, and the practical cost.

Most people think that the facsimile machine (the fax) is a modern invention. Actually its origin dates to 1926, when it was invented by Radio Corporation of America (Costigan 2). The first facsimile transmission was sent through radio waves in 1937 (6). Even though the military used early crude devices to transmit maps, orders, and weather charts (McCarroll 38), the machine did not gain public recognition until much later. In the early 1970s fax machines were used regularly by newspaper companies to transmit thousands of photos daily (Costigan 12). But the fax did not become commercially successful until the 1980s when Japanese companies developed better machines by replacing mechanical parts with

FIGURE 16.4 *Sample documented text in MLA style* (Courtesy of student Shawn Sabga)

sophisticated computer circuitry (McCarroll 38). This upgrade cut transmission time for a single page from six minutes to ten seconds, and also dramatically improved the quality of copy that was transmitted ("The Fax Revolution" 14).

The transmission process, though very complex, is now greatly improved. On the older fax machines, the process consisted of converting visual details of a document to analogous electric current, conditioning the current for transmission by wire or radio waves to a receiver, and restoring it to its original form at the receiver. These older fax systems used electromechanical scanning techniques to convert visual tonal variations for transmission to a receiver. The machines then recorded transmission by using transducers that came in contact with electromechanical scanners (Bonura 47–56). But modern fax machines send electronic copies of documents over ordinary telephone lines to a fax machine on the opposite end. These modern fax machines are equipped with innovative computer circuitry, and when contact is made, an electronic scanner is activated. As the scanner moves across the page, it converts the text, charts, and pictures into electrical pulses that are carried over the telephone line. On the receiving end the process is reversed (McCarroll 38, Bonura 57). The user does not have to worry about the difficulty of the procedures. The human part of the process is incredibly simple. First, . . .

FIGURE 16.4   *continued*

Content Endnotes

Content endnotes are optional. If you need to include notes about research problems, conflicts in the testimony of experts, interesting tidbits relating to your text but not crucial to supporting your thesis, or credits to people or sources not included in the Works Cited listing, follow these guidelines:

1. Place the content endnotes on a separate page(s) after the last page of the text. Title the list "Notes." Double-space throughout. Indent each separate note.

2. Place superscript numbers in numerical order in your text to refer to each note:

> The military has long delayed accepting responsibility for the effects of poisons on the soldiers in the Gulf War.[1] An interview with General E. D. Smyth revealed . . .[2]

3. Begin the endnote with the matching subscript:

> [1] Lately, the military is acknowledging that soldiers have suffered some poison gas reactions, but insists that only those soldiers who did not take proper precautions when destroying our own weapons depots before pulling out can be pinpointed. On this point see Glascoe (4) and Meyers (B8).
>
> [2] Funds to travel to Washington to interview General E. D. Smyth were provided by the DCC Foundation.

Final Works Cited List

The essential differences between the MLA style of referencing sources and the APA style and its variations are that MLA requires the full name of authors rather than initials and the standard use of capitalization for titles and places the date at the end. Remember when using the MLA style to employ hanging indentation; underscore book, periodical, and other major titles; and center the title "Works Cited." Include a page number in the upper, right-hand corner. For listings not included in the following examples, consult the *MLA Handbook*.

Books

Bequai, August. Computer Crime. Lexington, Mass.: Lexington Books, 1988.

Hsiao, David K., Douglas S. Kerr, and Stuart E. Madnick. Computer Security. 2nd ed. New York: Academic Press, 1979.

Shank, Roger C. Computer Models of Thought and Language. Ed. by Roger C. Shank and Kenneth Mark Colby. San Francisco: W. H. Freeman, 1973.

Management Information Corporation. <u>Computer Privacy.</u> Cherry Hill,
    N.J.: Management Information Corporation, 1982.

Cardoza, Juan. <u>Access Control of Computers.</u> Trans. by Alan Jameson.
    New York: New American Library, 1991.

Special Reference Works

Smythe, Charles John. "Computers." <u>Encyclopaedia Britannica.</u> 1993 ed.

"Computer Services." <u>Statistical Abstract of the United States,</u> 82
    (1993):72.

Journal

Barthelmew, Rosemary. "Computer Viruses Increase." <u>Journal of
    Communication</u> 42.2 (1993): 5–24.

Magazine

Taflich, Peter. "Opening the Trapdoor Knapsack." <u>Time</u> 25 Oct. 1992: 72.

Newspaper

"Computer-based Home Security System Sales Soar." <u>New York Times</u> 15
    Dec. 1993, natl. ed.: B7+.

Government Document

United States. Cong. Senate. Subcommittee on DNA Legislation. <u>DNA and
    Cloning: A Moral Issue.</u> 103rd Cong, 2nd sess. S. Hearing 709.
    Washington, D.C. USGPO, 1997.

Pamphlet (treat like a book)

Radio Shack, a Division of Tandy Corporation. <u>TRS-80 Model II Micro-
    Computer System.</u> U.S.A. 1991.

Interview

Larsen, Robert H. Personal interview. 11 Mar. 1995.

Television

"Should We Get On With Computer Literacy?" <u>The Firing Line.</u>
    Washington, D.C.: PBS-TV, 16 Oct. 1993.

Graphics

Buckley, Charles L. Accounting system flowchart. <u>Introduction to
    Accounting.</u> New York: Oxford University Press, 1997.

Slide Presentation

Marijuana: The Addition Question. Slide presentation. Developed by
　　　　Project Cork, Dartmouth Medical School, 7 July 1991. 55 slides.

Electronic Sources (the format is not yet standardized; consult your
most up-to-date sources. Again, one good source is **www.refdesk.com**
with a hyperlink to "Citing Electronic Citation Style for Internet
Sources.")

Sample Internet Sources

Abers, Roslyn. "Pharmacists Question Warnings." New York Times 20 May
　　　　1997, late ed.: C1. New York Times Online. Online. Nexis. 10 Feb.
　　　　1994.

Gerard, Donald. "Cloning." Encyclopedia Britannica. CD-ROM. Chicago:
　　　　Encyclopaedia Britannica, Inc., 1997. Art. 608.

Tenner, E. "Republicans Voice Concern over Cloning." Boston Globe 15
　　　　June 1997: 1. Online. DIALOG. 30 May 1997.

Jackson, Merrill. "Cloning Letter," 1996.
　　　　<http://sunset.backbone.clemiss.edu/~mjackson/pmla.htm> 13 May
　　　　1996.

Smelzer, Sally. <smelzers@hillwpos.hill.af.mil> "Gulf War Poisoning." 29
　　　　April 1996. Personal e-mail. 3 May 1996.

Figure 16.5 shows a Works Cited list for the MLA system. Notice that
entries by the same author employ three dashes followed by a period, and
that the works are then alphabetized by titles.

Figure 16.7 later in this chapter shows a complete paper documented in the MLA style.

Works Cited

"Architects." <u>Encyclopedia of Careers and Vocational Guidance.</u> 11th

ed. Ed. by William C. Swope. Chicago: I. G. Ferguson Publishing

Company, 1994.

Berger, Karen. <Karen@jobweb.org> "Careers in Architecture."

Personal e-mail. 29 Aug. 1997.

Daley, Thelma T., and others, eds. <u>4:</u> <u>Construction.</u> 5th ed. Encino, CA:

Glencoe Press, 1994.

Duckman, Winchell. <u>A Guide to Professional Careers.</u> New York:

Julian Messner, 1990.

McReynolds, C. M. "So You Want to Be an Architect." <u>Progressive</u>

<u>Architecture</u> June 1994: 55–60.

Szerdi, John, President of Szerdi and Associates, Fort Lauderdale,

Florida. Personal interview. 24 June 1997.

U.S. Department of Labor. <u>Occupational Outlook Handbook.</u>

Washington, D.C.: USGPO, April 1984.

Wright, John W. <u>The American Almanac of Jobs and Salaries.</u> New

York: Avon Books, 1996.

---. "Careers in Architecture." <u>Architectural Record</u> May 1997:

200–206.

FIGURE 16.5   *Sample MLA Works Cited page (The three dashes and a period*
[---.] indicate the work is by the previously named author;
alphabetization is then by title.)

CBE STYLES

The newest edition of The Council of Biology Editors' *The CBE Manual for Authors, Editors, and Publishers* (1994) advocates two citation and reference styles. The first style is based on a name and year system and is similar to the APA system but with slight variations quoted in the *CBE Manual* for specific disciplines. The CBE style is used in general biological and earth sciences (agriculture, anthropology, archaeology, astronomy, biology, botany, geology, and zoology). The second style is a number system used for the applied and medical sciences. Refer to the APA system earlier in this chapter for details of the first system (name and year). The following section in this chapter discusses the number system for applied and medical science papers.

In the name and year system a list of references for agriculture will be titled "References"; a list of references for anthropology and archaeology will be titled "References Cited"; references for astronomy and geology will be titled "Literature Cited"; and references for biology, botany, and zoology will be titled "Cited References." All entries are alphabetized.

NUMBER SYSTEM

The disciplines of the applied sciences (chemistry, computer science, engineering, mathematics, and physics) and the medical sciences (health, medicine, nursing, and bio-medicine) employ the Number System, but variations by discipline exist both in the in-text citations and in the entries of the list of references. The system requires an in-text number, rather than the year, and the references at the end of the paper are numbered, not alphabetized, to correspond to the in-text citations.

Internal References

In general, follow your summary, paraphrase, or direct quotation with a superscript number of the source or place the number in parentheses after the citation. These numbers dictate the order of the reference list at the end of the paper. Consult the *CBE Manual* or ask your supervisor or professor which style to use for a specific discipline. Place the number of the source in parentheses or in raised superscripts:

> Graves found that thermal contact resistance at the die-billet interface plays an important role in determining the level of the die and billet temperature profiles (8).

or

> Graves found that thermal contact resistance at the die-billet interface plays an important role in determining the level of the die and billet temperature profiles[8] while previous investigators, Kellow and others, analytically and experimentally obtained heat transfers by presetting the temperatures.[9]

If the text does not mention the authority's name, employ one of the following methods:

> It is known (1) that the DNA concentration . . .

> *or*

> It is known[1] that the DNA concentration . . .

> *or*

> Additional observations include alteration in the genes (Victor, 3).

> *or for a direction quotation*

> "The use of genetic removal and replication is crucial" (Victor, 3, p. 44).

Mathematics papers require the in-text citation number to be placed within brackets in boldface (use a wavy line under the number if you are not keyboarding the text):

> Additionally, it is known **[3]** that every d-regular Lindelof space is D-normal. Further results on D-normal space will appear while in preparation **[4–5]**.

Figure 16.6 shows a sample space of text documented in the Number System.

The paper reports the results of a study of heat transfer in a hot upsetting system where only the surface layers of a cylindrical billet undergo plastic deformation. This process, designated here as a coining type process, is achieved by maintaining the peak load at a level that ensures that the peak pressure at no time exceeds 80 percent of the yield strength. The process parameters, such as die preheat, machine speed, rate of deformation, dwell time, and initial temperature of the billet, influence the batch thermal characteristics. Although it is common knowledge that the die billet temperatures play a crucial role in determining forging efficiency, very little information is available on the subject of heat transfer in forging (1). Lahoti and Altan (2), Boer and Schroeder (3), Rebellow and Kobayashi (4), Dadras and Wells (5), and Oh et al. (6) have worked on estimation of die billet temperature in forging. While Dadras (5) takes into account the radiative heat loss from the billet only during its transfer from the furnace to die, other workers have observed the domination of conductive and convective modes of heat transfer.

Thermal contact resistence at die-billet interface plays an important role in determining the levels of die and billet temperature profiles (7, 8). Previous investigators, Kellow et al. (9), Klafts (10), and Semiatin et al. (11), have analytically and experimentally obtained heat transfer. . . .

FIGURE 16.6 *Sample documented text in Number System*

Final Reference List

The sources will be numbered and listed in order of reference within the text or alphabetically. In each of the following discipline samples, pay close attention to underlining, italics, and parentheses.

In chemistry, title the list "References" and number each entry in the order of the text reference. Place the numbers in parentheses before the source; omit journal article titles; boldface the date:

(1)   Gallidaut, M., et al. J. Biol Chem. **1994,**   268, 21666–21667.

(2)   Dean, J. L.; Keith, T. T. J. Organ. Chem. **1944,** 59, 2745.

In computer science, title the list "Works Cited" and number each entry in the order of the text reference. Use numbers followed by periods:

1.   Gregory, A. "Three ways to insert superscripts." P.C. Computing 33 (July 1994), 54–57.

2.   Swenson, K. "Future trends in templates." IEEE Trans. Knowledge and Data Eng. 2 (2 Mar. 1990), 37–39.

In engineering and physics, title the list "References" and use raised superscripts both in the in-text and reference list in the order of the text reference; underscore book titles but not journal titles; place a journal volume in boldface. Omit journal article titles:

[1]C. Reece. Physics Today **55,** 27–45 (1992).

[2]G. Farley and C. Naugle. Low-Noise Electronics (Wiley, New York, 1985), p. 18.

In health, medicine, nursing, and bio-medicine, title the list "References" and number the entries in the order of text reference:

1.   Cardiff, C., et al. A Study of Zocor. Am J Med Sci. 1995: 185: 222–335.

2.   Allman, D. New Findings in Cholesterol Medicines. JAMA 1994: 274: 52–67.

Figure 16.8 later in the chapter shows a complete paper documented in the Number System.

CHICAGO STYLE

The 14th edition of *The Chicago Manual of Style* (1993) sets the standards for the fine arts (art, dance, music, and theater) and for some fields in the humanities (history, philosophy, religion, and theology, but not literature). This system employs superscript numerals within the text and places documentary footnotes on the bottom of the corresponding pages. Although no "Works Cited" page is usually necessary because it would be redundant, some professors may ask for one at the end of the paper.

In-Text Citations

Use arabic numeral superscripts. Place the superscript numeral at the end of quotations or paraphrases with the number following immediately without a space after the final word or mark of punctuation, as in this sample:

> Dali advises: "Let us be satisfied with the immediate miracle of opening our eyes, becoming skillful in the apprenticeship of looking well."[11] For Dali photography exemplifies a new method of inquiry into physical appearances through which a simple operation like a change in scale can suggest unusual analogies.[12] Dali explains his position on Surrealism and the automatic process:
>
>> To know how to look at an object, an animal, through mental eyes is to see with the greatest objective reality. . . . To look is to invent. All this seems to me more than enough to show the distance which separates me from Surrealism.[13]
>
> In this way, rejecting Surrealism, automatism, and mimetic painting alike, Dali defines artistic invention as the process through which the physical world is framed as a representation.[14]

FOOTNOTES

Primary References

Place your footnotes (the corresponding bibliographical reference) at the bottom of the pages where superscripts occur. Separate the footnotes from the text with triple spacing or a 12-space bar line beginning at the left margin. Double-space between footnotes, but single-space within each entry. Indent the first line 5 spaces. Number the footnotes consecutively throughout the entire paper.

Book

1. Salvador Dali, *Dali on Modern Art: The Cuckolds of Antiquated Modern Art* (New York: Knopf, 1957), 78–79.

Book with Three or More Authors

2. Rafael I Torella et al., *La Miel* (New York: Appleton, 1967), 307–423.

Journal

2. Clement Greenberg, "Avant-Garde and Kitsch," *Art News* 99 (1994), 111.

Newspaper

> 3. William Rubin, "Dada and Surrealism," *The New York Times,* 18 September 1979, sec. C, pp. C1, C12.

Subsequent References

After the first full reference, you should shorten subsequent references to the same work by using only the author(s) name and the page number. When there are two works by the same author, use the author, shortened title, and the page number. In general, avoid Latin abbreviations (*loc. cit.* or *op. cit.*); however, whenever a note refers to the source in the immediately preceding note, use *Ibid.* with a page number. Do not italicize Ibid. in your note. Notice the footnotes 2, 4, and 5 following:

> 1. Salvador Dali, *Dali on Modern Art: The Cuckolds of Antiquated Modern Art* (New York: Knopf, 1957), 78.
>
> 2. Ibid., 79.
>
> 3. Clement Greenberg, "Avant-Garde and Kitsch," *Art News* 99 (1994), 111.
>
> 4. Dali, *Dali on Modern Art,* 79.
>
> 5. Ibid., 79.

Your professor may allow you to list all of your notes together at the end of your paper as endnotes because it simplifies your keyboarding of the notes on a computer. If you are allowed to group your notes at the end, entitle the page "Notes," centered and placed two inches from the top of the page. Triple-space to begin the first note. Indent each first line five spaces, and return to the left-hand margin for second and consecutive lines. Double-space the notes and between each note.

Final Reference List

As previously mentioned, a final reference list is redundant because you will have included complete documentation in your footnotes, but some professors may still ask you to prepare one. If you do use one, use a heading that best represents its contents, such as "Works Cited," "Selected Bibliography," or "Sources Consulted." Triple space between the heading and the first entry. Alphabetize the list by the author(s) last name and use the hanging indentation format (unlike your footnotes). Refer to the examples of works cited listings in the MLA system earlier. Figure 15.5 in Chapter 15 shows a paper in the APA style. Figure 16.7 shows a complete paper in the MLA style, and Figure 16.8 shows a complete paper in the Number System.

Shawn Sabga

Professor J. VanAlstyne

English Comp. 2210-04

May 1, 1988

The Phenomenal Facsimile Machine

Librarian Joan Rutledge helped a student locate a bibliography in a book in another library of a multi-campus university. In the past, Rutledge would have to telex a request to the other library to send the book to the first campus. But now she can request the other campus to send a copy of the bibliography immediately by the use of a facsimile machine. This process will save the student hours, if not days, in her research activities. The facsimile machine transmits documents electronically. All one needs are a telephone line and two parties with machines. Faxing popularity is growing to auspicious heights due to evolution of its capabilities, the simplicity of the transmission process, innovative features, and the practical cost.

Most people think that the facsimile machine (the fax) is a modern invention. Actually its origin dates to 1926, when it was invented by Radio Corporation of America (Costigan 2). The first facsimile transmission was sent through radio waves in 1937 (6). Even though the military used these early crude devices to transmit

1

FIGURE 16.7 *Sample paper documented in MLA style* (Courtesy of student Shawn Sabga)

Sabga 2

maps, orders, and weather charts (McCarroll 38), the machine did
not gain public recognition until much later. In the early 1970s fax
machines were used regularly by newspaper companies to transmit
thousands of photos daily (Costigan 12). But the fax did not become
commercially successful until the 1980s when Japanese companies
developed better machines by replacing mechanical parts with
sophisticated computer circuitry (McCarroll 38). This upgrade cut
transmission time for a single page from six minutes to ten seconds,
and also dramatically improved the quality of copy that was transmitted ("The Fax Revolution" 14).

The transmission process, though very complex, is now greatly
improved. On the older fax machines, the process consisted of converting visual details of a document to analogous electric current,
conditioning the current for transmission by wire or radio waves to
a receiver, and restoring it to its original form at the receiver. These
older fax systems used electromechanical scanning techniques to
convert visual tonal variations for transmission to a receiver. The
machines then recorded transmission by using transducers that
came in contact with the electromechanical scanners (Costigan
47–56). But modern fax machines send electronic copies of documents over ordinary telephone lines to a fax machine on the opposite end. These modern fax machines are equipped with innovative

FIGURE 16.7 *continued*

Sabga 3

computer circuitry, and when contact is made, an electronic scanner is activated. As the scanner moves across the page, it converts the text, charts, and pictures into electrical pulses that are carried over the telephone line. On the receiving end the process is reversed (McCarroll 38). The user does not have to worry about the difficulty of the procedures. The human part of the process is incredibly simple. First, the sender feeds the copy through the paper feeder; then in a matter of seconds a copy of the document reaches the receiving fax to be printed exactly as it was sent. The receiving person does not have to be present when the document arrives because the document will feed automatically on the receiving fax (39). This, of course, makes it perfect for transmitting vital data to a time zone out of sync with one's own. Financial centers in New York, London, and the Far East fax material before and after business hours ("The Fax Revolution" 14).

Another reason for the sudden boost in popularity is the innovative features of the modern fax machines. Engineer Robert Johns points out that a standard top-of-the-line fax comes with a photocopy machine, an answering machine, a telephone and the fax system. Once considered too bulky and costly to be practical, Johns reports that fax machines have shrunk to half the size of personal computers. This transformation in size has been taken to the limit.

FIGURE 16.7   *continued*

Sabga 4

Mitsubishi Electric has introduced a fax unit that fits under a car
dashboard and connects to the cellular car phone (McCarroll 38).
Already New York-based Medbar Enterprises Inc. is selling a
Japanese-made model targeted for executives on the go. It is battery
operated and can be used from a plane or train as long as a telephone line is available (Gelfond 59).

    The fax has not only shrunk in size but also in price. In 1971
fax transceivers sold for as much as $10,000 and up (McCarroll
38). Later in the 1970s prices dropped to $3,000 to $5,000 ("The
Fax Revolution" 16). In the 1980s faxes the size of a portable typewriter entered the market for as low as $1,500 to $2,000. The
1990s see machines available from $500 to $800 (16). If these
prices do not attract the buyer's eye, there is still one more cost factor to consider: the amount of money that could be saved when using
a fax. Federal Express charges about $15.00 to deliver a one-page
letter overnight. The same letter can be faxed in a matter of seconds
for less than fifty cents. To telex a document, a keyboard operator
must retype it on a computer terminal before sending it to its desti-
nation. This process can require more than an hour and cost about
$5.00 for fifty words (McCarroll 38). Mark Winther, an electronics
analyst at Manhattan-based LINK Resources, says, "The growth of
fax is coming out of the hides of Federal Express and Western

FIGURE 16.7 *continued*

Sabga 5

Union. Fax poses a serious threat to overnight mail, and it could

make telex obsolete" (qtd. in Costigan 227). As a result of the cost

analysis, sales have soared from some 200,000 units sold in 1986 to

over 700,000 units in 1988. By the mid-1990s sales should hit one

million (McCarroll 38 and "The Fax Revolution" 14). The fax offers a

host of benefits:

> It eliminates the need for a typist to send something, as
>
> you do when you telex. There are no anxieties about typing
>
> errors—especially worrisome if you're sending critical
>
> financial numbers. You can maintain some degree of confi-
>
> dentiality if the document goes directly from you to a pro-
>
> tected receiver. In contrast to electronic mail, you don't
>
> have to be computer literate to send or receive ("The Fax
>
> Revolution" 16).

Fascination with communicating by fax is international. "It's <u>the</u>

electronic boom," declares John A. Widlicka, who is in charge of mar-

keting fax machines in the United States for Sharp Electronics

Corporation (qtd. in Gelfond 59).

FIGURE 16.7 *continued*

Sabga 6

Works Cited

Costigan, Daniel. FAX: The Principles and Practice of Facsimile

Communication. Philadelphia: Chilton Books, 1971.

"Fax Revolution, The." Travel & Leisure Oct. 1988: 14–16.

Gelfond, Susan. "Will There Be a FAX in Every Foyer?" PC Magazine

23 June 1987: 59.

Johns, Robert, Engineer with Canon Products, Chicago. Personal

interview on facsimile machines. 1 Dec. 1989.

McCarroll, Thomas. "Just the FAX, Ma'am." Time 31 Aug. 1987: 38.

FIGURE 16.7   *continued*

Cathy Venci

Professor M. Minnassian

English Comp. 2210-09

December 12, 1989

Hospice Versus Hospital: An Examination of Contrasting Approaches to Patient Care

Medicine and health care have made enormous life-saving advancements in this age of technology. Such procedures as organ transplants and laser surgery are now used routinely. Due to scientific research, many diseases that once were life threatening no longer pose a serious threat to our health. However, despite scientific advancements, there comes a point when technology reaches its limits. This is when the health-care professional must face the reality of the terminally ill patient. This patient has very special needs that go beyond medical treatment. The hospice movement is meeting the needs of these dying patients. Mosely's Medical Dictionary defines hospice as "a program offering continuous supportive care to dying people enabling them to live out the final days of their lives comfortably and as fully as possible" (1). The hospice concept of care differs from the traditional hospital in the approach to the treatment of the patient and his/her family, the role of the health-care professional, and the development for care in the future.

FIGURE 16.8 *Sample paper documented in the Number System style*
(Courtesy of student Cathy Venci)

Venci 2

A brief history of hospice care will further define the philosophy behind the modern hospice care movement. In the Middle Ages,
the hospice was a place of refuge for the traveler. The religious sects
that operated the hospices offered food and shelter to the poor. In
the middle 1800s these shelters developed into retreats for people
who were dying from incurable diseases such as tuberculosis (2).
According to Hamilton and Reid, the advancement of hospice care
came in 1967 when Cicely Saunders opened St. Christopher's Hospice
in London (3). They state that St. Christopher's Hospice was formed
with the purpose of helping the terminally ill patient to remain comfortable and relatively free from pain without any artificial means to
prolong dying (3). These same principles are the foundation of the
hospice concept in America today.

The hospice program has taken a different approach to the
treatment of the patient. Although the hospital and hospice are both
committed to medical needs of the patient, the hospital's emphasis is
on the care of victims of acute illness. The goals of the hospital,
according to Michael Hamilton and Helen Reid, authors of <u>A Hospice
Handbook</u>, are to diagnose and cure disease through modern technology (3). Another hospice expert, Kenneth P. Cohen, states that since
technology is the main source of treatment, it is used to the extent

FIGURE 16.8 *continued*

Venci 3

of prolonging life, even when there is no known cure (4). He adds
that life support systems and aggressive treatment are normal procedures in the hospital setting (4). Health reporter Jane Toot, notes
that the treatment of the terminal patient includes

- Intravenous fluids
- Forced feedings
- Nasogastric tubes
- Laboratory examination (5)

Cohen says that the most important treatment the dying patient needs
is pain control therapy (4). According to Cohen, hospital procedures
have a fixed routine of medication every four hours or so. The patient
may be experiencing serious pain before relief is given (4).

There are differences between hospital and hospice attention to the
dying patient's family. According to Toot, the hospital is not equipped to
deal with the family (5). The patient is often in an acute-care ward that
has very strict visiting regulations, which tend to create a sense of iso-
lation for both the patient and the family (5). Toot notes that the family is often confused over the treatment of the patient. She states that
this confusion is due to the hospital excluding the family in making
decisions concerning the treatment of the patient (5).

FIGURE 16.8 *continued*

Venci 4

On the other hand, the hospice is developed to accommodate the terminally ill patient. The emphasis is on caring for the patient's symptoms and not curing his/her disease (5). Hamilton and Reid refer to the hospice as "a place for dying" (3, p. 48). Accordingly there are no life support systems in the hospice program, there are no last-minute life saving attempts (3). Toot notes the major treatment of the patient is pain control; pain-killing drugs, such as morphine, are given to keep the patient free from pain (5). Other forms of treatment, according to Toot, consist of

- Massage for relaxation
- Heat therapy for circulation
- Exercise for flexibility (5)

Family participation is encouraged by hospice staff. Patients go home frequently to spend time with the family (5). Further, the hospice gives the patient access to

- Unlimited visiting hours for family
- Personal belongings and pets
- Around-the-clock medical care and pain relief
- Counseling and spiritual guidance for both the patient and the family (5)

FIGURE 16.8   *continued*

This daily care emphasizes the quality of life at all times (5). The whole hospice program can be summarized in one statement: "Dying patients are human beings. These people have real needs and cares that need to be attended to" (6). A survey (see attached instrument with numbered tabulations) of 100 hospice patients in Moore County reveals the value patients place on hospice services (7). Figure 1 shows the number one ratings of four services by percentages:

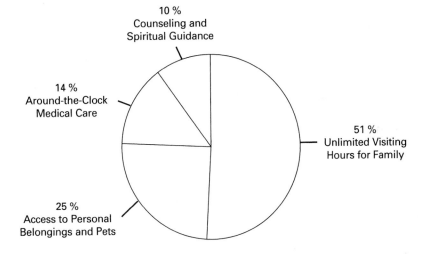

10 %
Counseling and
Spiritual Guidance

14 %
Around-the-Clock
Medical Care

51 %
Unlimited Visiting
Hours for Family

25 %
Access to Personal
Belongings and Pets

Figure 1 Number one ratings for four hospice services (7)

In addition, 41 percent selected access to belongings and pets as the second most highly valued. Thirty percent rated around-the-clock medical care as third highest, and 40 percent rated counseling and

FIGURE 16.8 *continued*

Venci 6

spiritual guidance least highly. Clearly being in their own homes with family, belongings, and pets nearby is the premier desire of terminally ill patients (7).

Another area of contrast between hospice and hospital care concerns the role of the health-care professional. Cohen notes that the hospital staff is increasingly involved in the challenge of technical care and diagnosis; he states

> Hospital staff do not necessarily ignore the dying patient, but they give priority to patients for whom they can provide life-saving measures. This is not unreasonable, considering their training and emphasis on curative functions rather than the caring functions needed by the dying (4).

Table 1 shows the contrast of involvement in hospital and hospice staff members:

Table 1   Contrast between staff members in hospice and hospital

Hospital staff	Hospice staff
Hierarchy system of care	Team approach to sharing
physician	responsibility
nurses	
aides	

FIGURE 16.8 *continued*

Table 1 Continued

Hospital staff	Hospice staff
Rotation of wards and shifts by staff members	Daily care of same patient
Little communication concerning patient cases	Daily discussions and meetings concerning patient cases
Few volunteers on staff	Many volunteers on staff

Based on Toot (5)

The cost of hospital and hospice care differs. The hospital trend will continue to focus on technology and research. Due to the high cost of these components, the cost of hospital care will continue to increase. Another study (8) states that the cost of a bed space in a hospital is now $240 and upward a day (8). The study also notes that this same bed space in a hospice costs as little as $34 per day (8). This difference in medical expenses is one of the major factors in hospice growth (8). A New York Times reporter (9) points out that another factor in hospice care growth is the age of the population. With an aging population, there is a greater number of people who need the kind of care hospice programs offer (10). Two-thirds of hospice patients are over sixty-five years of age (10).

In the past the majority of hospice funding came from donations by individuals, civic groups, and church organizations (9). Medicare

FIGURE 16.8   *continued*

is now paying some of the expenses (10). Fifteen years ago there were only 200 hospices in the United States (10). There are now thousands of groups, and with the help of Medicare funding there will be even greater numbers (10).

In conclusion, hospitals are doing an excellent job in providing a cure for the victim of acute illness and accident. The hospital should continue to do everything possible to advance research and technology to meet the needs of the patient who has a chance to return to society. On the other hand, with the increase of the age of our population, it is apparent there will be a greater need for the care and counseling that hospice provides to those in their final days of living.

FIGURE 16.8 *continued*

Venci 9

Hospice Services Survey

The purpose of this survey is to determine the value you place on benefits received through hospice care. Results will be included in a brief paper examining hospital versus hospice care.

Please rate the following benefits of hospice care in your opinion. (Place a 1 before the service you deem most valuable, a 2 before the second most valuable service, a 3 before the third most valuable service, and a 4 before the least valuable).

_____ a.   Unlimited visiting hours for family
_____ b.   Access to belongings and pets
_____ c.   Around-the-clock medical care
_____ d.   Counseling and spiritual guidance

Tabulated Survey Results

Most valuable selections:	Second most valuable:
51 selected a.	22 selected a.
14 selected b.	41 selected b.
25 selected c.	10 selected c.
10 selected d.	27 selected d.

Third most valuable:	Fourth most valuable:
17 selected a.	10 selected a.
30 selected b.	15 selected b.
30 selected c.	35 selected c.
23 selected d.	40 selected d.

FIGURE 16.8   *continued*

Venci 10

References

1. Mosley's Medical Dictionary, 1993 ed.

2. Hospice care. Encyclopedia Americana, 1993 ed.: 436–438.

3. Hamilton, M., Reid, H. A Hospice Handbook. Grand Rapids: William B. Eerdmans Publishing, 1980.

4. Cohen, K. P. Hospice: Prescription for Terminal Care. Germantown, Md.: Aspen Publications, 1996.

5. Toot, J. Physical Therapy and Hospice. Physical Therapy Journal 64 (1994): 665–670.

6. Coor, C., Coor, D. Hospice Care Principles and Practices. New York: Springer Publishing Company, 1983.

7. Venci, C. Survey on Value of Hospice Services. Moore County, 1997.

8. Mudd, P. High Ideals and Hard Cases. Hastings Center Report (April 1982): 11–14.

9. Boundy, D. Growth of Hospice Programs Is Cited. The New York Times 20 May 1994, sec. 22:6.

10. Friedland, S. Hospice Benefit off to Slow Start. The New York Times 22 Nov. 1994, sec. 11: 1, 7–8.

FIGURE 16.8   *continued*

CHECKLIST

Documenting a Research Paper

GENERAL

❏ **1.** Have I selected the appropriate documentation system for the field of research I am pursuing?

❏ **2.** Have I correctly referenced by name and year each summary, paraphrase, and quotation?

❏ **3.** Have I completed taking and organizing all of the notes necessary to prove my thesis in a developed paper?

APA SYSTEM

❏ **1.** Have I clarified the directions for a working draft or manuscript for publication?

❏ **2.** Have I completed all the parts of the paper?
   ❏ Title page?
   ❏ Outline (optional)?
   ❏ Abstract?
   ❏ In-text documented paper?
   ❏ List of references?

❏ **3.** Have I double-spaced and numbered each page correctly?

❏ **4.** Have I used only last names in the text?

❏ **5.** Have I reported the findings in the past tense?

❏ **6.** Have I correctly referenced by name and year each summary, paraphrase, and quotation (and page)?

❑ **7.** If I am writing a working draft, have I adhered to all of the regulations?

    ❑ Courier or Times Roman 12 type?

    ❑ Underscored titles?

    ❑ Unjustified right margins?

    ❑ Paragraph indentation for the reference list?

❑ **8.** If I am writing a paper for publication, have I adhered to all of the regulations?

    ❑ Optional fonts and other document design features?

    ❑ Justified right margin?

    ❑ Hanging indentation for the reference list?

MLA SYSTEM

❑ **1.** Have I completed all parts of the paper?

    ❑ Title page?

    ❑ Outline (optional)?

    ❑ Abstract (optional)?

    ❑ In-text documented paper?

    ❑ Content endnotes (optional)?

    ❑ The Works Cited listing?

❑ **2.** Have I double-spaced and numbered the pages correctly?

❑ **3.** Have I reported the research in the present tense?

❑ **4.** Have I correctly referenced summaries, paraphrases, and quotations?

❑ **5.** If I am including content endnotes, have I placed superscripts in the text and before the corresponding notes in the "Notes"?

CBE STYLES

❑ **1.** Have I distinguished the appropriate system of the two according to the subject field?

❑ **2.** Have I prepared all in-text citations and the appropriate (by field) list of references?

❑ **3.** Have I assembled my paper correctly?

NUMBER SYSTEM

❑ **1.** Have I chosen a consistent in-text documentation system for the discipline?

❑ Superscript numbers?

❑ Parenthesized numbers?

❑ Boldface numbers in brackets?

❑ **2.** Have I provided either a numbered and alphabetized or a numbered listing in order of reference in the text?

❑ **3.** Have I, at minimum, included a titled first page of text, consecutively numbered pages, and a final list of sources?

CHICAGO STYLE

❑ **1.** Have I used consecutive, superscript numbers within the text for all summaries, paraphrases, and quotations?

❑ **2.** Have I used footnotes to explain each source at the bottom of the page corresponding to the reference, or has my professor allowed a complete list of notes at the end of the paper entitled "Endnotes"?

❑ **3.** Have I used the correct double-spacing for text plus a line to separate text and footnotes allowed by triple-spacing before the first entry?

❑ **4.** Are my footnotes single-spaced and each first line indented five spaces?

❑ **5.** If I use endnotes, is each entry double-spaced and listed with hanging indentation?

EXERCISES

1. **Bibliography Card.**   What is missing on the following bibliography card?

Tulving, E., & Schacter, Priming and human memory Systems: Science, 247, 301–305
—case studies of unconscious mental Processes

---

2. **APA Documentation.**   Write the following information in the APA system for a paper on psychological aspects of communication:

    a.   An article, "How We Learn to Communicate Feelings," published in the 1997, vol. 83 edition of Psychological Review, written by R. P. Langstroth and M. E. French on pages 41–45.

    b.   An Appleton-Century-Crofts publication in New York of a book, Communication Theory, by Theodore Conover, M. D. in 1995.

    c.   An article, "Do Men and Women Think Alike?" by M. P. Chadwick in the December 1997 edition of Psychology Today on pages 72–76.

3. **MLA Documentation.**   Rewrite the following information in the MLA style for a "Works Cited" listing:

    a.   An article in Technical Communication by C. R. Miller entitled Some Thoughts on Document Design: on pages 108–111 of Volume 37 in July 1997.

    b.   Can the Public Understand Scientific Terms? by Dawn Funder, published by Madsen Publishers, Inc. in 1995 in Peoria, Illinois.

    c.   A Time magazine article entitled Technical Talk Is Tricky, on pages 37–39 of the February 17, 1996 issue.

    d.   An Encyclopaedia Britannica article on CD-ROM published in 1995, written by T. S. Stephan.

4. **Number System Documentation.**   Rewrite the following information in the Number System for an article on computer science:

    a.   A book, The Design of analysis of computer algorithms, written by M. E. Grasso, Seth Stephens, and J. D. Ulmann, published by Addison-Wesley in Reading, Massachusetts, in 1995.

    b.   An article, Word Perfect 8.0 Sales Soar, by Manning G. Abrams in Computer, on pages 189–199 of volume 27 in September, 1997.

    c. Computers and word processing, on a July 15, 1997 posting on the Internet at <http://www.computers.com/wdprcing/html>.

WRITING PROJECTS

1. **Documented Research Paper.** Write a documented paper (2,000–3,000 words) on a new development in your field or another specific field that interests you. The field will determine whether you should use the APA (working draft or paper for publication), MLA, CBE (two styles for variant fields), Number, or Chicago Style system. Where optional, determine from your professor whether to include a separate title page, an outline, an abstract, and other optional choices. Select a subject from the following suggestions or choose your own from your field of study:

autism	vocal computer problems
aardvark mating rituals	zocor
new food source	ozone deletion
nitric oxide	black holes
new surgical glues	robotics
new laser surgeries	new weapon(s)
new fuels	digital cameras
tomography	new chemotherapy procedures
cloning human body parts	changes in Medicare
computer scanners	magnetic resonance imaging
new heart medicine(s)	religious cults
arthroscopic knee surgery	last year's crime rate in your
changes in welfare	state
changes in life expectancy	effects of weightlessness
facts on UFOs	trends in transportation
future of boxing	a new Ice Age
Mars discoveries	new building material(s)
weather satellites	computer-influenced slang
an electronic marvel	U.S. relations with China
future space travel	American Catholicism
Hong Kong today	athletes and steroids
Indonesian economic	changes in the Democratic
development	National Party
new boundaries in Croatia/	an aspect of cyberspace
Serbia	changes in nursing studies
low cholesterol diet(s)	new tobacco industry
homosexuality	legislation
e-mail in the business	abuse treatment centers
world	digitized TV
suction lipectomy	911 abuses

other?

PART *five*

Presentation Strategies

CHAPTER *17*

Verbal Communications

CATHY   by Cathy Guisewite

S K I L L S

After studying this chapter, you should be able to

1. Identify potential professional situations in which oral communication skills are necessary.

2. Name three considerations in preparing for a successful information-gathering interview.

3. Prepare for and role play an information-gathering interview with appropriate strategies.

4. Name and practice eight strategies for productive telephone communications.

5. Name and practice three strategies for reaching group decisions and understand the drawbacks of each.

6. Name and practice six strategies for worthwhile appraisals.

7. Name and practice four strategies for productive reprimands.

8. Name five strategies for a convincing persuasive oral report.

9. Present both persuasive and informative oral reports.

10. Discuss the significance of at least five nonverbal messages that are unconsciously sent and received.

INTRODUCTION

Effective verbal communications are important in every work setting. Every organization requires its employees to develop competent and productive speaking skills. Initially to gain employment you must be able to speak persuasively about your own potential, and to reveal yourself to be knowledgeable about the organization that has granted you an interview (see Chapter 11).

Once you are employed, you will be expected to demonstrate careful and useful interpersonal communication skills in a variety of tasks including information-gathering interviews, telephone transactions, group decision-making discussions, and informal and formal speeches. You should even be aware of the nonverbal communications you purposefully and inadvertently send and receive.

INTERVIEWS

Employment Interviews

In addition to covering the writing strategies of the resume and cover letter necessary to obtain an employment interview, Chapter 11 covered employment interview strategies. Review those strategies now because they offer valuable information about all of the verbal communications that this chapter discusses.

Information-Gathering Interviews

You may often find yourself acting as an interviewer in order to obtain information from your co-workers, superiors, and people in other organizations. You need to determine in advance the exact purpose of the interview. In addition you need to preplan by jotting down the topics that need to be covered and organizing them logically. Be prepared to take notes or to tape the interview with the permission of the interviewee.

PURPOSE

Write out your purpose statement so that you identify the overall purpose and the subcategories of information that you are seeking. You are not going to read this purpose, but you are going to state it clearly both in seeking the interview appointment and in beginning the actual interview.

Sample Purpose Statement 1

The purpose of my interview with you, the twelve members of my department, is to obtain from each of you your opinions and answers to twenty questions regarding our needs for updating computers, software, and printers in order to draft a proposal for necessary improvements.

Sample Purpose Statement 2

The purpose of my interview with you, John Avery, Manager of the Media and Technology Communications Center of Motorola Corporation, is both to discuss the changing role of the industrial technical writer and to obtain sample user guides and other tiers of materials for publication in the fourth edition of my textbook.

The next step is to establish the appointment by phone call, memo, letter, or personal contact. At this time state the purpose of the interview and set a specific date giving the interviewee(s) plenty of time to prepare for the actual interview.

CONTENT PREPARATION

Decide on the specific questions you are going to ask and organize them logically. You may want to prepare a very brief topic outline to serve as your visual guide during the interview, but your most effective interview will be realized without much reference to personal notes (unless you are actually using a survey or questionnaire). Your questions should be specific, open-ended, and unbiased. Give the person(s) plenty of time to respond. Some sample questions for the interview on the changing roles of industrial technical writers may include the following:

1. In the past, the technical writer was given only a month or so after the development of a new product to write a user manual. I understand that technical writers now operate within a different time frame and that their tasks involve much more than just the written document. Would you please tell me at what point the writer is now involved and describe for me these new tasks?

2. Why has the time schedule changed?

3. Does your staff do all of the document design and desktop publishing or do you use professional printers? If so, please differentiate those tasks you undertake and those you job out.

4. You mention preparation of online materials. Would you please elaborate on the nature and purpose of these materials?

Continue along this line, taking your cues for further questions from the responses given to you. If you do not fully understand a response, rephrase your question or ask for amplification. Avoid questions that can be answered with a simple *yes* or *no* or are too vague, such as, "What do you do?" Do not ask biased questions, such as, "Would you agree that technical writers are underpaid?" Ask, "In view of the increased number of tasks, are modern technical writers being adequately compensated?" Be prepared to ask follow-up questions, such as, "Why? Why not? Do you foresee even more tasks being required? What recommendations would you make to cope with the increasing tasks of the writer?"

Your conduct should be similar to that you exhibit when you are the interviewee. Review the tips in Chapter 11 about being on time, dressing professionally, shaking hands and making eye contact, showing your personality and being courteous. Remember that the actions that will nullify the value of your interview include "winging it," over familiarity, nervousness, offensive language, failure to respond to questions, and abrupt endings. In addition, consider these suggestions for a successful information-gathering interview:

1. Thank your respondent at the beginning of the interview for granting you the time.

2. Restate your purpose and elaborate on how you intend to use the interview responses.

3. Let your interviewee do most of the talking. Don't "lead" the respondent or state your own opinions.

4. Do not tape an interview without the explicit consent of the interviewee. If you take notes, keep them to a minimum, jotting down words and phrases, statistics, and other specific information to jog your memory when you write up the proceedings.

5. If you interview more than one person on the same subject, standardize your questions and ask them in the same order to each respondent.

6. Offer to provide a copy of any reports, documents, or other materials that will incorporate the interviewee's answers and comments when your study is complete.

7. Arrange to write out your interview results very soon after the interview in order to remember clearly what transpired.

When interviewing a number of people about the same subject, you may want to devise a survey or questionnaire. Review the material on surveys in Chapter 14.

TELEPHONE CONVERSATIONS

At all times you are a representative of your organization. Many transactions are conducted by telephone both within and outside your company. Your telephone etiquette reflects not only your own verbal skills but also the company's image. Certain basic strategies will help you to receive messages in a professional manner. Talking on the phone involves both talking and listening skills. Consider these tips:

- Pay attention to the volume, tone, and clarity of your voice
- Use a warm and pleasant tone
- Do not shout or whisper
- Be sure that your mouth is about an inch or two from the speaker—no closer and no farther
- Enunciate clearly; avoid slurring of "Hullo," "Whajasay?," "Yeah," and the like
- Avoid distracting background noises, such as another conversation, computer printers, radios, tapping, blowing, or chewing
- Be prepared to take notes; have a notepad, pens, and pencils near your telephone
- Have on hand, too, any reports, letters, or other printed matter that may need to be referenced during your conversations; do not make your listener wait while you locate such materials

Placing a Call.   When placing a call, identify yourself by a courteous "Hello" and state your name and other identifying comments, such as, "Hello, this is Judy Withrow, the training consultant with Writing Skills Management. I spoke to you last Friday about the possibility of conducting a seminar for your midlevel management people." After an acknowledgment, proceed to state the purpose of your call, such as, "I'm calling to determine if you received the brochures I mailed to you and to set a date for an interview." After transacting your business, specify what it is you want your listener to do: call you back in an hour or a week, mail you material, or confirm details in writing. If it is appropriate, thank the listener for the information, interest, or time.

Receiving a Call.   When receiving a call, say "Hello" courteously and identify yourself, your department, or your organization: "Hello, Tom Brown speaking," "Hello. Personnel Department, Miss Shipley speaking," or "Ace Company. This is Bob Martin, Sales Manager. May I help you?" End the conversation courteously, too. It may be appropriate to say, "Thank you for calling," "I'll send the materials to you in today's mail," or "I hope I've assisted you." Say "Goodbye" in response to the caller's closing, and hang up carefully. A receiver that is banged down or dropped on its cradle may unduly irritate the caller who has not yet hung up.

If you answer the telephone for someone else, be doubly courteous. Let the caller know that the right number has been reached: "Hello, Ms Steven's office; this is Jerry Thomas, her assistant, speaking. May I help you?" Do not make lame excuses for another's absence, such as, "She's busy right now" or "I don't know where she is." State that the individual is engaged, in conference, or out to lunch and offer to take a message. Do whatever you can to have the call returned.

OTHER VERBAL COMMUNICATIONS

As a professional worker you will be expected to take part in a variety of verbal information situations. These may include group or panel discussions, appraisals and reprimands, as well as oral reports.

Group Discussions

From time to time you will be part of a decision-making group—a committee, a department, or an even larger group. Large groups tend to operate through parliamentary procedure to reach decisions, whereas small groups operate less formally. We are concerned here only with small task groups.

A number of factors affect group processes:

- Purpose for which the group exists
- Personal goals of its members (which may be in conflict with the overall purpose)
- Permanency of the group
- Power and relationships of the members of the group
- Methods (majority rule, compromise, or consensus) used for decision making

The most effective group is one with a clear purpose that is understood and supported by all of its members. Further, the longer a group can work together, the less it tends to be dominated by one or two people. Finally, a group that knows its decisions will be accepted by others will be more successful than a group that perceives an outside threat or an overriding decision maker. There are three major methods for reaching a group decision. Each has its drawbacks.

Majority Rule.   Arriving at a decision by majority rule involving a vote is a common method, but not always the best. A vote tends to polarize the camps, making the losers less committed or even antagonistic to the decision.

Compromise.   In a compromise decision both sides give up a little to gain a little. Collective bargaining typically uses compromise decision-making methods. Compromise is effective if members of the group have to answer to larger constituencies. The constituents can believe they won something and were not total losers, yet since nobody is 100 percent pleased with the decision, those involved may not work very hard to implement the decision.

Consensus.   This method requires that all members agree on the decision. It is the most difficult way to decide but the most effective because all members are satisfied with the outcome. Consensus should lead to better productivity and commitment to implementation of the decision. For a group to reach consensus decisions, follow these steps:

1. All members should participate in clarifying a goal and establishing a procedure.
2. Draw up an agenda of group procedure and appoint an initial leader.
3. Let each member talk freely about the goal and suggest procedures for the group to follow.
4. Hold a group discussion to determine the status and causes of the problem or situation.
5. Establish criteria for evaluating each possible solution or decision.

6. Brainstorm solutions in a creative say to consider all possible solutions:

   a. Have all members offer as many different solutions as possible with no interrupting criticism or evaluation.

   b. List all of the suggestions on a chalkboard or flip chart.

   c. Weigh each potential solution against the criteria to determine which is best in terms of effectiveness, cost, simplicity, possible implementation, and acceptability to the group.

   d. Finally, make a group decision (a consensus) on how to implement the solution with give and take on delegating and accepting responsibilities.

Appraisals and Reprimands

You will probably meet with your supervisor from time to time to assess your job performance. Alternatively, as you move up the career ladder, it may become your responsibility to appraise the performance of your subordinates. Giving and receiving reprimands are also inevitable transactions in any organization. Both appraisals and reprimands tend toward emotional communication, so strategies should be aimed toward objectivity.

Appraisals. Certain strategies will lead to an objective and effective interchange, including the following:

1. Schedule an appraisal interview in advance so that both sides may prepare.

2. Limit the interview to a specific time period—15 to 30 minutes—to avoid repetitious harangues.

3. Have both parties prepare separate written assessments of the performance and exchange them in advance of the interview.

4. During the appraisal, concentrate on task-related factors rather than personalities.

5. Offer (or seek) explanations for poor performance.

6. Overall, center the discussion on specific goals and methods for improvement.

These strategies should lead to good will, improved performance, and group cohesiveness.

Reprimands. Objectivity is also the key to the effective reprimand situation. A clear company procedure should be in place to handle these touchy interviews. The following steps should help to turn a reprimand into a maturation process:

1. The company should offer written policies and/or procedural manuals to spell out the expected activities of its employees. These written materials will also substantiate that a violation has occurred.

2. The supervisor should attempt to determine if the violation was due to lack of information or was willful. Clear communication is essential.

3. The supervisor must ask questions, use feedback, and be sensitive to the feelings of, and possible penalty to, the violator.

4. The violator should be prepared to ask questions about appropriate recourse, appeals, and corrective actions.

Both the supervisor and the violator must be honest and concerned and work toward the goal of amelioration.

ORAL REPORTS

You will be called upon in your career to make any number of informal or formal oral reports. These may be reports at meetings of your peers and supervisors, presentations at training seminars, speeches at conventions or before civic groups, or presentations to explain proposals and other projects. Oral reports are classified as

- Impromptu speeches
- Memorized speeches
- Manuscript speeches
- Extemporaneous speeches

The impromptu speech is an off-the-cuff presentation likely to occur at a meeting where either you are asked about a project or you decide to present your views on an agenda item. A memorized speech may be appropriate for material which must be communicated many times, but it is difficult to avoid sounding wooden and mechanical in memorized speeches. A manuscript speech, one which is read, may be appropriate to convey very technical or detailed information but calls for exacting practice to avoid a monotone delivery and lack of eye contact. The extemporaneous report is the most widely used and effective oral presentation.

Extemporaneous Reports

The extemporaneous report requires careful audience analysis, clear purpose, logical organization, supportive visual materials, sufficient rehearsal, and skillful delivery.

Audience Analysis

Previous chapters have stressed that an analysis of your audience is essential for effective written reports. The public speaker must also consider the audience who will listen to the oral report. Besides thinking about how much the audience already knows about your subject, what level of technical language is appropriate, and your relationship to the group, you should seek to discover in advance the average age of the group, political persuasions, religious or ethnic affiliations, rural or urban interest, and sex. Knowledge of these factors will help you to infer how the listeners will receive your information. These factors should indicate the degree of formality appropriate to your talk.

It is equally important to continue analyzing your audience during your speech through feedback. You can gauge your audience's reactions to your speech by noting facial expressions, postures, applause, and the like. The alert speaker will make adjustments in delivery based on this feedback.

Purpose

It is important to have a clear purpose in mind: to entertain, to persuade, or to inform. We are not concerned here with the entertaining speech, but as a professional many of your reports will persuade or inform your audiences.

Persuasive Purpose. Actually, you call upon the strategies of persuasion not only in most of your verbal communications but in written communications, as well. Every time you ask for action in a memo, letter, brief report, proposal, interview, group discussion, phone call, or verbal presentation, you consciously or unconsciously employ persuasive strategies. These have been discussed in almost every chapter of the book. Even graphics and visuals help to persuade an individual or a group to your way of thinking by presenting in condensed visual form irrefutable facts, relationships, and abstract concepts.

You need to employ persuasive strategies to bring about overt action or to change beliefs and attitudes. Examples of overt actions are the purchase of products, the election of officers, the adoption of policies, or the alteration of procedures. Persuasion may also change workers' attitudes toward minority and women workers, alleviate difficult working situations, or promote pride in organizational membership. Five strategies will help you to deliver a compelling persuasive speech:

1. **Audience comfort.** Arrange to give your speech in a setting that is comfortable for the audience. Consider whether you should take an authoritative stance up front or arrange the group into a roundtable mode.

2. **Identification with the group.** Arrange to be introduced by a person who is well liked by the group and who will stress your qualifications to speak on the subject. Establish that you are part of the group by the way you dress and act or by actually expressing similarities in ideas, beliefs, or experiences.

3. **Goals and rewards.** State your goals in an honest, friendly, yet assertive manner emphasizing the rewards (anything that meets the needs and desires of the group) that persuaded listeners will receive.

4. **Credibility.** Use accurate statistics and other evidence along with the *reasons* for believing that the evidence is relevant to your conclusions. Cite the experts whose claims can be readily accepted by the group. If you cite authorities whose expertise is not fully accepted by the group, you will lose credibility.

Above all remember that change is very difficult for some people to accept. Some members of the group may believe their hierarchal position or very job is threatened; others may believe they are going to have to work harder or fall behind. If you are the spokesperson for the company, you must let everyone know how each will benefit from the change.

Informational Purpose.   The informative report, or in this consideration, the informative speech, is the most common among professionals. Such speeches may be patterned along the same lines as written reports, for example, analytical and evaluation reports, instructions, analysis of processes, descriptions of mechanisms, and so forth. Do not lose sight of your primary goal, to impart information.

Organization of Speeches

Your evidence or data must be logically organized. A listener is not a reader. A listener cannot back up to review information or skip ahead to the conclusion. Organize your report with the listener in mind.

If your purpose is to persuade, you may consider the problem-solution organizational approach or the advantages-disadvantages approach. Figure 17.1 illustrates the organization of these approaches.

If your purpose is to inform, you will organize along the lines of the type of report which you are presenting orally. Figure 17.2 illustrates an organizational approach to your speech.

Next, prepare an outline of your actual speech and gather your information. Other chapters in this book review outlines and organization for specific types of reports.

Finally, using your outline, prepare 3 × 5 in. or larger note cards on the main topics of your speech. Print large enough so that you can read the materially easily. Underline key points in red, and indicate by asterisks where you plan to use your visual materials. Use only one side of the note cards, and do not overload a card.

PROBLEM-SOLUTION APPROACH

   I. Approach
     A. Gain attention and goodwill.
     B. Develop credibility, if necessary.
     C. Orient receiver to subject and purpose.

  II. Body
     A. Develop problem.
        1. Explain symptoms or results (problem description).
        2. Explain size and/or significance.
        3. Explain cause.
     B. Develop solution.
        1. Explain solution.
        2. Explain how solution eliminates problem.

 III. Conclusion
     A. Appeal for action or desired belief.
     B. Allow for discussion.

ADVANTAGES-DISADVANTAGES APPROACH

   I. Introduction
     A. Gain attention and goodwill.
     B. Develop credibility, if necessary.
     C. Orient receiver to subject and purpose.

  II. Body
     A. Explain disadvantages of present situation.
     B. Explain advantages of new idea, proposal, policy, or situation.
     C. Explain that advantages cannot be obtained without proposed changes.

 III. Conclusion
     A. Appeal for action.
     B. Allow for discussion.

FIGURE 17.1   *Persuasive oral report approaches*

INFORMATIVE SPEECH APPROACH

I. Introduction
   A. Gain attention and goodwill.
   B. Orient receiver to subject and purpose.

II. Body
   A. Present data in logical organization.
   B. Repeat key terms and provide verbal transitions between parts.
   C. Employ visual aids.

III. Closing
   A. Summarize.
   B. Emphasize.
   C. Allow for questions and/or discussion.

FIGURE 17.2   *Informative oral report approach*

Visuals and Graphics

Visuals and graphics will help you add impact to both persuasive and informational speeches by clarifying and emphasizing your information. The size of the room or auditorium, the kind of people in your audience, the available monies, your particular talents, and the available equipment (chalkboards, projectors, flip chart stands, televisions, computers, etc.), and the nature of your speech are all factors for consideration in planning visual material.

Chapter 3 discussed the value and conventions of graphics in written reports. These same types of graphics can be presented in oral reports. Chapter 18 discusses at length a wide variety of oral report visual props from simple flip charts to video and computer presentations.

Rehearsal

Rehearse your speech a number of times before you actually give it. Practice before a mirror, and use a tape recorder. If possible, give your speech to a small group of friends. Ask them to assess your poise, eye contact, voice, gestures, and rate.

Your voice should be conversational, confident, and enthusiastic. Avoid a monotonous sound by varying the pitch, intensity, volume, rate, and quality of your voice.

Delivery

All of the preceding steps should prepare you for an effective delivery. How your audience perceives you and your information depends on skillful delivery.

General Appearance.   Dress appropriately. Approach the podium confidently and maintain good posture. Gesture naturally. Gestures may be larger in a large room than in a small room. Be animated and maintain eye contact. Do not be ramrod stiff nor clutch the podium. Do not jingle keys or make distracting adjustments to your hair, glasses, or clothes.

Audience Interaction.   Pause before you begin your speech to gain the listener's attention. Begin forcefully and engagingly. Keep tabs on your audience. Your listeners will be confirming or contradicting what you say through nonverbal messages. Allow for questions and discussion at the end of your report.

Notes and Visual Usage.   Use your notes and visuals with ease. Place these items in comfortable positions. Do not fidget with your cards or pointers. Stand aside when referring to visual materials and maintain eye contact as you make your points about the materials.

Voice.   Try for vocal variation. Let your voice exude warmth and sincerity. Pronounce your words clearly and distinctly. Do not vocalize "uhs," "ums," and other nervous sounds. Pause when appropriate.

ORAL SPEECH RATING SHEETS

At the end of the chapter are speech rating blanks for your classmates to use to evaluate speeches. Figure 17.3 (see p. 581) is a general speech rating sheet useful for any extemporaneous persuasive or informational speech. Figures 17.4 through 17.6 (see pp. 582–588) may be used to assess oral descriptions of a mechanism, instructions, and analyses of a process, all of which are good subjects for informational speeches. Use these rating sheets in addition to the chapter checklist on preparing verbal communications.

NONVERBAL COMMUNICATIONS

Verbal communications involve more than the words being spoken. Our senses of sight, hearing, touch, and smell are also receiving messages that can color our own and others' perceptions of the spoken word. These perceptions are called *nonverbal messages*. A number of message carriers have already been mentioned—posture and position, voice modulation, eye contact, handshaking, background noise, and the like. For effective communications you should be alert both to the nonverbal messages you send and those you receive. Ideally, nonverbal messages will be consistent with verbal messages, but such consistency is often violated.

Overt Actions.   Actions often do speak louder than words. Obvious actions, such as jabbing with a finger, pacing back and forth, waving papers around, and pounding on desks quite obviously signal aggression and anger. But subtle body actions, such as slouching, staring at the ceiling, standing with crossed arms, and the like, convey messages, too. Slouching conveys boredom; staring at a ceiling conveys disinterest; and crossed arms convey coldness or aggression. Open arm and body positions convey trust, warmth, and sincerity.

Covert Actions.   Some actions, such as blushes, tight jaws, smiles, and the like, reveal how a person is feeling. Who has not been confused by the person with furrowed brow and a tight-mouthed smile asserting, "No, I'm not mad at you"? Be alert to these signals in interpersonal communications, and learn to control your own covert actions to convey consistent verbal and nonverbal messages.

Eye Contact.   Eyes are the primary conveyors of nonverbal messages. Use your eyes to your advantage. An eyebrow flash by both parties often occurs within seconds of making eye contact with another person. Both you and the other person's eyebrows will lift briefly in a visual handshake. If you do not receive the warm eyebrow flash, consider whether the person is signaling hostility or is just shy, anxious, or even nearsighted. Intermittent three-second, direct eye contact conveys interest and receptivity, whereas reduced eye contact may convey concentration, disinterest, or deceit. Between communicators eye contact regulates interactions and monitors the verbal messages. Breaking eye contact sideways conveys distraction or lack of interest and generates discomfort in the other person. Looking up will also throw the other person off balance.

Appearance.   Physical appearances also convey messages although often inaccurately. We tend to stereotype persons according to three basic body types: athletic, frail, or obese. We expect the athletic person to be strong, mature, or self-reliant, whereas we expect the frail person to be indecisive, bookish, or pessimistic. Examine your personal tendency to stereotype people by their body type and actively rethink the validity of these prejudices. Be aware of how your own body type influences the perceptions of others. Changeable factors in appearance—length of hair, mustaches, eyeglasses, uniforms, jewelry, cosmetics, hem lengths, and so on—convey impressions, too. We attribute skills, personality traits, and abilities based on how people look.

Space.   Space is also an important nonverbal message carrier. The arrangement of office furniture or conference facilities can enhance or detract from open communication. Also, individuals have a sense of territory; to encroach on another's space is perceived as a threat or an aggression. Women are generally more comfortable conversing face-to-face, while men tend to prefer a side-on position that moves into a more frontal one. If you remain standing while your conversation partner is seated, you signal dominance, which may make the seated party uncomfortable. Even the distance that we set between ourselves and those with whom we are communicating conveys messages. Arabs and Latins tend to communicate at much closer distances (18 to 36 inches) than do North Americans or other Westerners who unconsciously prefer 30 to 48 inches. Arabs may perceive Western colleagues as remote and unfriendly, whereas the Westerners feel intimidated as their hosts keep moving in closer.

**Extemporaneous Speech
Rating Sheet**

Speaker _____ Subject _____ Evaluator _____

ITEMS	COMMENTS	SCORE
ORGANIZATION: Clear arrangement of ideas? Introduction, body, conclusion? Pattern of development adapted to ideas and audience?		
LANGUAGE: Clear, accurate, varied, vivid? Appropriate standard of usage? In conversational mode?		
MATERIAL: Specific, valid, relevant, sufficient, interesting? Properly distributed? Adaptive to audience? Personal credibility? Use of evidence?		
DELIVERY: Poised, at ease, communicative, direct? Eye contact? Aware of audience reaction to speech? Do gestures match voice and language?		
ANALYSIS: Approach to subject original, interesting? Central idea, purpose clear, divided into significant, interesting, subordinate ideas?		
VOICE: Pleasing, adequate, distracting? Varied or monotonous in pitch, intensity, volume, rate, quality? Expressive of logical emotional meanings?		

TOTAL _____

SCALE:

10	7	4	1
Superior	Average	Inadequate	Poor

FIGURE 17.3 *Extemporaneous speech rating sheet*

Mechanism Description Rating Sheet

Speaker _____ Mechanism _____ Evaluator _____

Did the speaker	Yes	Somewhat	No	Comment
1. Make necessary preparations before starting?	___	___	___	_____
2. Define intended audience?	___	___	___	_____
3. Name mechanism precisely?	___	___	___	_____
4. Define and/or state purpose of mechanism?	___	___	___	_____
5. Provide an overall description?	___	___	___	_____
6. Discuss the operational theory?	___	___	___	_____
7. State by whom, when, and where the mechanism is operated?	___	___	___	_____
8. Provide a list of the main parts?	___	___	___	_____
9. Describe the parts in the order listed?	___	___	___	_____
10. Define and/or state purpose of each part?	___	___	___	_____
11. List subparts of assemblies?	___	___	___	_____
12. Describe each part adequately?	___	___	___	_____
13. Avoid wordiness?	___	___	___	_____
14. Assess the advantages and disadvantages?	___	___	___	_____
15. Describe optional uses?	___	___	___	_____

FIGURE 17.4 *Rating sheet for oral description of a mechanism*

Did the speaker	Yes	Somewhat	No	Comment
16. Compare the mechanism to other models?	—	—	—	————
17. Discuss the cost and availability of mechanism?	—	—	—	————
18. Avoid reading the presentation?	—	—	—	————
19. Use adequate, well organized notecards?	—	—	—	————
20. Show evidence of rehearsal?	—	—	—	————
21. Maintain effective eye contact?	—	—	—	————
22. Speak clearly so all could hear?	—	—	—	————
23. Use appropriate graphics?	—	—	—	
a. Number and title clearly?	—	—	—	————
b. Print legibly?	—	—	—	————
c. Keep graphics simple and uncluttered?	—	—	—	————
d. Lay out logically?	—	—	—	————
e. Credit sources of graphics?	—	—	—	————
24. Handle graphics with ease?	—	—	—	————
25. Do you feel you can judge the reliability, practicality, and efficiency of the mechanism?	—	—	—	————

SUGGESTED GRADE ——————————

FIGURE 17.4   *continued*

Instructions Rating Sheet

Speaker ——————— Instructions for ————— Evaluator —————

Did the speaker	Yes	Somewhat	No	Comment
1. Make necessary preparations before starting?	___	___	___	___
2. Define intended audience?	___	___	___	___
3. Provide a specific, limiting title?	___	___	___	___
4. State the instructional or behavioral objective?	___	___	___	___
5. Stress the importance of the instructions?	___	___	___	___
6. Define key terms?	___	___	___	___
7. State preliminary warnings or cautions?	___	___	___	___
8. Divide the process into main steps?	___	___	___	___
9. Provide a precise list of tools, materials, etc.?	___	___	___	___
10. Divide the main steps into substeps?	___	___	___	___
11. Provide warnings and notes for individual steps?	___	___	___	___
12. State each step as a command?	___	___	___	___
13. Avoid combining steps and substeps?	___	___	___	___
14. Express all steps in parallel, grammatical terms?	___	___	___	___
15. Avoid wordiness?	___	___	___	___

FIGURE 17.5 *Rating sheet for oral instructions*

Did the speaker	Yes	Somewhat	No	Comment
16. Avoid reading presentation?	—	—	—	————
17. Use adequate, well-organized notecards?	—	—	—	————
18. Show evidence of having rehearsed?	—	—	—	————
19. Maintain effective eye contact?	—	—	—	————
20. Use appropriate graphics?	—	—	—	————
a. Number and title clearly?	—	—	—	————
b. Print legibly?	—	—	—	————
c. Keep graphics simple and uncluttered?	—	—	—	————
d. Lay out logically?	—	—	—	————
e. Credit sources of graphics?	—	—	—	————
21. Handle graphics with ease?	—	—	—	————
22. Invite questions from the audience?	—	—	—	————
23. Do you feel you could perform the set of instructions accurately and efficiently?	—	—	—	————

SUGGESTED GRADE ————————————

FIGURE 17.5   *continued*

Process Analysis Rating Sheet

Speaker _____ Process _____ Evaluator _____

Did the speaker	Yes	Somewhat	No	Comment
1. Make necessary preparations before starting?	—	—	—	_____
2. Define intended audience?	—	—	—	_____
3. Name process precisely?	—	—	—	_____
4. Select a process primarily involving human action?	—	—	—	_____
5. Define or state purpose of process?	—	—	—	_____
6. Define or explain terms?	—	—	—	_____
7. Explain theory on which process is based?	—	—	—	_____
8. Indicate by whom, when, and where the process is performed?	—	—	—	_____
9. Indicate special conditions, requirements, preparations, precautions which apply to the entire process?	—	—	—	_____
10. Precisely list materials, tools, and apparatus?	—	—	—	_____
11. Divide process into five or six main stages?	—	—	—	_____
12. Arrange steps in numbered, chronological order?	—	—	—	_____
13. Provide a flow chart of main steps?	—	—	—	_____
14. Discuss steps in order listed?	—	—	—	_____
15. Define or state purpose of each main step?	—	—	—	_____

FIGURE 17.6   *Rating sheet for oral analysis of a process*

Did the speaker	Yes	Somewhat	No	Comment
16. Divide main steps into sub-steps?	—	—	—	————
17. Describe special conditions for each step?	—	—	—	————
18. Explain theory which applies only to one step?	—	—	—	————
19. Adequately analyze each main step with emphasis on why, to what degree, to what extent, etc.?	—	—	—	————
20. Avoid excessive detail?	—	—	—	————
21. Evaluate effectiveness of process?	—	—	—	————
22. Discuss advantages/ disadvantages?	—	—	—	————
23. Discuss importance of process?	—	—	—	————
24. Evaluate results of process?	—	—	—	————
25. Explain how process is part of larger process?	—	—	—	————
26. Compare process to similar processes?	—	—	—	————
27. Assess cost and time factors?	—	—	—	————
28. Avoid reading presentation?	—	—	—	————
29. Use adequate, well-organized notecards?	—	—	—	————
30. Show evidence of rehearsal?	—	—	—	————
31. Speak clearly so all could hear?	—	—	—	————
32. Avoid shifting to instructions?	—	—	—	————

FIGURE 17.6   *continued*

Did the speaker	Yes	Somewhat	No	Comment
33. Use appropriate graphics?	—	—	—	————
a. Number and title clearly?	—	—	—	————
b. Print legibly?	—	—	—	————
c. Use color effectively?	—	—	—	————
d. Keep graphics simple and uncluttered?	—	—	—	————
e. Lay out logically?	—	—	—	————
34. Handle graphics with ease?	—	—	—	————
35. Invite questions from the audience?	—	—	—	————
36. Do you feel you can judge the reliability, practicality, and efficiency of the process?	—	—	—	————

SUGGESTED GRADE ——————————

FIGURE 17.6   *continued*

CHECKLIST

Verbal Communications

INFORMATION-GATHERING INTERVIEW

❏ **1.** Have I reviewed the **do's** and **do not's** of all interviews (see Chapter 11)?

❏ **2.** Have I determined a clear-cut purpose that I can present to the interviewee(s)?

❏ **3.** Have I decided on the questions I am going to ask?

  ❏ Are these questions organized logically?

  ❏ Specific?

  ❏ Open-ended?

  ❏ Unbiased?

❏ **4.** Have I reviewed, so that I can practice the courtesies of interviews?

  ❏ Will I remember to thank the respondent in advance for granting me the time?

  ❏ Will I inform the respondent how I intend to use the information?

  ❏ Am I prepared to listen rather than to "lead" the respondent?

  ❏ Have I received the respondent's permission to tape the interview?

  ❏ Am I prepared to take notes?

  ❏ Have I prepared to offer a copy of my report, document, or other material that will incorporate the interviewee(s) responses?

  ❏ Have I prepared a survey or questionnaire for multiple respondents?

TELEPHONE CONVERSATIONS

❏ **1.** Have I practiced modulating the volume, tone, and clarity of my voice?

❏ **2.** Am I prepared to enunciate all words correctly?

❏ **3.** Am I aware of the need to avoid distracting background noises?

❏ **4.** Have I provided a note pad, pencils, and pens by my telephone?

❏ **5.** Will I need any reference materials for a particular telephone conversation?

GROUP DISCUSSIONS

❏ **1.** Have I assessed the group that will participate in the discussion?

  ❏ Have I made clear the purpose of the discussion?

  ❏ Have I evaluated the members' personal goals that might be in conflict with the overall goal?

  ❏ Have I considered the permanency or impermanency of the group?

  ❏ Have I considered the power and relationships of the group members?

❏ **2.** Have I decided on the method (majority rule, compromise, or consensus) that will guide the group's decisions?

  ❏ Will a vote of the majority suffice?

  ❏ Will the members be able to compromise to reach decisions?

  ❏ Have we the time for a consensus decision?

❏ **3.** If I am looking for a consensus, am I prepared to give the group the power to establish a consensus?

  ❏ Has the group collectively established a goal and a procedure?

❏ Has the group devised an agenda?

❏ Have I allowed time for each member to contribute freely?

❏ Have I planned for a group discussion to determine the status and causes of the problem or situation?

❏ Has the group established criteria for evaluating each possible solution or decision?

❏ Will the group employ the four strategies of brainstorming?

   ❏ Suggestions without criticism or interruption?

   ❏ Listing of all the suggestions so that everyone can see them?

   ❏ Weighing the potential solutions against the criteria?

   ❏ Making a group decision and delegating and accepting responsibilities?

APPRAISALS

❏ **1.** Am I prepared to participate in a performance appraisal?

❏ **2.** Is the appraisal interview scheduled far enough in advance for both parties to prepare?

❏ **3.** Has a specific time limit been established?

❏ **4.** Am I ready to concentrate on task-related factors rather than personalities?

❏ **5.** Am I ready to seek or offer explanations for poor performance?

❏ **6.** Will I be able to concentrate on steps for improvement?

REPRIMANDS

❏ **1.** Am I prepared to participate in an objective reprimand situation?

❏ **2.** Are there written policies and/or procedural manuals for the type of behavior to be discussed?

❏ **3.** Am I prepared to evaluate if violations are due to lack of information or willfulness?

❏ **4.** Am I prepared to discuss the violations with questions, feedback, and sensitivity?

❏ **5.** Am I prepared to ask or to answer questions about recourse, appeals, and corrective actions?

ORAL REPORTS

❏ **1.** Am I prepared to give an impromptu, memorized, manuscript, and extemporaneous report?

❏ **2.** Have I assessed my audience?

    ❏ Have I considered their prior knowledge of the subject?

    ❏ Have I considered their average age, sex, and rural or urban interests?

    ❏ Have I assessed their political and religious persuasions as well as their ethnic affiliations?

❏ **3.** Have I organized my report logically?

❏ **4.** Have I outlined my speech?

❏ **5.** Have I prepared note cards that are brief, readable, and highlighted for key points and visuals?

❏ **6.** Have I prepared supporting visuals and made the necessary preparations for handling them?

❏ **7.** Have I rehearsed my speech?

❏ **8.** Have I considered my appearance, audience interaction, and voice?

❏ **9.** Have I considered strategies for a **persuasive** speech?

    ❏ Have I arranged for the audience's comfort?

    ❏ Have I arranged for an introduction to stress my qualifications?

    ❏ Have I considered my dress, actions, and other possible expressions of similarities with the group?

    ❏ Have I stated my goal emphasizing the groups' rewards should such emerge?

> ❏ Have I included convincing statistics and the statements of acceptable authorities?
>
> ❏ Have I considered my nonverbal communications?
>
> ❏ **10.** Have I considered strategies for an **information** speech?
>
> ❏ Besides employing all the above strategies, will my audience be able to judge the reliability, practicality, and efficiency of the subject of my speech?

EXERCISES

1. **Information-Gathering Interview.**   Working in groups of two or three, devise interview questions about a specific subject that will affect your classmates or the entire student body. Possible subjects are the need for computer lab changes, the establishment of snack booths on campus, class registration procedures, campus bookstore policies, cafeteria procedures, and bike racks. Be sure to have a clear purpose, logical organization, and specific, open-ended, and unbiased questions. Review the relevant interview strategies discussed in Chapter 11 and the seven suggestions in this chapter. Have each member of the group participate either as the interviewer, the respondent, or the critic who will assess the verbal and nonverbal communication skills of the other two. Encourage the rest of the class to participate in a critique.

2. **Telephone Communications.**   Working in pairs, role play a business telephone transaction. Devise your own transaction or try the following situation: The caller has mailed a cover letter and resume to the director of personnel. The caller is trying to find out if the letter arrived and seeks to arrange an interview for the next week. The callee, the director of personnel, is eager to arrange an interview but will ask a few questions. Both should be prepared to take notes. The class will assess the skills of both the caller and the callee.

3. **Appraisals.**   Working in pairs, devise a performance appraisal situation between a supervisor and a subordinate. Each will provide a written assessment to be exchanged before the role playing in class. The supervisor will discuss five elements of the subordinate's work: timeliness, productivity, attention to detail, enthusiasm, and

peer relationships. The supervisor will discuss two areas needing improvement. The class will assess the skills of both the supervisor and the subordinate in terms of objectivity, concentration on task-related factors, explanations for negative appraisal, and discussion on goals and improvement.

4. **Reprimands.**   Working in pairs, devise a reprimand situation between a supervisor and a subordinate. The pair should devise a brief, written policy or procedural statement that, presumably, the company already had in writing. The reprimand may be over repeated tardiness, low productivity, lack of attention to detail (as in a written report), general demeanor, or peer relationships. The class will assess the skills of both the supervisor and subordinate in terms of determination of cause, questions, feedback, sensitivity, and agreement on appropriate recourse, appeal, or corrective action.

5. **Group Discussion for a Consensus Decision.**   Role play a group discussion among five or six classmates. The group is to reach a consensus on one of the following situations:

   a.   A donor has offered $5,000 to your school for a scholarship in a professional field. The group consists of one member each from the computer department, the nursing department, the architectural department, and the dentistry department (or if these are separate schools in your university, the members are from each of the schools). The group is to decide how to implement the scholarship. Following the discussion the class will assess the verbal and nonverbal skills of the participants.

   b.   A department in a business has $5,000 budgeted for employee travel to professional conferences and seminars around the country for the fiscal year. There are 10 members in the department; each member has two or three conferences he or she would like to attend. Expenses would range from $200 for short, in-county seminars to $2,000 for week-long, out-of-state conferences. The group must decide on the criteria for spending the money. Following the discussion the class will assess the verbal and nonverbal skills of the participants.

6. **Extemporaneous Oral Reports.**   Prepare a 6- to 10-minute extemporaneous oral report with visual materials for one of the following situations:

   a.   *A persuasive speech.* Choose a controversial subject (four-day work week, class registration procedures, bookstore policies, degree requirements in your field, etc.). Prior to your speech, pass out index cards to the class and ask each member to write opinions on this subject. Deliver your speech. Each classmate will evaluate your presentation by rating you according to the rating sheet in Figure 17.3. Following the speech, ask your classmates to read their opinion cards and discuss how you have or have not changed their opinions.

    b. *An informative speech.* Refer to Chapters 7, 8, and 9 on instructions, descriptions of mechanisms, and analysis of a process. Select one of the report strategies; prepare and deliver an informative oral report. Each classmate will assess your presentation by filling out the appropriate rating sheet in Figures 17.4, 17.5, or 17.6.

7. **Nonverbal Communications.** Select an employer, a co-worker, a professor, or a friend. Tell that person you desire just a five-minute interview about a problem you are having. Devise a situation. Hold the informal interview paying close attention to the nonverbal messages the interviewee sends, such as overt actions, covert actions, eye contact, appearance, and space. Be mindful of your own nonverbal messages during this interchange. Following the interview, quickly write down your observations. Write a two- or three-paragraph paper about your observations. Do not use the actual name of the interviewee.

NOTES

CHAPTER 18

Visual Communications

MR. FUMBLE   by PeJay Ryan

By permission of PeJay Ryan.

S K I L L S

After studying this chapter, you should be able to

1. Identify potential professional situations in which visual aids are necessary.

2. Name eleven visual communication aids along with their individual strengths and weaknesses.

3. Understand and practice effective design, emphasizing simplicity, unity, emphasis, balance, and legibility.

4. Practice the attribution of sources for your visuals.

5. Handle visuals properly and with ease.

6. Construct at least six visual aids from among the following: handouts, demonstrations or exhibits, chalkboards, flip charts, posters, transparencies, slides, filmstrips, video tapes, and computer presentations.

7. Understand the value and production steps of videotaped employment interviews and job performances.

8. Participate in a videoconference.

INTRODUCTION

Throughout this textbook the power of visual communications has been stressed. As you have devised page designs, incorporated emphatic features such as boldface and italic type, explored various fonts and type sizes, and devised graphics and other visuals for your documents, you have experienced first hand the value of visuals. They can be further put to use by incorporating them into your oral presentations by the following methods:

printed or photocopy handouts	transparencies
demonstrations/exhibits	slides
chalkboards	filmstrips
flip charts	videotapes
posters	computer presentations
videoconferencing	

Each of these tools may be incorporated into your formal presentations alone or in combination with any other visual aids. The key to success is to select and to place them carefully and intentionally where they best fit the situation. Obviously, then, you cannot design and incorporate visuals until you have planned your verbal presentation and considered your audience, purpose, space, and the equipment available to you.

EFFECTIVE DESIGN

Review the design materials in Chapters 3 and 4, which include document page design information, graphics construction, drawings, illustrations, maps, photographs, text art, clip art and icons (including tips on the amount of detail, clarity, texture, color, size, and orientation on a page). Besides clip art, you may find many visuals that are useful to your presentations by searching online by keywords, narrowing the field, and finding the right link.

Some key considerations for visual communications are

- Simplicity
- Unity
- Emphasis
- Balance
- Legibility

If you are designing visuals from scratch or using other sources to produce large-scale visuals for use with verbal communications, consider each of these carefully.

1. **Simplicity.** Fewer elements are more pleasing to the eye than a hodgepodge of detail. Bold key detail has more impact than does complex art. If the material contains words, limit them to 15 or 20. Figure 18.1 illustrates both a simple design and a complex design:

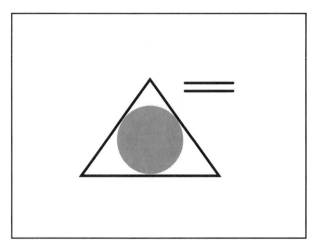

Simple, pleasing design

FIGURE 18.1   *Simplicity of design versus complexity of design*

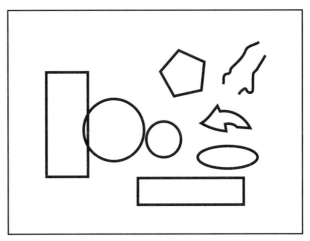

Too complex design

FIGURE 18.1 *continued*

The circle within the triangle with two parallel lines indicating text exemplifies simplicity of design, whereas the multiple figures with squiggly text lines illustrate a complexity of design that is difficult on the eye (and the nervous system).

2. **Unity.** Unity may be achieved by overlapping elements and arrows, and by a conformity of shapes and sizes on any one visual. Random layout reduces unity. Figure 18.2 illustrates unity:

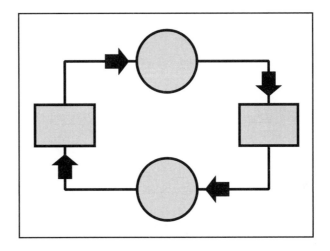

FIGURE 18.2 *Unity of design*

The double circles, double rectangles, and connecting arrows produce the unifying result.

3. **Emphasis.** Use color, bold sizes, and white space to achieve emphasis.

4. **Balance.** Balance may be formal or informal. Formal balance is achieved when an imaginary axis divides the design into two mirror halves horizontally or vertically. Informal balance is asymmetrical but logical. It is more dynamic and more attention-getting. Figure 18.3 illustrates some principles of balance:

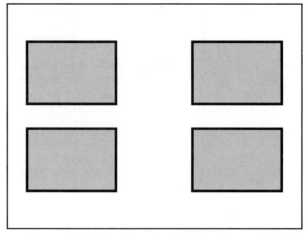

Symmetrical design

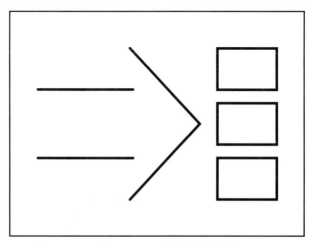

Assymmetrical design

FIGURE 18.3   ***Formal balance and informal (asymmetrical) designs***

The first design is formal with the halves mirrored both horizontally and vertically while the second design is informal having asymmetrical elements. Both are effective, but, perhaps, the informal design is more sophisticated.

5. **Legibility.** Wording on visual materials should be minimal. Sanserif or gothic letters are easier to read than script and other fancy fonts. Figure 18.4 illustrates a variety of fonts.

TIMES NEW ROMAN   **FUTURA**

CLAIRVAUX   GOUDY

FIGURE 18.4   *Illustration of plain and fancy fonts*

Use capital letters for short titles and labels, but use a combination of upper- and lowercase letters for verbal content of six or more words. Figures 18.5 and 18.6 show examples of all capital titles and upper- and lowercase combination titles.

SHORT

TITLES

FIGURE 18.5   *All capital letters used for short titles*

Combinations of Upper and Lowercase Titles Are Easier to Read

The maximum viewing distance is generally accepted as being eight times the horizontal dimension of the graphic. Thus, a four-foot-wide poster has a maximum viewing distance of 32 feet. Minimum lowercase letter sizes are 1-in. high on standard posterboard and ¼-in. high (35-point font) for transparencies. Capital letters are correspondingly larger. Thick letters are more legible than thin letters.

MATERIALS

Other than computer-generated art and lettering, basic supplies for visual aid construction are available in college bookstores and art and paper supply shops and include

- Felt-tip pens in various widths and colors
- Grease pencils in a variety of colors
- Color projection pens and pencils for transparencies
- Prepared cut-a-color and shading sheets both translucent and transparent
- Chart pax tapes both translucent and transparent
- Lettering templates
- Letter, symbol, and shape stencils
- Dry transfer letters
- Adhesive letters in black and colors

Poster boards of various qualities, transparencies (acetate and thermals), flip charts, photocopiers, blank videotapes, and all of the other possible materials you may consider are also readily available.

Each of the design principles—simplicity, unity, emphasis, balance, and legibility—applies equally to all forms of visual presentation to be discussed in this chapter.

PLAGIARISM

Plagiarism is as much a concern with visual materials used in oral presentations as it is with written documents. Thus, your visual presentations need to be credited to their sources. These credits may be printed directly on the visual aid, but this is not always possible. Therefore, you should *always* verbally acknowledge the source of a graphic when using it in a presentation.

HANDLING VISUALS

Visuals will lose all of their inherent power if you fail to use them carefully. Follow these tips:

1. Ensure ahead of time that your facility has the required equipment (easels, videocassette recorders, slide projectors, overhead projectors, screens, microphones, and pointers), electrical outlets, and accessible light switches. Check the furniture (desks, podiums, tables) and seating arrangement and rearrange if necessary.

2. As a rule limit your exhibits, flip chart pages, posters, and transparencies to no more than four or five in a 20-minute presentation to avoid a circus atmosphere.

3. Try not to display your visuals until you refer to them in your presentation in order not to distract your audience.

4. Do not get in the way of your visuals. Stand aside and avoid turning your back to your audience.

5. Interpret or discuss each visual. Do not just let it "speak" for itself.

6. After discussing a visual, remove it or step away from it and refocus attention on yourself.

7. Distribute handouts *after* a presentation in order to focus your audience's attention on you. Sometimes you will want to distribute handouts ahead of time to help the audience to follow your agenda.

TYPES OF VISUALS

Print or Photocopy Handouts

Computer-printed or photocopied handouts are very easy to use, especially with larger audiences that may have trouble seeing the physically distant chalkboard, exhibit, flip chart, or poster. If you prepare too much material on the page, your audience will tend to read it as you speak, thereby missing part of what you say.

Keep the art and words to a minimum. Make your handouts clear and concise as well as relevant to the material you are discussing in your presentation. Handouts offer the advantages of being prepared in advance, being simple to transport, and giving your audience something to take home for review.

Demonstrations/Exhibits

Occasionally your purpose will be to demonstrate a technique, a piece of equipment, or a process, such as coronary pulmonary resuscitation, a blood pressure cuff, or a laboratory experiment. You may consider using exhibits such as architectural models, products, and equipment. These visuals must be easy to use to avoid long and complicated pauses during your presentation and large enough for your audience to see clearly.

Limit the use of demonstrations and exhibits to small audiences and facilities, such as a conference room or small classroom. Do not use demonstrations and exhibits unless everyone in the room can clearly see the exhibit plus your use or manipulation of it. You may require your audience to cluster around you or the exhibit.

Chalkboards

Most classrooms and many conference rooms contain chalkboards. You may wish to present an outline, simple text, graphics, and the like on the chalkboard; however, if you do plan to use a chalkboard, prepare the material ahead of time and use multicolored chalk. Your audience will be greatly distracted if it has to wait while you write and draw. Consider whether everyone in the facility will be able to see and to read all of your material.

Flip Charts

Use flip charts (large, ring-bound pads) on an easel for informal presentations. You can prepare some pages of graphics or other visuals in advance. Write in any color felt-tip marker and flip the paper over the top to move on to the next sheet. Flip charts are very handy if you are going to be listing topics in a situation such as a brainstorming session. New electronic flip charts are now on the market that will allow you to write on their surface, print out the messages, or download information to a personal computer.

Flip charts have advantages over chalkboards. You may easily transport them and place them closer to your audience; a chalkboard is usually fixed and generally behind you. Flip charts, like exhibits, should be used in small rooms and with small audiences of fewer than 25 people.

Posters

Posters are very useful but are limited in scope. The complicated posters, with typewritten pages and small diagrams, used by high school students for a science fair exhibit are functionally useless for an oral presentation. They cannot be seen clearly from more than a few feet away. The rule is simple: your poster must be clearly visible and easy to read from the farthest seat in the room and from all angles. This requirement limits the amount of information you can include on the poster.

Masking tape will hold posters firmly on to a blackboard without marking the slate. Posters should not be propped up against the podium or on chairs; they are sure to fall. Use easels if at all possible to hold them. Both posters and easels are highly transportable.

Greatly enlarged photographs, usually mounted on foam core board along with brief text, are widely used for seminar and sales table displays. Use photographs of text, persons, situations, or graphics. Two drawbacks to the technique of visual presentation can be high cost and loss of picture quality.

For regular posters use sturdy white or pale-colored poster material. If you use more than one poster, use the same color for each; multiple color poster displays are distracting. Sketch artwork and lettering directly onto the poster. Letter stencils or paste-on letters are useful. You may, of course, also use felt-tip pens, crayons, grease pencils, translucent color tapes and color sheets, and many other art supplies.

Transparencies

Overhead transparencies are simple sheets of clear plastic that can be written on by hand, prepared by photocopying printed or art material, or printed directly on your computer printer. Transparencies are cheap and allow you to apply artwork and lettering directly onto the plastic for quick, temporary use. You may also prepare your material on plain white paper and photocopy it onto a transparency. The most popular method of preparation, however, is to prepare your material on a computer using at minimum 35-point type size, a variety of fonts, art, and color. If you have an inkjet or laser printer, you may print directly onto transparencies. Copies of your transparencies can be made to use as handouts in conjunction with your presentation.

Ensure ahead of time that your facility has an overhead projector and a white wall or screen of the appropriate size for your group.

Slides

Anyone with a knowledge of photography may produce slides. Effective presentations can be created using only a 35-mm camera and a small amount of ingenuity. You may photograph people, scenery, and artwork; journal, magazine and book illustrations; graphics; and computer-printed clip art, icons, and lettering.

Once your slides have been developed and returned, you need only to select those you intend to use in your presentation, place them in the order you intend to use them, and arrange them in the tray of a projector.

Two projectors are often used simultaneously to make comparisons and contrasts. Determine ahead of time, of course, if a slide projector(s) and screen(s) are available in the facility you intend to use. A light-projection pointer will help you to direct the audience's attention to specific detail within the slide. Many facilities provide pointers but you may decide to buy your own reasonably priced, pen-sized pointer to carry with you.

Filmstrips

Filmstrips are extremely useful because of their ease of handling; however, they are difficult to make, expensive to purchase, and are becoming difficult to find. More versatile slide presentations and videotapes are replacing filmstrips.

Videotapes

A videotape is without a doubt the most popular currently used visual aid. It is used by executives, teachers, lawyers, students, and others. There are countless videos available for purchase, or you may prepare your own using a video camera, or even copy material from a television program. Some understanding of photography, cinematography, and videotape editing will enable you to produce professional videos with voice and other sound effects. Your camera user manual will include helpful information.

You will require a VCR in your presentation facility. Remember that the size of the screen will dictate the acceptable size of your audience.

VIDEOTAPED EMPLOYMENT INTERVIEWS AND JOB PERFORMANCES

In addition to those tapes that you prepare for an oral presentation, some companies may request a video of you expressing your qualifications for employment or performing the duties of your job. You may prepare a video that shows you either talking about yourself and your goals or actually working at your job. This will allow potential employers to get a much clearer image of what you are like as a person and how you present yourself to the public.

Interview and job performance videos may be created by you or by professional videographers (ask about the possibility in your learning resource center) with a minimum of equipment. The following elements need to be addressed if you decide to produce your own video:

- Designing the presentation
- Preparing the script
- Designing and preparing the set
- Rehearsing the presentation
- Recording and editing the interview
- Copying the presentation

In designing the presentation, consider your strengths and weaknesses and anticipate any questions that an interviewer may ask about the information you have offered. For the job performance video, tape yourself actually doing your work and interacting with others. Prepare a script not to exceed 20 minutes. Design a set that reflects your field of work. If the position involves desk work, use a desk with books, possibly a computer, and other appropriate paraphernalia arranged on it. If you are looking for an architectural or engineering position, stand or sit next to a drafting table and carry out some actual work. Let the setting demonstrate that you are a part of the profession and comfortable within its surroundings.

If you design the set, consider color, lighting, and camera placement. Avoid clashing colors or strong patterns that may detract from you, the main focus. Use soft light coming directly toward your face to eliminate the blemishes, wrinkles, and shadows that overhead lights may underscore. Determine the best side of your face and place the camera about 30 degrees to the opposite side, which will allow you to turn your head slightly, giving a more relaxed look and allowing the front light to highlight the best side of your face. The incidental shadow falling on the opposite side of your face provides modeling and dimension.

Rehearse your presentation on tape so that you can retain what looks good and eliminate what looks bad. Check the sound and lighting. Tape your final presentation and make two or three initial copies from the master, saving the master for further use. Copies from copies tend to lose their contrast.

Mailing boxes specifically designed for videocassettes are available from postal supply stores. Be sure to label both the tape and the box appropriately, include the correct postage, and include your cover letter and resume in an envelope attached to the package.

Computer Presentations

With the right software you can create drawings, graphics, and text on slides for a computer presentation. You may also create your own materials, incorporate clip art and on-line downloads, or use a scanner to import materials from other sources. Templates and tools allow you to select background and layout layers. You may add titles and text in any size type or font within the presentation and add arrows, banners, and textures, and even pictures within pictures. With a few clicks you can alter the spacing around objects. You may easily change the order of slides, add more slides, or import existing ones from another presentation. Further, you can turn any drawing or text into a 3-D object or reshape text or objects to special images such as waves, bow ties, and pennants. You can even rotate (or flip) a selected object to change its orientation. In addition, animation, movie files (digitally encoded live action), and unique transition effects can be added, making the slide seem to roll to the side or to open out of the center of the screen. You can open a page on the Internet, play a sound file, or skip to another part of the slide show. The possibilities are almost endless.

To create an effective presentation, follow these steps:

1. **Familiarize yourself with the presentation options.** Practice developing presentations until you are able to work with all of the options.

2. **Evaluate your purpose.** Do you want to entertain, motivate, sell, or instruct? How much does your audience already know about your subject? How much detail will you need to include in any one presentation?

3. **Write down your objectives.** Your computer will help you to develop presentation notes, make copies for your audience to take with them, and print out what you want your audience to learn at the end of the presentation.

4. **Develop and organize your slides.** Consider three main sections: the *opener*, which should begin with a title and a statement about the presentation; the *main body*, which should present just

one concept to a slide in logical order; and the *conclusion,* with a bullet chart to summarize and reinforce keypoints.

5. **Practice, review, and edit.** Practice your oral presentation along with the computer presentation until it is smooth. Evaluate each slide to see if it relates to your point and objective. Pace the slides to your spoken words. Use the highlighter to emphasize detail. Ensure that every slide is eye catching.

Practice and experience will enable you to incorporate dynamic visual communications into your oral presentations.

Videoconferencing

Videoconferencing is becoming increasingly popular because it saves thousands of hours lost in travel time. In the videoconference, you share live video, audio, and data over wired or wireless telephone lines—meeting face-to-face on television or computer monitors and by video telephones. You can even transmit graphics, slides, or videotape during the meeting. If you do not have the capability for a personal computer videoconference or an on-site videoconference facility, you may use the videoconferencing rooms and state-of-the-art equipment in hundreds of Kinko's and other multiservice office centers around the world.

In the videoconference you may share any number of other visual aids already discussed, such as demonstrations, exhibits, flip charts, posters, slides, or videotapes. You may also distribute premeeting materials, such as handouts, agendas, and the like. You can even hold multi-point conferences (more than two sites).

Set the conference date and time well in advance so that all participants can plan to be on hand. Dress conservatively and avoid white, plaids, prints, and stripes, which tend to vibrate or glare. Set up your props where they are accessible and can be easily seen on camera. At the beginning of the conference have each location chair introduce all participants even if they are off camera at the time. Review your meeting agenda and goals. Be yourself and speak naturally and clearly. Identify yourself as you speak from time to time in multipoint conferences. Pause slightly after finishing a comment to allow for a brief lag in the video image transfer. Avoid coughing into the microphone, drumming your fingers, shuffling papers, or holding side conferences. Allow time to review the decisions made and end on time.

CHECKLIST

Visual Communications

GENERAL

❑ **1.** Have I selected appropriate visuals for the content of my presentation and the facilities in which I will present them?

❑ **2.** Have I considered handouts, chalkboard material, flip charts, posters, transparencies, slides, filmstrips, videotapes, and computer presentations?

❑ **3.** Will I be using these visuals in a face-to-face situation or in a videoconference?

❑ **4.** Have I considered for each aid the principles of effective design?

  ❑ Simplicity?

  ❑ Unity?

  ❑ Emphasis?

  ❑ Balance?

  ❑ Legibility?

❑ **5.** Have I collected the appropriate art materials?

❑ **6.** Do I need to attribute verbally the sources of my visuals?

❑ **7.** Have I practiced handling my visuals effectively?

HANDOUTS

❑ **1.** Have I limited my art work and words to a minimum?

❑ **2.** Have I used thick enough lettering and appropriate fonts and art work to distinguish this from a normal page of print?

❏ 3. Will I distribute my handouts before or after my presentation?

DEMONSTRATIONS/EXHIBITS

❏ 1. Should I consider giving a demonstration?

❏ 2. Should I use an item as an exhibit?

❏ 3. Can I handle a demonstration and exhibit with ease and without long pauses?

❏ 4. Will everyone in the room be able to see my demonstration or exhibit?

CHALKBOARDS

❏ 1. Will I be able to prepare my chalkboard material in advance?

❏ 2. Do I have multicolored chalk for emphasis?

❏ 3. Will everyone in the room be able to see and to read my material?

FLIP CHARTS

❏ 1. Would a portable flip chart be more appropriate than a chalkboard?

❏ 2. Should I prepare some of the graphics or print ahead of time?

❏ 3. Do I have color felt-tip markers for emphasis?

❏ 4. Will everyone in the room be able to see and to read my material?

POSTERS

❏ 1. Would posters be appropriate for my visual aids?

❏ **2.** How will I display my posters—by taping to a wall or chalkboard or using an easel?

❏ **3.** Would enlargements of photographs be appropriate for my posters?

❏ **4.** Will all of my posters be uniform in color and size?

❏ **5.** Will everyone in the room be able to see and to read my posters?

TRANSPARENCIES

❏ **1.** Should I create transparencies?

     ❏ By hand?

     ❏ On plain white paper for photocopying?

     ❏ By computer printout?

❏ **2.** Shall I make copies to distribute as handouts?

❏ **3.** Do I have the necessary acetates, reproduction equipment, and projection equipment?

SLIDES

❏ **1.** Do I have the equipment and photographic ability to produce slides?

❏ **2.** Have I organized my slides into a logical order?

❏ **3.** Do I have the necessary equipment (slide projector and tray, light-projection pointer, and wall or screen large enough for my projections)?

FILMSTRIPS

❏ **1.** Is it possible to purchase a filmstrip that will serve as an appropriate visual aid?

❏ **2.** Do I have access to the machinery to produce a filmstrip?

❏ **3.** Do I have the necessary projection equipment and screen?

VIDEOTAPES

❏ **1.** Do I have the necessary equipment to use a video-tape (video camera, VCR, television screen, and blank tapes)?

❏ **2.** Should I add voice and other sound effects?

❏ **3.** Will everyone in the room be able to see the screen?

❏ **4.** Should I consider making a video of myself for an employment interview or job performance?

    ❏ Have I carefully designed the presentation?

    ❏ Have I prepared a script?

    ❏ Have I arranged an appropriate set?

    ❏ Have I rehearsed my presentation?

    ❏ After recording, have I carefully edited the tape?

    ❏ Have I made copies and saved the master tape?

    ❏ Have I obtained the appropriate mailing box?

    ❏ Should I include my resume, cover letter, or other written material?

COMPUTER PRESENTATIONS

❏ **1.** Does my facility have a computer and screen?

❏ **2.** Do I have the software to develop a computer presentation?

❏ **3.** Do I have the know-how to use the software?

    ❏ Have I practiced incorporating clip art, on-line downloads, and scanned material into the presentation?

    ❏ Have I considered backgrounds and layout layers?

    ❏ Have I used appropriate fonts, type sizes, arrows, banners, and textures?

    ❏ Have I spaced around the material artfully?

❏ Should I turn any object into a 3-D object or reshape text into special images?

❏ Should I change the orientation of any of my slides?

❏ Would animation, movie files, rolled slides, or center-opening slides add to the presentation?

❏ Should I incorporate words and music into my presentation?

❏ **4.** Have I carefully planned the purpose, objectives, and organization of my presentation?

❏ **5.** Have I practiced, reviewed, and edited my presentation?

VIDEOCONFERENCING

❏ **1.** Would a videoconference presentation be appropriate?

❏ **2.** Do I have access to videoconferencing facilities?

❏ **3.** Should I develop visual aids for the presentation?

❏ **4.** Have I set the date and time well in advance with my groups?

❏ **5.** Have I arranged for the appropriate introductions at each site?

❏ **6.** Does everyone understand the agenda and goals?

❏ **7.** Have I practiced speaking naturally and clearly and avoiding distractions?

❏ **8.** Have I dressed conservatively and avoided white, plaids, prints, and stripes?

❏ **9.** Have I paused after my comments to allow for the brief lag in image transfer?

❏ **10.** Have I encouraged the participants to ask questions and add comments?

EXERCISES

1. Prepare a handout plus a chalkboard, flip chart, poster, and transparency for a subject of your choice.

2. Orally evaluate a classmate's visuals used in exercise 1.

3. For each of the following, determine the best type of visual aid for the conditions presented:

   a. A stockholder's meeting with an audience of 500

   b. A board meeting with 10 participants

   c. A classroom oral report on describing a mechanism

   d. A demonstration on emergency care to a group of 25

   e. A sales presentation in a person's home with 6 attendees

ORAL/VISUAL PROJECTS

1. Prepare an oral presentation for class and include at least three visual aids (handouts, demonstration/exhibit, chalkboard material, flip charts, posters, or transparencies) in your presentation.

2. Prepare a slide, videotape, or computer presentation on a subject of your choice to present in class. Any one of these projects will earn you extra credit because they are more difficult than the aids assigned in project 1.

NOTES

Conventions of Construction, Grammar, and Usage

INTRODUCTION

Effective communication depends not only on content and format but also on precise adherence to the conventions of sentence construction, grammar, and usage. This appendix will provide you with a brief handbook, exercises, and reference tables of professional and technical writing conventions.

The conventions that govern written English are not prescriptive rules. Rather, they are patterns developed by careful writers over decades and accepted by publishers, professional and technical writers, educators, and the public. Because English is a living language, these conventions alter over periods of time. For instance, it has long been conventional to use a comma before *and* in a written series of three or more items (i.e., *He assembled the nuts, bolts, and rivets*). In recent years some publishers, businesses, and industries have agreed to omit the comma (i.e., *He assembled the nuts, bolts and rivets*). Those who are concerned with language classify the conventions as *Standard English* and *General English*. The majority of professional and technical writers adopt Standard English conventions; therefore, this appendix covers the more formal conventions. Use these conventions to edit your texts and to troubleshoot your writing problems.

THE SENTENCE

Main Sentence Elements

A sentence is a group of words containing a subject and a verb and expressing a complete thought. English sentences are arranged in many word order patterns which convey meaning. Lewis Carroll's nonsense sentence

T'was brillig and the slithy toves did gyre and gimble in the wabe.

reveals patterns which make sense to English-speaking people. We recognize a subject and verb *(T'was)*, the conjunctions *(and)*, a helping verb *(did)*, a preposition *(in)*. Using the pattern we can develop any number of sentences:

It was Monday, and the electronic technicians did assemble and test in the plant.
It was noon, and the construction workers did toil and sweat in the sun.

There are five basic sentence patterns in English. Subjects, objects, and complements are nouns or pronouns. Verbs are words which express action *(jump)* or state of being *(is, seem)*.

Pattern 1: Subject + Verb
    S + V
    Water boils.

Pattern 2: Subject + Verb + Direct Object
    S   + V + DO
    Technicians repair computers.

Pattern 3: Subject + Linking Verb + Subjective Complement
    S +   V + SC
    Helium is a gas.
            (A subjective complement renames the subject.)

Pattern 4: Subject + Verb + Indirect Object + Direct Object
    S   + V   +   IO + DO
    The company gave Ms. Barnes a plaque.
            (An indirect object is the receiver of the direct object.)

Pattern 5: Subject + Verb + Direct Object + Objective Complement
    S + V + DO + OC
    The union elected him president.
            (An objective complement renames the direct object.)

There are actually many patterns beyond these basic five and many ways to invert or expand a sentence with single, one-word adjectives and adverbs or multiword phrases and clauses. Consider that a Pattern 2 sentence

S + V + DO

The man has a compass.

may be inverted to ask a question

Has the man a compass?

or expanded to

adjective clause
The man who shares my drafting cubicle has

adjective   **prepositional phrase**
a metric compass in his hand.

Recognition of the basic elements allows you to construct sentences which make sense and to punctuate the elements according to the conventions.

Secondary Sentence Elements

Secondary sentence elements are typically used as modifiers; that is, they describe, limit, or make more exact the meaning of the main elements.

Adjectives and Adverbs.   Single words used as modifiers are related to the words they modify by word order. Adjectives modify nouns or pronouns and usually stand before the word modified.

ADJ   ADJ
He is a precise, concise writer.   (modifies writer)

Adjectives may stand after the modified noun or pronoun or follow a linking verb:

N   ADJ
The temperature made the metal brittle.   (modifies metal)

N   ADJ
The metal was brittle.   (modifies metal)

Adverbs, which modify verbs, adjectives, or other adverbs, are more varied in position. They usually stand close to the particular word or element modified.

ADV

He worked late. (modifies the verb <u>worked</u>)

ADV

He worked quite late. (modifies the adverb <u>late</u>)

ADV

He was rather late. (modifies the adjective <u>late</u>)

ADV

He had never been late. (modifies verb phrase <u>had been</u>)

ADV

Unfortunately, he was always late. (modifies whole sentence)

Edit your writing so that the modifiers are clearly related to the words or statements they modify.

Ambiguous:	The plasterers have finished the wall <u>almost</u>. (<u>Almost</u> seems to modify <u>wall</u>.)
Clear:	The plasterers have <u>almost</u> finished the wall.
Ambiguous:	I <u>only</u> need a few seconds.
Clear:	I need <u>only</u> a few seconds.
Ambiguous:	The red brick doctor's office is closed.
Clear:	The doctor's red, brick office is closed.

Do not place an adverb within an infinitive phrase (<u>to read</u>, <u>to plan</u>, <u>to construct</u>).

Split infinitive:	We have to <u>further</u> plan the assembly.
Clear:	We have to <u>plan</u> <u>further</u> the assembly.

EXERCISE

Add adjectives or adverbs within each sentence. If more than one position is possible, explain what change of emphasis would result from shifting the modifier.

1. **Add <u>only</u>.** Last week I ordered a computer package with a spelling checker.

2. **Add <u>definitely</u>.** I am convinced that you are the one to complete the audit.

3. **Add <u>hardly</u>.** Although I adjusted the lever, I could hear the bass resonate.

4. **Add <u>engineer's</u>.** The well-worn report lay on the shelf.

5. **Add <u>carefully</u>.** The committee has to read all of the reports.

Phrases as Modifiers.   A phrase is a group of related words without a subject or verb. It cannot stand alone. A phrase is connected to a sentence or to one of its elements by a preposition or a verbal.

PrepositionalPhrases.   A prepositional phrase consists of a preposition (*in, at, by, from, under,* etc.) followed by a noun or pronoun plus, possibly, modifiers. It functions as an adjective or adverb, depending on what element it modifies.

> He entered <u>from the door</u> (modifies the verb <u>entered</u>) <u>of my office</u> (modifies the noun <u>door</u>).

Verbal Phrases.   A verbal phrase consists of a participle, gerund, or infinitive (verb forms without full verb function) plus its object or complement and modifiers. A participial phrase functions as an adjective; a gerund phrase as a noun; and an infinitive phrase as either a noun, adjective, or adverb.

Participial phrase:	Circuit boards <u>containing any defects</u> should be scrapped. (modifies <u>boards</u>)
Gerund phrase:	<u>Drafting the blueprint</u> was the next task. (functions as noun subject)
Infinitive phrase:	<u>To conduct a market survey</u> (functions as noun subject) is the easiest way <u>to determine the cost.</u> (functions as adjective modifying <u>way</u>)

Place modifying phrases next to the word modified. Participial and infinitive phrases will give you the most trouble.

Misrelated:	He distributed notebooks to the trainees <u>bound in plastic.</u> (The participial phrase seems to modify <u>trainees.</u>)
Revised:	He distributed notebooks <u>bound in plastic</u> to the trainees.
Misrelated:	The man who was lecturing <u>to emphasize a point</u> pounded the podium. (The infinitive phrase seems to modify <u>lecturing.</u>)
Revised:	The man who was lecturing pounded the podium <u>to emphasize a point.</u>
Dangling participial phrase:	<u>Looking up from my desk,</u> Jane gave me the report. (The phrase seems to modify <u>Jane.</u>)
Revised:	<u>Looking up from my desk,</u> I accepted the report from Jane.

Clauses.   A clause is a group of words that contains a subject and verb plus modifiers. An **independent** (main) **clause** is a complete expression which could stand alone as a sentence. A **dependent**

(subordinate) **clause** also has a subject and verb but functions as part of a sentence. It is related to the independent clause by a connecting word which shows its subordinate relationship either by a relative pronoun (*who, which, that,* etc.) or a subordinate conjunction (*because, although, since, if,* etc.).

Independent clauses:	The water tastes brackish because it is contaminated.
	The laser beam penetrated the metal plate, and the plate glowed red.
Dependent clauses:	If your engine is hot, add antifreeze.
	The drive belt which slipped shredded.
	After I attached the heat sink, the rectifier cooled.
	We will make a final test because a dry run was never completed.

Sentence Classification

Sentences may be classified according to the kind and number of clauses they contain as simple, compound, complex, or compound-complex.

Simple Sentences.   A simple sentence contains an independent clause and no dependent clauses. It may contain any number of modifiers or compound elements.

The capsule exploded.

The tiny, white, plastic capsule expanded and exploded due to the high temperature in the storage bin.

Compound Sentences.   Compound sentences contain two or more independent clauses and no subordinate clauses. They may be joined by coordinating conjunctions (*and, or, but,* etc.), semicolons, or conjunctive adverbs (*nevertheless, therefore, however,* etc.).

A dot matrix printer is acceptable, but a daisy wheel printer produces easier-to-read copy.

A Radio Shack computer is flexible; it allows you to print hard copy of graphics.

A personal computer is expensive; nevertheless; it is a practical tool for the professional writer.

Complex Sentences.   A complex sentence contains one independent clause and one or more dependent clauses.

Because it is raining, the slump test will be postponed.

The engineer who originally specified seven pilings changed her mind when she considered the sand content of the soil.

Compound-Complex Sentences.   A compound-complex sentence contains two or more independent clauses and one or more dependent clauses.

Because the text is illustrated with tables and sample materials, it is an indispensable guide for technical writers, and it may be used by students and professional writers in business and industry.

Complex sentences are used more frequently than simple, compound, or compound-complex sentences in published writing today. Complex sentences allow for more variety than simple sentences and allow the writer to manipulate emphasis. Recognition of independent and dependent clauses is necessary also for adding conventional punctuation.

EXERCISE

1. Combine these simple sentences as directed.

   a. Combine into a compound sentence.
      A v-t voltmeter measures sine waves. An oscilloscope measures nonsinusoidal voltage.

   b. Combine into a complex sentence.
      Pressure-sensitive tapes serve the electrical industry well. They insulate all manner of equipment and last longer than other tapes.

   c. Combine into a compound-complex sentence.
      Magnetic memories can store more digital information on a par with optical disks. Optical disks can recreate visual images at a lower cost than magnetic memories. Optical disks can recreate visual images at a higher speed.

2. Rewrite these complex sentences to shift the emphasis as directed.

   a. Emphasize the idea that he disliked his car rather than the idea that it uses too much gas.
      He disliked the car because, as far as I could determine, it used too much gas.

   b. Emphasize the idea that soaking will separate the components rather than the idea that you can separate them in acids.
      If you like, you can separate the components by soaking them in acids.

BASIC SENTENCE ERRORS

Sentence Fragments

If a group of words is written as a sentence, but the group lacks a subject or a verb or cannot stand alone independently, it is called a sentence fragment. A fragment may be corrected by adding a subject or verb, joining it to another sentence, or rewriting the passage in which it occurs.

Fragment: The siphons, which were described earlier.

Revision: The siphons, which were described earlier, <u>must be ordered</u>. (verb added)

Fragment: The company continues to lose money. Although production has increased.

Revision: The company continues to lose money although production has increased.   (joined to another sentence)

Fragment: Effective writing requires many skills. For example, a command of conventional grammar and the application of correct mechanics.

Revision: Effective writing requires many skills, such as a command of conventional grammar and the application of correct mechanics.   (rewritten and combined into a sentence)

EXERCISE

Rewrite or combine these fragments into complete sentences.

1. Although we are trying to please the personnel of each department by presenting a variety of training programs.
2. The late arrivals having been named in this report along with the reasons for their tardiness.
3. The arrangements should be checked immediately. Registrants to be confirmed immediately.
4. Each month more than $1,500 can be saved if the department buys a copy machine. No increase in quality if the machine is top quality.
5. I recommend we use copper. Not zinc.

Sentence Parallelism

Parallel structure involves writing related ideas in the same grammatical constructions. Adjectives should be parallel with other adjectives, verbs with verbs, phrases with phrases, and clauses with clauses.

Nonparallel:	Tungsten steel alloys are <u>tough</u>, <u>ductile</u>, and <u>have strength</u>. (adjective, adjective, verb plus noun)
Parallel:	Tungsten steel alloys are <u>tough</u>, <u>ductile</u>, and <u>strong</u>. (all adjectives)
Nonparallel:	He must <u>learn</u> the language and <u>to be knowledgeable</u> about his computer. (verb and infinitive)
Parallel:	He must <u>learn</u> the language and <u>become</u> knowledgeable about his computer. (both verbs)
Nonparallel:	He will be hired <u>if he has the required training</u>, <u>if he has three years experience</u>, and <u>by being recommended by his former employer</u>. (dependent clause, dependent clause, phrase)
Parallel:	He will be hired if he has the required training, if he has three years of experience, and if he is recommended by his former employer. (all dependent clauses)

Edit your writing to ensure that parallel ideas are expressed in parallel structures.

EXERCISE

Rewrite the nonparallel structures.

1. The computer is inexpensive, compact, and it is easy to use.
2. Before studying architecture, you should assess whether you have design ability and if you are exacting with numbers.
3. To write well one must be able to organize materials, have a flexible vocabulary, and one should know grammatical and mechanical conventions.
4. He was well liked and had training in management skills.
5. She attached the sphygmomanometer by positioning the patient, placing the cup above the elbow, and the clasp was secured.

Run-on and Comma-spliced Sentences

A run-on (also called a fused) sentence occurs when two or more independent clauses are written as one sentence without appropriate punctuation. A run-on sentence may be corrected by separating the fused clauses with a period or semicolon or by rewriting the sentence.

Run-on:	Ace Company revised its maternity policy men are now eligible for child-rearing leave.
Correction:	Ace Company revised its maternity policy; men are now eligible for child-rearing leave.

Ace Company revised its maternity policy by allowing men to be eligible for child-rearing leave.

A comma-spliced sentence occurs when two or more independent clauses not joined by a coordinating conjunction or conjunctive adverb are written with only a comma between them.

Splice:	Employees are entitled to eight sick days per year, they may be concurrent.
Correction:	Employees are entitled to eight sick days per year; they may be concurrent.

or

Employees are entitled to eight sick days per year. They may be concurrent.

Splice:	The restaurant requires a deposit for our annual dinner engagement, therefore, we must send a $50.00 check.
Correction:	The restaurant requires a deposit for our annual dinner engagement; therefore, we must send a $50.00 check.

Using your understanding of clause structure, edit your writing to avoid run-on and comma-spliced sentences.

EXERCISE

Correct these run-on and comma-spliced sentences.

1. This guide is a product of months of research, compilation was done by specialist Jim Smith.
2. Adult education opportunities are plentiful, moreover, all classes may be offered at our training facility.
3. After three straight days of bargaining, the talks broke down they will resume on Monday.
4. The union refused to consider benefit reductions but it did express willingness to negotiate increased work hours.
5. Mr. Larsen has participated in other civic activities in addition to his involvement in public schools he is a member and past officer of the Chamber of Commerce.

AGREEMENT

The most common grammar errors are subject-verb disagreements and pronoun-antecedent disagreements. For instance, if a subject is plural, then its verb must be plural as well, and if a pronoun antecedent is singular, then its pronoun must be singular as well.

Subject and Verb Agreement

Verbs change form from the singular to the plural. Consider:

Singular	Plural
I am	We are
He is	They are
She was	They were
He edits	They edit

A verb must agree with its subject in number. Generally, English-speaking people make these alterations automatically, but problems arise in a variety of structures.

In a sentence with compound subjects joined by *and* the verb is plural unless the subjects are considered a unit.

The <u>technician</u> and <u>engineer</u> **are** consulting.   (two persons)

The <u>accountant</u> and <u>auditor</u> **reviews** my books monthly.   (one person)

In a sentence with compound subjects joined by *or, nor, either . . . or, neither . . . nor* the verb usually agrees with the closest subject.

The <u>diodes</u> or the <u>transistor</u> **is** faulty.

Neither the <u>transistor</u> nor the <u>thermistors</u> **are** operating.

A singular subject followed by a phrase introduced by *as well as, together with, along with, in addition to* ordinarily takes a singular verb.

The <u>president</u> as well as the vice-president **was held** responsible for the mismanagement.

Collective nouns (*committee, jury, crowd, team, herd,* etc.) usually take a singular verb.

The <u>committee</u> **is meeting** Tuesday.

The <u>jury</u> **is arguing** with the judge.

When a collective noun refers to members of the group individually, a plural verb is used.

The <u>jury</u> **are arguing** among themselves.

Expressions signifying quantity or extent (*miles, years, quarts,* etc.) take singular verbs when the amount is considered as a unit.

Ten <u>dollars</u> **is** too much to pay for a tablet.

Six <u>hours</u> **is** too long to work without lunch.

A singular subject followed by a phrase or clause containing plural nouns is still singular.

The highest <u>number</u> of diesel trucks **is** produced in Europe.

The <u>nurse</u> who tends the heart patients **finds** them to be grateful.

When a sentence begins with *there is* or *there are,* the verb is determined by the subject which follows.

There **are** an estimated 100 <u>employees</u> in this building.

There **is** a conflicting <u>opinion</u> over capital punishment.

A verb agrees with its subject and not with its complement.

Our chief <u>trouble</u> **was** (not <u>were</u>) malfunctions in the testing equipment.

EXERCISE

Select the appropriate verb to agree with its subject.

1. Three semesters *(is, are)* not enough to master French.
2. The bulk of our tax dollars *(go, goes)* to defense spending.
3. He is one of those people who *(is, are)* always willing to help.
4. The best benefit *(is, are)* the vacations.
5. Either the men or the woman *(assist, assists)* me with the payroll.

Pronoun Agreement

A pronoun is a word which takes the place of a noun, such as *he, who, itself, their, ourselves,* etc. (A complete list of pronouns is reviewed in Table A.1.)

The <u>technician</u> checked **his** circuit boards.

The <u>woman</u> **who** trained me could assemble the parts **herself.**

A pronoun must agree in number with the word for which it stands, its antecedent.

Faulty:	<u>Ace Company</u> is furloughing 20 of **their** employees.	
Correct:	<u>Ace Company</u> is furloughing 20 of **its** employees.	
Faulty:	Send the receipts to the <u>bookkeeping department</u>. **They** will issue the refunds.	
Correct:	Send the receipts to the <u>bookkeeping department</u>. **It** will issue the refunds.	
Faulty:	<u>Anyone</u> can take **their** accrued sick leave when necessary.	
Correct:	<u>Anyone</u> can take **his** (or **his or her**) accrued sick leave when necessary.	

TABLE A.1 *Pronouns*

	Subject	*Object*	*Possessive*
PERSONAL PRONOUNS:			
First person			
Singular	I	me	my, mine
Plural	we	us	our, ours
Second person			
Singular & plural	you	you	your, yours
Third person			
Singular			
masculine	he	him	his
feminine	she	her	her, hers
neuter	it	it	its
Plural	they	they	their, theirs
RELATIVE PRONOUNS:	who	whom	whose
	that	that	
	which	which	whose, of which
INTERROGATIVE PRONOUNS:	who	whom	whose
	which	which	whose, of which
	what	what	
REFLECTIVE AND INTENSIVE PRONOUNS:	myself	herself	ourselves
	yourself	itself	yourselves
	himself	oneself	themselves
DEMONSTRATIVE PRONOUNS:	this	that	
	these	those	
INDEFINITE PRONOUNS:	all	everybody	no one
	another	everyone	nothing

any	everything	one
anybody	few	other
anyone	many	several
anything	most	some
both	much	somebody
each	neither	someone
each one	nobody	something
either	none	such

RECIPROCAL PRONOUNS: each other   one another

NUMERAL PRONOUNS: one, two, three . . .
first, second, third . . .

When a pronoun's antecedent is a collective noun, the pronoun may be either singular or plural, depending on the meaning of the noun.

The <u>committee</u> planned **its** next meeting.   (the unit)

The <u>committee</u> gave **their** reports.   (the individual members)

Usually a singular pronoun is used to refer to nouns joined by *or* or *nor*.

Neither <u>Jane</u> nor <u>Judy</u> did **her** share.

When the antecedent is a common-gender noun (*customer, manager, instructor, student, supervisor, employee,* etc.), the traditional practice has been to use *he* and *his* as in

A <u>manager</u> routinely evaluates **his** employees.

However, writers who are sensitive to sexist elements of our language are more prone to use both *his* **or** *her* if the gender of the antecedent is not known.

If *his* **or** *her* must be repeated frequently, the cumbersome usage may be avoided by changing the singular antecedent to a plural construction.

<u>Managers</u> routinely evaluate **their** employees.

Some indefinite pronouns (*some, all, none, any,* etc.) used as antecedents require singular or plural pronouns, depending on the meaning of the statement.

Everyone, everybody, anyone, anybody, someone, no one, and *nobody* are always singular.

Everyone must turn in **his or her** timesheet.

Somebody erased **his or her** floppy disk.

All, any, some, or *most* are either singular or plural, depending on the meaning of the statement.

All of the employees received **their** payroll deduction forms.   (All refers to employees and is plural; all is the antecedent of their.)

All of the manuscript has been typed, but **it** has not been proofread.   (All refers to manuscript and is singular; all is the antecedent of it.)

In Standard English usage *none* is usually singular unless the meaning is clearly plural.

Standard:	None of the men finished **his** work.
General:	None of the men finished **their** work.
Clearly plural:	None of the new computers **are** as large as their predecessors. (The sentence clearly refers to all new computers.)

When a pronoun is used, it must have a clearly identified antecedent.

Ambiguous:	The CRT fell on the keyboard and broke **it**. (It could refer to CRT or keyboard.)
Clear:	The CRT broke when **it** fell on the keyboard.
Ambiguous:	The consultants recommended a new method of shipping parts. This is the company's best alternative for the future. (It is not clear if this refers to method or to the implied word recommendation.)
Clear:	The consultants recommended a new method of shipping parts. This recommendation is the best alternative for the future.

EXERCISE

Correct these misused pronouns.

1. The company had high hopes for the new research program, but they encountered financial problems.
2. Neither the supervisor nor the laborers want his pay reduced.
3. Everybody supported their union.
4. An instructor should encourage his students to ask questions.
5. Electrical engineering is an interesting field, and that is what I want to be.

Pronoun Case

Review Table A.1. The personal pronouns and the relative or interrogative pronoun *who* have three forms depending on whether the pronoun is used as a subject, an object, or a possessive. Writers frequently encounter a few problems in proper case usage.

The object form of a pronoun is used after a preposition.

Incorrect:	The work was divided between **he** and **I.**
Correct:	The work was divided between **him** and **me.**
Incorrect:	That is the data processor about **who** I have spoken.
Correct:	That is the data processor about **whom** I have spoken.

In written English *than* is considered a conjunction, not a preposition, and it is followed by the form of the pronoun that would be used in a complete clause, whether or not the verb appears in the construction.

I am more experienced than **she** [is].

I like him better than [I like] **her.**

In general usage many educated people say "It is me" or "This is her," but Standard English usage prefers the subject form after the linking verb *be.*

It is **I.**

This is **she.**

That is **he.**

It will be **I** who fail.

Although the distinction between *who* and *whom* is disappearing in oral communications, Standard English usage prefers the distinction in writing. *Who* is the standard form when it is the subject of a verb; *whom* is the standard form when it is the object of a preposition or the direct object.

That is the professor **who** taught me chemistry.   (Who is the subject of the verb taught.)

That is the woman **whom** I recommended for promotion.   (Whom is the direct object of recommended.)

To **whom** are you speaking?   (Whom is the object of the preposition to.)

Reflexive and Intensive Pronouns

The reflexive form of a personal pronoun is used to refer back to the subject in an expression where the doer and recipient of an act are the same.

He had only **himself** to blame.

I timed **myself** typing.

The same form is sometimes used as an intensive to make another word more emphatic.

The raise was announced by the president **himself.**

Safety **itself** is crucial.

In certain constructions writers mistakenly consider *myself* to be more polite than *I* or *me,* but in Standard English the reflexive forms are not used as substitutes for *I* or *me.*

Faulty: Mr. Jones and **myself** attended the meeting.
Correct: Mr. Jones and **I** attended the meeting.

Faulty: The work was completed by Ms Burns and **myself.**
Correct: The work was completed by Ms Burns and **me.**

EXERCISE

Select the correct pronoun in each of the following sentences.

1.   From *(who, whom)* will we receive the instructions?
2.   The Director of Training assigned the project to Jones and *(I, me).*
3.   It is *(we, us)* who were to leave early.
4.   She was later than *(I, me).*
5.   Smith, White, and *(I, myself)* were assigned to conduct the survey.

USAGE GLOSSARY

Many words in the English language are so similar that they cause confusion. Following is a brief glossary and exercise of such terms for you to review. (Words that are asterisked are corruptions of correct usage; they are not to be used in formal writing.)

accept, except

Accept means "receive" or "agree to."

Community colleges accept a wide variety of students.

As a preposition, *except* means "other than."

I did all of the work except your report.

As a verb, *except* means "exclude," "omit," "leave out."

If you except Mr. Jones, no other president has owned his own Lear Jet.

advice, advise

Advice is a noun meaning "guidance."

If I wanted it, I would ask for your advice.

Advise is a verb meaning "counsel," "give advice to," "recommend," or "notify."

I advise you to exercise stock options.

affect, effect

Affect means "change," "disturb," or "influence."

The rising cost of fuel has drastically affected the trucking industry.

It can also mean "feign" or "pretend to feel."

Although she knew she was to be promoted, she affected surprise when notified.

As a verb, *effect* means "bring about," "accomplish," or "perform."

She effected a perfect word-processed report.

As a noun, *effect* means "result" or "impact."

Her extra work had no effect on the vice-president.

aggravate, irritate

Aggravate means "make worse."

The recent charge of police brutality has aggravated racial hostilities.

Irritate means "annoy" or "bring discomfort to."

His endless twitching began to irritate me.

allude, refer

Allude means "call attention indirectly."

When the candidate spoke about the Mid East, he alluded to the past administration's involvement in secret arms sales.

Refer means "direct attention specifically."

He referred to the second paragraph which stated the fact.

all ready, already

Use *all ready* when *all* refers to things or people.

At noon the secretaries were all ready to lunch.

Use *already* to mean "by this time" or "by that time."

I have already typed that report.

all right, *alright

All right means "completely correct," "safe and sound," or "satisfactory."

My answers to the interviewer were all right.   (The meaning is that *all* of the answers were *right*.)

Despite a few cuts, I was all right.

Do not use *all right* to mean "satisfactorily" or "well."

*Do not use *alright* anywhere; it is a misspelling of *all right*.

almost, *most all

Almost means "nearly."

> By the time we reached Miami, the tank was almost empty.

*In formal writing, do not use *most all*; use *almost all* or *most*.

> Almost all (or most) of the employees were eligible for vacation.

a lot, *alot

A lot of and *lots of* are colloquial and wordy. Use *much* or *many*.

*Do not use *alot* anywhere; it is a misspelling of *a lot*.

among, between

Use *among* with three or more persons, things, or groups.

> There are no secrets among my friends.

Use *between* with two persons, two things, or two groups.

> Voters must choose between the Democrat and Republican candidates.

amount, number

Use *amount* when discussing uncountable things.

> No one knows the amount of good a female President might do.

Use *number* when discussing countable things or persons.

> A number of us are thinking of a third candidate.

anyone, any one

Anyone is an indefinite pronoun (like *everyone, everybody*).

> *Anyone* who desires to run for office must file by Wednesday.

Any one means "any one of many."

> I could not find any one who could fix my tire.

as, as if, like

Use *as* or *as if* to introduce a clause.

> As I drove into the parking lot, my tire blew out.
>
> He looked as if he had worked all night.

Use *like* to mean "similar to."

> The logo looked like ours.

assure, ensure, insure

Assure means "state with confidence to."

> I assure you that he will be hired.

Ensure means "make sure" or "guarantee."

There is no way to ensure that every policy is understood.

Insure means "make a contract for payment in the event of specified loss, injury, or death."

He insured the package for $100.00.

bad, badly

Bad is an adjective meaning "not good," "sick," or "sorry."

We paid a lot for the movie, but it was bad.

In spite of the medicine, I still felt bad.

She felt bad about losing her job.

Badly is an adverb meaning "not well."

She starts races well but ends badly.

Used with *want* or *need, badly* means "very much."

The diver badly wanted a gold award.

I need some money badly.

Do not use *bad* as an adverb.

*He swam bad for the first two laps.

Revised: He swam badly for the first two laps.

censor, censure

As a verb, *censor* means "exercise censorship."

Some groups want the NEA to censor certain art.

As a noun, *censor* means "one who censors."

Rap groups should act as censors of their own lyrics.

As a verb, *censure* means "find fault with" or "reprimand."

The Prime Minister was censured by Parliament.

As a noun, *censure* means "disapproval" or "blame."

Ridicule often hurts more than censure.

complement, compliment

As a verb, *complement* means "bring to perfect completion."

His red tie complemented his blue suit.

As a noun *complement* means "something that makes a whole when combined with something else" or "the total number of persons needed."

Practice is the complement of learning.

Without a full complement of workers, we cannot complete the task.

As a noun, *compliment* means "expression of praise."

He complimented the appearance of my report.

continual, continuous

Continual means "going on with occasional slight interruption."

At the office there is a continual humming of typewriters.

Continuous means "going on with no interruptions."

We have 24-hour guard service; surveillance is continuous.

council, counsel, consul

Council means "groups of persons who discuss and decide certain matters."

The city council authorized a new park.

As a verb, *counsel* means "advise."

My stockbroker counseled me not to sell my bonds.

As a noun, *counsel* means "advice" or "lawyer."

However, I didn't follow his counsel and sold short.

A *consul* is a government official working in a foreign country to protect the interest of his or her country's citizens there.

If you lose your passport in France, you will need to see the American consul.

criterion, criteria

A *criterion* is a standard by which someone or something is judged.

The criterion for selecting the car was color.

Criteria is the plural of *criterion*.

The criteria for selecting the car were color, factory air conditioning, and horsepower.

data

Data is the plural of Latin *datum* (literally, "given"), meaning "something given"—that is, a piece of information; however, *data* may be treated as singular.

The data stored in my computer is readily accessible.

different from, *different than

*Do not use *than* after different. Use *from*.

His management style is different from mine.

due to, because of

Due to means "resulting from" or "the result of."

The leaking roof was due to broken tiles.

Because of means "as a result of."

Animosities persist in the Middle East because of Iraq's nuclear capability.

eminent, imminent, immanent

Eminent means "distinguished," "prominent."
  An eminent astronomer has questioned the "big bang" theory.
Imminent means "about to happen."
  The dark sky and distant rumbling indicated the storm was imminent.
Immanent means "inherent," "existing within."
  The Trobrianders believe that evil is immanent in stone, sand, and water.

farther, further

Farther means "a greater distance."
  The road went farther into the forest.
As an adjective, *further* means "more."
  He needs a further education beyond high school.
As a conjunctive adverb, *further* means "besides."
  I dislike carrots; further, I detest asparagus.
As a transitive verb, *further* means "promote" or "advance."
  How much has the mayor done to further tourism in our town?

few, fewer, little, less

Use *few* or *fewer* with countable nouns.
  We have fewer employees than we did a year ago.
Use *little* or *less* with uncountable nouns.
  I have less experience than you on the word processor.

frightened, afraid

Frightened is followed by *at* or *by*.
  Many people are frightened by thunder.
  I was frightened at the thought of gaining weight.
Afraid is followed by *of*.
  I am afraid of heights.

good, well

Use *good* as an adjective, but not as an adverb.
  The proposal for staggered work hours sounded good to many employees.
Use *well* as an adverb when you mean "in an effective manner," "ably."
  He did so well on the project that he was promoted.
Use *well* as an adjective when you mean "in good health."
  She hasn't looked well since her operation.

hanged, hung

Hanged means "executed by hanging."
  The rustler was hanged at dawn.

Hung means "suspended" or "held oneself up."

> I hung my hat on the rack.

> I hung on the side of the cliff until I could get a foothold.

hopefully

Use *hopefully* to modify a verb.

> She looked hopefully at her boss as he scanned her proposal.

*Do not use *hopefully* when you mean "I hope that," "we hope that," or the like.

> **Incorrect:**   Hopefully, the company will make a profit this quarter.

> **Revised:**   The stock holders hope that the company will make a profit this quarter.

immigrate, emigrate

Immigrate means "enter a country in order to live there permanently."

> Those Haitians decided to immigrate to Florida.

Emigrate means "leave one country in order to live in another."

> My grandfather emigrated from Cuba.

in, into, in to

Use *in* when referring to a direction, location, or position.

> Everyone wants to live in the Sun Belt.

Use *into* to mean movement toward the inside.

> He dove into the water.

Use *in to* when the two words have separate functions.

> I walked in to see the movie.

its, it's

Its is the possessive form of *it*.

> I like this company because of its location and its benefits.

It's means "it is."

> It's evident that we are an expanding company.

kind of, sort of

In formal writing, do not use *kind of* or *sort of* to mean "somewhat."

> **Faulty:**   When I moved my head, I felt sort of queasy.

> **Revised:**   When I moved my head, I felt rather queasy.

lie, lay

Lie (lie, lying, lies, lied) means "tell a falsehood."

> She sometimes lies about her age.

> She has lied about her age for 20 years.

Lie (lie, lying, lay, lain) means "rest," "recline," or "stay."

I love to lie in bed late.

Lying on the sand, I watched the water sparkle.

After I lay on the sand for two hours, I was sunburned.

That book has lain on the table for two months.

Lay (lay, laying, laid) means "put in a certain position."

A good education lays the foundation for a good life.

The workers were laying tile on the floor.

They laid the tile into stacks.

The hurricane struck after the tile had been laid.

likely, apt, liable

Likely indicates future probability.

It is likely that inflation will continue.

Apt indicates a usual or habitual tendency.

Most cars are apt to rust after a few years.

Liable indicates a risk or adverse possibility.

Apartments which are left unlocked are liable to be burgled.

*not very, none too, *not too*

In formal writing, use *not very* or *none too* instead of **not too.*

Faulty: Not too pleased with the weather, we changed our plans.

Revised: None too pleased with the weather, we changed our plans.

precede, proceed, proceeds, proceedings, procedure

To *precede* is to come before or go before in place or time.

A dead calm often precedes a hurricane.

To *proceed* is to move forward or go on.

The boat could not proceed because the bridge would not open.

When the bridge opened, the boat proceeded.

Proceeds are funds generated by a business deal or a money-raising event.

The proceeds from the garage sale will be used to pay the rent.

Proceedings are formal actions, especially in an official meeting.

The stenographer transcribed the proceedings of the meeting.

A *procedure* is a standardized way of doing something.

The correct procedure for moving a block of type is in the manual.

*regardless, *irregardless*

Use *regardless,* not *irregardless.*

Faulty: Irregardless of the potential circumstances, Scott decided to bungee jump.

Revised: Regardless of the potential circumstances, Scott decided to bungee jump.

stationary, stationery

Stationary means "not moving."

The typewriter was on a stationary table.

Stationery means "writing paper."

We had to order more stationery from the printing department.

such a

Do not use *such a* as an intensifier unless you add a result clause beginning with *that*.

Weak: The political rally was such a bore.

Better: The political rally was such a bore that I fell asleep.

EXERCISE

Supply the correct word. You may have to change the tense of the verbs.

1. Everyone has _____ the invitation _____ Sam.
    (accept, except)

2. I _____ you to follow your instructor's _____.
    (advise, advice)

3. The malfunctioning air conditioning _____ our tempers.
    The manager _____ a defiant look.
    The _____ of nuclear fallout are under study.
    (affect, effect)

4. Finally, the reports were xeroxed, and we were _____ to begin the board meeting.
    The President had _____ left for lunch when I reported for our interview.
    (already, all ready)

5. It is _____ with me if you use correction tape.
    (alright, all right)

6. The technician looked _____ he were tired.
    (as, as if, like)

7. I _____ you that we can _____ your right to strike.
    (ensure, assure, insure)

8. If there are _____ members, it means _____ work for the secretary.

   (fewer, less)

9. The departmental members work _____ together, and I feel _____ about their cooperation.

   (good, well)

10. Ace Company is an ideal employer because of _____ benefits, and _____ improving _____ stock option program each year.

    (its, it's)

APPENDIX *B*

Punctuation and Mechanical Conventions

INTRODUCTION

The professional or technical writer must be conscious and demanding of punctuation and mechanical conventions to prevent vagueness and misreading. Two practices of punctuation are prevalent today. A few businesses and industries adopt an open punctuation system, which favors only essential marks and omits those that can be safely omitted, such as the comma before *and* in a series (screws, nuts, and bolts). The majority of professional writers use standard, or close, punctuation conventions because these conventions promote greater accuracy. This appendix reviews close punctuation conventions.

One of the major characteristics of professional and technical writing is the extensive use of abbreviations, numbers, symbols, and other mechanics. This appendix reviews the general conventions that govern mechanics.

Both the punctuation and mechanical conventions are arranged in alphabetical order to help you edit your writing quickly.

PUNCTUATION

1.0   Apostrophe

1.1   Use an apostrophe to indicate the possessive case of the noun. (If the noun ends in *s*, use an apostrophe after it, but do not use an additional *s*.)

The company's product
Jack and Bob's office (joint possession)
Bill's or Jack's car (individual possession)
his sister-in-law's law practice
The Jones' car crashed.

1.2   Use an apostrophe to indicate the possessive case of indefinite pronouns:

another's tools	neither's wrench
anybody's desk	one's customers
each one's station	somebody's computer

Do *not* use an apostrophe to indicate the possessive case of personal pronouns:

his schedule
Ours is the newest model.
The mistake was hers.
Its handle is steel.

1.3   Use an apostrophe to indicate the omission of letters in contractions:

I'm	o'clock
he'll	we're
can't	you're

Do not confuse *they're* with *their* or *there*.

Their supervisor knows they're there.

Do not confuse *it's* (it is) with *its* (a possessive).

It's demonstrating its graphic function.

1.4   Use an apostrophe to indicate the plural symbols and cited words:

You use too many *and*'s.

The *&*'s are broken on all of the typewriters.

1.5 Do not use an apostrophe to form the plural of term in all capital letters:

The *CD-ROMs* are by the computer.

The university awarded eight *Ph.D.s* in engineering.

1.6 Do not use an apostrophe to form the plural of letters and numbers unless confusion would result without one:

Your *rs* look like your *ls.*

Two IOUs

the 1980s

nine l's.

2.0 Brackets

2.1 Use brackets within a quotation to add clarifying words that are not in the original.

Mr. Roberts stated, "They [computers] have revolutionized his business."

2.2 Use brackets within a quotation to enclose the Latin word *sic* ("so," "thus") which indicates that a misspelling, grammatical error, or wrong word was in the original:

He wrote, "Your [sic] selected to head the committee."

3.0 Colon

3.1 Use a colon after a formal salutation:

Dear Ms Benson:

Good morning:

3.2 Use a colon to introduce a phrase or clause which explains or reinforces a preceding sentence or clause:

Food processing consists of three main steps: selecting the blade, measuring the ingredients, and processing at the appropriate speed.

The position sounds attractive: the salary is high and the opportunities for advancement are excellent.

3.3 Use a colon when a clause contains an anticipatory expression *(the following, as follows, thus, these)* and directs attention to a series of explanatory words, phrases, or clauses:

The requirements for the position are as follows:

1. A master's degree
2. Three years experience
3. Willingness to relocate

3.4 Use a colon to express ratios, to separate hours and minutes, and to indicate other relationships.

3:1	signal:noise
A:B	8:25 P.M.
Acts: 14:7	12:101–104 (volume 12, pages 101–104)

3.5 Use a colon between the main title and subtitle of a book:

Technical Writing: An Easy Guide

4.0 Comma

Refer to Table B.1 for a review of comma and semicolon usage in compound and complex sentences.

4.1 Use a comma to separate independent clauses joined by a coordinating conjunction.

TABLE B.1 *Comma and Semicolon Review for Compound and Complex Sentence Construction*

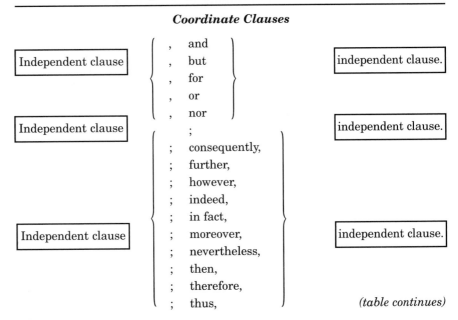

Coordinate Clauses

Independent clause	, and / , but / , for / , or / , nor	independent clause.
Independent clause	;	independent clause.
Independent clause	; consequently, / ; further, / ; however, / ; indeed, / ; in fact, / ; moreover, / ; nevertheless, / ; then, / ; therefore, / ; thus,	independent clause.

(table continues)

Subordinate Clauses

After
Although
As
Because
Before   [dependent clause]   ,   [independent clause.]
If
Since
Until
While

The cursor shows where you are typing, and it moves across the screen as you type.

4.2   Use a comma after an introductory dependent clause:

If you have a two-drive computer system, you place your program diskette in drive A.

4.3   Use a comma after a conjunctive adverb introducing a coordinate clause:

This system is easy to use; however, we suggest that you read the directions carefully.

4.4   Use a comma to separate a nonrestrictive word, phrase, or clause from the rest of the sentence:

You should, however, make copies of your original diskette for safekeeping.

His goal, to become computer literate, is easy.

The cursor, which is a blinking square, shows where the entry will appear on the screen.

4.5   Use a comma to separate items in a series:

These instructions will teach you how to create, edit, or proof a file.

The Personnel Department will receive all letters of application, forward them to the appropriate search committee, and handle all correspondence.

The comma is often omitted in company names:

Jones, Smith and Scully

4.6   Use a comma to separate a series of adjectives or adverbs not connected by a conjunction:

Store your thin, sensitive, magnetic disks out of direct sunlight.
The computer blinked haphazardly, noisily.

4.7   Use a comma in a date to separate the day and year:

June 9, 1997
the April 15, 1998, deadline

Do *not* use a comma in the military or international form:

9 June 1987

4.8   Use a comma to separate city and state, state and county, county and state:

Fort Lauderdale, Broward County, Florida, has 27 electronics firms.

4.9   Use a comma to separate titles from names and to set off appositives:

Juan Murillo, M.D., prescribed aspirin.
Julie Koenig, attorney-at-law, tried the case.
K. D. Marshal, Jr., ran for office.
Robert H. Larsen, president, hired a computer guru.
Jerry Noosinow, the Delta pilot, checked the KAL log.

4.10   Use commas to group numbers into units of threes in separating thousands, millions, and so forth:

3,845

74,763

9,358,981

4.11   Use a comma after the salutation of an informal letter and after the complimentary close of most letters:

Dear Sally,

Very truly yours,

Cordially,

4.12   Use a comma to separate corporation abbreviations for company names:

Trolleys, Ltd.

Jones and Scully, Inc.

Ace Company, Inc.

4.13 Use a comma after expressions that introduce direct quotations:

President Dan Barker said, "We must double our profits."

If the phrase interrupts the quotation, it is set off by two commas:

"We must," said President Dan Barker, "double our profits."

5.0 Dash

5.1 Use dashes to set off emphatic and abrupt parenthetical expressions:

The idea of this program—it has been tested thoroughly—is to simplify spelling correction.

5.2 Use a dash to mark sharp turns in thought:

He was an arrogant man—with little to be arrogant about.

5.3 Use dashes to separate nonrestrictive material which contains commas from the rest of the sentence:

These manuals—*Guide to Operations, Disk Operating System, Word Proof,* and so on—are protected by copyright.

6.0 Ellipsis

6.1 Use the ellipsis (three spaced periods) to indicate any omission in quoted material:

Martin stressed, "The technical writer . . . must master punctuation." (The words *as well as the professional writer* have been omitted.)

6.2 Use four spaced periods to indicate the ellipsis at the end of a sentence.

The consultant stressed, "Write carefully. . . ." (The words *and edit endlessly* have been omitted.)

7.0 Exclamation Point

7.1 Use an exclamation point at the end of an exclamatory sentence to show emotion or force:

Shred the files!
We will not file bankruptcy!

7.2 Use an exclamation point at the end of an exclamatory phrase:

What a disaster!

8.0 Parentheses

8.1 Use parentheses to enclose an abruptly introduced qualification or definition within a sentence:

You may place your program diskette in drive A and your storage diskette (the one with your file on it) in drive B.

8.2 Use parentheses to enclose a cross-reference within a sentence:

The ellipsis (see Section 6.0) is used in direct quotations.

8.3 Use parentheses to enclose figures or letters to enumerate points:

To use this program (a) insert your DOS diskette in drive A, (b) turn on your computer, (c) type in the date, and (d) press the Enter key.

9.0 Period

9.1 Use a period to signal the end of declarative or imperative sentences:

Diskettes are sensitive to extremes of temperature.
Do not try to clean diskettes.

9.2 Use a period with certain abbreviations (see pages 653–654 for exceptions):

Dr.	Ph.D.
A.M.	Nov.
P.M.	J. C. Lewis
Jr.	etc.
Mr.	

9.3 Use a period before fractions expressed as decimals, between whole numbers and decimals, and between dollars and cents:

.10	$3.50
3.6	$0.92 (or 98¢ or 93 cents)

9.4 Use a period after number and letter symbols in an outline:

I.

    A.

    B.

        1.

        2.

10.0 Question Mark

10.1 Use a question mark at the end of an interrogative question:

Do you own a personal computer?

Do *not* use one after an indirect question.

He asked me if I owned a personal computer.

10.2 Use a question mark in parentheses to indicate there is a question about certainty or accuracy:

This is the best (?) computer.

11.0 Quotation Marks

11.1 Use quotation marks to set off direct speech and material quoted from other sources:

Dr. William Haskell writes, "Before 1960 the thing rarer than a marathoner was a health professional trained to care for one. Most doctors," he points out, "forbade post-cardiac patients to do anything more vigorous than walk to the refrigerator."

11.2 Use quotation marks to indicate nonstandard terms, ironic terms, and slang words:

This is a "gimmick."

His "problem" was his genius IQ.

He "got his act together."

11.3 Use quotation marks to indicate titles of articles, essays, short stories, chapters, short poems, songs, television and radio programs, and speeches:

I read the article "Ten Years of Sports Medicine" in *Runner's World*.

Do *not* use quotation marks around quoted material which requires more than four lines in your paper. Display a long quote by indenting it 10 spaces from your regular margins and omitting the quotation marks.

Commas and periods are always placed *inside* the closing quotation mark:

"Yes," Ms Gloss said, "we have a swine flu epidemic."

He said "electrons," but meant "electronics."

Semicolons and colons are placed *outside* the closing quotation mark:

He said, "The trapped air bubble will leave honeycombs"; honeycombs are sections of little indentations.

He said, "The trapped air bubble will leave honeycombs": little indentations.

Question marks, exclamation points, and dashes are placed inside *or* outside the final quotation mark, depending upon the situation:

He asked, "Is the oscillator connected to the mixer?"

Did he say, "The assembly is constructed of heavy-gauge stainless steel"?

12.0   Semicolon

Refer to Table B.1 for a review of comma and semicolon usage in compound and complex sentences.

12.1   Use a semicolon between coordinate clauses not connected by a conjunction:

The new system will use low-powered transmitters; it is called a cellular radio.

12.2   Use a semicolon before a conjunctive adverb including a coordinate clause:

Retort pouches are like cans; however, they do not dent.

12.3   Use a semicolon before a coordinating conjunction introducing a long or loosely related clause:

Niobium, which is used primarily as an alloy, is a metallic element that resists heat and corrosion and hardens without losing strength; and it is widely available in Canada and South America.

12.4   Use a semicolon in a series to separate elements containing commas:

J. D. Smyth, member of the board; Carol Winter, president; Glenn Morris, committee chairperson; and I attended the conference.

13.0   Virgule (Slash)

13.1   Use a virgule to indicate appropriate alternatives:

Define and/or use the words in sentences.

13.2   Use a virgule to represent *per* in abbreviations:

17 ft/sec   12 mi/hr

13.3   Use a virgule to separate divisions of a period of time:

the April/May report

the 1996/97 academic year

MECHANICAL CONVENTIONS

1.0   Abbreviations

1.1   Avoid the overuse of abbreviations. (See Table B.2 for a list of common technical abbreviations.)

1.2   Explain an abbreviation the first time you use it:

He has worked for the Department of Transportation (DOT) and the Office of Mental Health (OMH).

1.3   Omit most internal and terminal punctuation in abbreviations of technical measurements:

BTU	lb
psi	ft
DNA	rpm

1.4   If the abbreviation forms another word, use the internal and terminal punctuations to avoid confusion:

in.	A.M.
gal.	No.

1.5   Use uppercase (capital) letters for acronyms and degree scales:

NASA   (National Aeronautics and Space Administration)

VHF   (very high frequency)

OEM   (original equipment manufacturer)

C   (Centigrade)   F   (Fahrenheit)

TABLE B.2 *Common Technical Abbreviations*

ac	alternating current	**kw**	kilowatt
amp	ampere	**kwh**	kilowatt-hour
A	angstrom	**l**	liter
az	azimuth	**lat**	latitude
bbl	barrel	**lb**	pound
BTU	British Thermal Unit	**lin**	linear
C	Centigrade	**long**	longitude
Cal	calorie	**log**	logarithm
cc	cubic centimeter	**m**	meter
circ	circumference	**max**	maximum
cm	centimeter	**mg**	milligram
cps	cycles per second	**min**	minute
cu ft	cubic foot	**ml**	milliliter
db	decibel	**mm**	millimeter
dc	direct current	**mo**	month
dm	decimeter	**mph**	miles per hour
doz	dozen	**No.**	number
dp	dewpoint	**oct**	octane
F	Fahrenheit	**oz**	ounce
f	farad	**psf**	pounds per square foot
fbm	foot board measure	**psi**	pounds per square inch
fl oz	fluid ounce	**qt**	quart
FM	frequency modulation	**r**	roentgen
fp	foot-pound	**rpm**	revolutions per minute
fpm	feet per minute	**rps**	revolutions per second
freq	frequency	**sec**	second
ft	foot	**sp gr**	specific gravity
g	gram	**sq**	square
gal.	gallon	**t**	ton
gpm	gallons per minute	**temp**	temperature
gr	gram	**tol**	tolerance
hp	horsepower	**ts**	tensile strength
hr	hour	**v**	volt
in.	inch	**va**	volt ampere
iu	international unit	**w**	watt
j	joule	**wk**	week
ke	kinetic energy	**wl**	wavelength
kg	kilogram	**yd**	yard
km	kilometer	**yr**	year

1.6   Use lowercase (small) letters for abbreviations for units of measure:

> gph   (gallons per hour)
> cc   (cubic centimeters)
> rpm   (revolutions per minute)
> mph   (miles per hour)
> bps   (bits per second)

1.7 Write plural abbreviations in the same form as the singular:

> 17 in.
> 47 lb
> 5 hr
> 30 gph
> 10 cc

2.0 Capitalization

2.1 Use Standard English conventions.

2.2 Begin all sentences with a capital letter.

2.3 Capitalize all proper nouns (proper names, titles which precede proper names, book and chapter titles, languages, days of the week, months, holidays, names of organizations and groups, races and nationalities, historical events, names of structures and vehicles, and so forth):

John Doe	Ace Construction Company
Professor Jane Doe	American Federation of Labor
Introduction to Nursing	Caucasian
French	Jewish
Monday	the Korean War
October	the Statue of Liberty
Labor Day	a Ford Mustang

2.4 Capitalize adjectives derived from proper nouns:

> English
> Elizabethan

2.5 Capitalize words like *street, avenue, corporation,* and *college* when they accompany a proper name:

> Elm Street
> Forty-second Avenue
> Ace Company, Inc.
> Yale University
> Broward Community College

2.6 Capitalize *north, east, midwest, near east,* and so on when the word denotes a specific location:

the South
the Midwest
the Near East
101 Northwest First Street

2.7 Capitalize brand names:

Kleenex tissues	Xerox photocopies
Scotch tape	a Frigidaire
a Formica counter	the Astro-Turf field
a Polaroid camera	Sanforized

3.0 Hyphenation

3.1 Use a hyphen between some compound names for family relations:

Hyphenated: brother-in-law's company
One word: my stepmother's portfolio
Two words: my half brother was graduated

3.2 In nontechnical prose use a hyphen in compound numbers from twenty-one to ninety-nine and in fractions:

thirty-seven cartons
forty-third year
four-fifths of the book
one-eighth inch

3.3 Use a hyphen after the prefixes *all-, ex-, self-,* and before the suffix *-elect:*

all-American
ex-president
self-contained
president-elect

3.4 Use a hyphen in some compound nouns:

kilowatt-hour
dyne-seven
foot-pound

3.5 Use a hyphen in compound adjectives when the latter precedes the word it modifies:

alternating-current motor

closed-circuit television

high-pressure system

easy-to-build model

3.6 Use a hyphen between a number and a unit of measure when they modify a noun:

6-inch ruler

12-volt charge

a 3-week-old prescription

4.0 Italics

4.1 Use italics (underline in handwritten or typed material) to indicate the names of books, magazines, newspapers, and other complete works published separately:

the book *Introduction to Nursing*

the magazine *Newsweek*

the movie *The Right Stuff*

Dante's *Devine Comedy*

Word Proof: A Manual

4.2 Use italics to indicate the names of ships and planes:

the H.M.S. *Ark Royal*

the U.S.S. *Independence*

4.3 Use italics to indicate words, symbols used as words, and foreign words which are not in general English usage:

The prizewinning orchids were *Alleraia* Ocean Spray, *Bloomara* Jim, and *Guantlettara* Noel.

coup d'etat

deus ex machina

The word *thrombosis* is derived from the Greek word *thrombos,* which means "a clot" or "a clump."

Your *9*s look like 7s.

Do not italicize foreign expressions that are established as part of the English language, such as:

a priori	bona fide	habeas corpus	pro tem
ad hoc	carte blanche	laissez faire	resume
ad infinitum	etc.	per annum	status quo
	ex officio	pro rata	

5.0 Numbers

5.1 Handle numbers consistently in any one report.

5.2 Write out single digit numbers from zero through nine when the number modifies a noun:

five disks	two printers
three word processors	nine keyboards

5.3 Use numerals for zero through nine when the number modifies a unit of measure, time, dates, pages, chapters, sections, percentages, money, proportions, tables, and figures:

2 inches	section 9
3-second delay	2 percent
5 gph	a 4% increase
9 years old	$50
2:40 A.M.	$0.05 or .05 cents
June 9, 1996	1:9
9 June 1996	4 to 2 odds
page 7	Figure 2
Chapter 1	Table 6

5.4 In technical prose use numerals for decimals and fractions:

0.6	1/4 or 0.25
3.341	7/16 in.
3/5 or 0.6	6½ lb

5.5 In technical prose use numerals for any number greater than nine:

10 psi	237 lb
97 employees	101,400 people

5.6 Write out numbers which are approximations:

a half cup of coffee

a quarter of a mile farther

a fifth of the energy

approximately three times as often

5.7   Place a hyphen after a number of a unit of measure when the unit modifies a noun:

7-in. handle

6-inch-diameter circle

10½-lb box

27-gal. capacity

5.8   When many numbers, both smaller than and greater than nine, are used in the same section of writing, use numerals:

Buy 4 sheets of 8 x 11½-in. paper, 15 sheets of 8 x 20-in. paper, and 7 manila envelopes.

Exception: If none of the numbers are greater than nine, write them all out:

The office contains eight desks, seven chairs, six file cabinets, and seven typewriters.

5.9   When one number appears immediately after another as a part of the same phrase, avoid confusion by writing out the shorter number:

nine 50-watt bulbs

two 4-inch wrenches

thirteen 20-pound packages

twenty-two 2,500-component circuit boards

5.10   Place a comma in numbers in the thousands:

1,000

17,276

427,928

5.11   Write numbers in the millions in one of two ways:

2,700,000	*or*	2.7 million
16,000,000	*or*	16 million
$1,500,000	*or*	$1.5 million
72,110,427		

5.12   Write numbers in the billions, trillions, quadrillions, and so on in numerals:

2,700,000,000

47,337,426,104,900

5.13   Do not begin a sentence with a numeral:

Fifteen inches of rain fell.

not

15 inches of rain fell.

6.0   Symbols

    6.1   Use symbols sparingly.

    6.2   Define symbols in your text. (See Table B.3 for common technical symbols.)

TABLE B.3   *Common Technical Symbols*

Symbol	Word	Symbol	Word
%	percent	=	equals
°	degree	F	Fahrenheit
&	and	C	Centigrade
'	feet	Rx	take (on prescriptions)
"	inches	☉	the Sun, Sunday
$	dollar	£	pound
¢	cents	#	number, or leave space
@	at (12 @ $2.00 each)	Hb	mercury (the element)
+	plus	☿	Mercury (the planet)
−	minus	X	snow
×	times	↑	gas
÷	divide	Ω	ohm
‖	greater than or derived from	S	Silurian soil

7.0   Spelling

    7.1   Use a dictionary when in doubt about the proper and preferred spelling of a word. (See Table B.4 for a list of frequently misspelled words.)

TABLE B.4 *Frequently Misspelled Words*

accidentally	comparative	heroes	prominent
achievement	competitive	humorous	propaganda
acquaintance	consensus	immediately	psychology
amateur	contemptible	indispensable	pursue
analysis	convenience	irrelevant	questionnaire
anonymous	courageous	irresistible	receive
anxiety	criticism	knowledge	rhythm
appreciate	definitely	laboratory	schedule
arctic	descent	leisure	scissors
athletics	desirable	lieutenant	secretary
auxiliary	despair	lighting	seize
awkward	disappear	loneliness	separate
bachelor	discipline	maneuver	sergeant
beggar	efficient	meant	siege
beginning	eighth	medieval	similar
believe	eligible	minimum	sophomore
benefited	equipped	mortgage	souvenir
bookkeeper	exaggerate	necessary	subtle
breath	exercise	ninth	succeed
bulletin	exhausted	noticeable	successful
bureau	existence	occasionally	surprise
business	familiar	occurred	synonym
calendar	fascinating	omitted	thoroughly
campaign	fatigue	opportunity	tragedy
caricature	fiery	parallel	twelfth
catastrophe	foreign	paralysis	unforgettable
cemetery	forty	pastime	unmistakable
colonel	government	possibility	vacuum
coming	guarantee	privilege	vengeance
committee	height	procedure	weird

WORKS CONSULTED

CHAPTER 1

Anson, Chris M., and L. Lee Forsberg. "Moving Beyond the Academic Community. Transitional Stages in Professional Writing." Written Communication 7 (1990): 200–31.

Bass, John, Manager, Motorola, Inc. Media and Communication Technology Center Radio Products Group. Personal interview. 12 Mar. 1997.

Beauchamp, Tom L., and Norman E. Bowie. Ethical Theory and Business. 2nd ed. Englewood Cliffs, NJ: Prentice Hall, 1983.

Boisjoly, Roger M. "Ethical Decisions—Morton Thiokol and the Space Shuttle Challenger Disaster." Journal of Mechanical Engineering Science 87 (April 1987): 92–95.

Corbett, Jan. "From Dialog to Praxis: Crossing Cultural Borders in the Business and Technical Communication Classroom." Technical Communication Quarterly 5.4 (Fall 1996): 411–24.

Cross, Mary. "Aristotle and Business Writing: Why We Need to Teach Persuasion." The Bulletin of the Association for Business Communication 54.1 (March 1991): 3–6.

Finsberg, Stephen. "Dilbert Tackles Ethics Dilemmas." Sun-Sentinel 17 (April 1997): D1–2.

Havelock, Eric. The Muse Learns to Write: Reflections on Orality and Literacy from Antiquity to the Present. New Haven: Yale University Press, 1986.

Hunt, Peter. "The Teaching of Technical Communication in Europe: A Report from Britain." Technical Communication Quarterly 2.3 (Summer 1993): 319–30.

Lorek, L. A. "Device Lets TV Surf Web." Sun-Sentinel 13 (Dec. 1996): D1.

Ruggiero, Vincent R. The Art of Critical and Creative Thought. 4th ed. New York: HarperCollins Publishers, 1995.

Thrush, Emily A. "Bridging the Gaps: Technical Communication in an International and Multicultural Society." Technical Communication Quarterly 2.3 (Summer 1993): 271–83.

University of Michigan. Engineering: Master of Science in Technical Information Design and Management. Ann Arbor, MI: University of Michigan, n.d.

CHAPTER 2

Lay, Mary M., and William M. Karis, eds. Collaborative Writing in Industry: Investigations in Theory and Practice. Amityville, NY: Baywood Publishing Company, Inc., 1991.

Pinker, Steven. "Chasing the Jargon." Time 13 (Nov. 1995): 30–31.

Roundy, N., and D. Mair. "The Composing Process of Technical Writers: A Preliminary Study." Journal of Advanced Composition 3 (1982): 89–101.

CHAPTER 3

Bikle, David, Lucent Technologies, Manager, Public Relations. Letter to the author. 15 Apr. 1997.

"Color Graphics: The Results of Our Tests." Consumer Reports Oct. 1996: 58–59.

Corel Corporation Limited. Corel Word Perfect Suite 7 User's Manual, Version 7.01. Ottawa, Ontario, Canada: Corel Corporation, n.d.

Pinker, Steven. "Chasing the Jargon." Time 13 Nov. 1995: 30–31.

Tritt, Merrill, Graphics Designer. Personal interview. 10 Feb. 1997.

White, Jan. V. Color for the Electronic Age: What Every Desktop Publisher Needs to Know About Using Color Effectively in Charts, Graphs, Typography, and Pictures. New York: Watson Guptill Publications, 1990.

XACT Development System. XEPLD Schematic Design Guide. San Jose, CA: XILINX, Dec. 1994.

CHAPTER 4

Bernhardt, Stephan A. "Seeing the Text." College Composition and Communication 37 (1986): 66–78.

Kostelnick, Charles. "Supra-textual Design: The Visual Rhetoric of Whole Documents." Technical Communication Quarterly 5.1 (Winter 1996): 9–23.

Kramer, Robert, and Stephen A. Bernhardt, "Teaching Text Design." Technical Communication Quarterly 5.1 (Winter 1996): 35–60.

Lay, Mary. "The Non-Rhetorical Elements of Design." Technical Writing, Theory and Practice. New York: Modern Language Association, 1989.

Markel, Mike. "Using Design Principles to Teach Technical Communication." Journal of Business and Technical Communication 9 (1995): 206–18.

CHAPTER 5

Bass, John, Manager, Motorola, Inc. Media and Communication Technology Center Radio Products Group. Personal interview. 12 Mar. 1997.

Weiss, Amy. Manager, Motorola, Inc. Integrated Technical Communications Paging Products Group. Personal interview. 7 Apr. 1997.

Weiss, Edmond H. How to Write a Usable User Manual. Philadelphia, PA: ISI, 1985.

Whitaker, Ken. A Guide to Publishing User Manuals. New York: Wiley, 1995.

CHAPTER 7

Braunwald, E., ed., Heart Disease. 4th ed. 2 vols. Philadelphia, PA: W. B. Saunders, 1991.

Columbia University College of Physicians and Surgeons. "How the Heart Works." Complete Home Medical Guide. Rev. ed. New York: Crown Publishers, Inc., 1989.

Frazier, O. H. "The Human Heart." Grolier Multimedia Encyclopedia, 1996. Online. 23 Apr. 1997.

"Jargon: The Time Digital Dictionary." Time 10 Mar. 1997: 48.

Tierney, L. M. Current Medical Diagnosis and Treatment. 34th ed. Stamford, CT: Appleton & Lange, 1997.

CHAPTER 8

Clement, David E. "Human Factors, Instructions and Warning, and Product Liability." IEEE Transactions on Professional Communication 30.3 (1987): 149-56.

Spyridakis, Jan H., and Michael J. Wenger. "Writing for Human Performance: Relating Reading Research to Document Design." Technical Communication 39.2 (1992): 202–15.

CHAPTER 9

Baker, David, and Dean Weinlaub. "Fossil Fuels." Sun-Sentinel 1 May 1994: G6.

"Geography Facts." Online edition of The Universal Almanac 8 Mar. 1994.

Krantz, Michael. "A Tube for Tomorrow." Time 14 Apr. 1997: 69.

"Making Electricity and Steam Together." The Lamp of Exxon Corporation 78.4 (Winter 1996): 4–5.

"Science Explained, The Way Nature Works." Online edition of Grolier's Academic American Encyclopedia 7 Mar. 1997.

CHAPTER 10

Blake, Gary. Quick Tips for Better Business Writing. New York: McGraw-Hill, Inc., 1995.

CHAPTER 11

Barthel, Brea, and Amanda Golrick-Jones, Renssalaer Polytechnic Institute. "Preparing a Resume." Online. 9 June 1997.

Forman, Ellen. "Judgement Day: Job Interview Survival Guide Advises Applicants to be Prepared for Any Question." Sun-Sentinel 31 Mar. 1997: B3.

Forman, Ellen. "On-line Bulletin Boards Aiding Job Hunters." Sun-Sentinel 12 May 1997: B2.

Grappo, Gary Joseph. "Selling 'Me Inc." Sun-Sentinel 24 Feb. 1997: B3.

Half, Robert. How to Get a Better Job in This Crazy World. New York: NAL/Dutton, 1993.

Hanson, Amy. "What Ever Happened to the Traditional Resume: Preparing Students for an Electronic Job Search. ATTW Bulletin 6.2 (Spring 1996): 4–6.

Harcourt, Jules, A. C. Krizan, and Patricia Merrier. "Teaching Resume Content: Hiring Officials' Preference Versus College Recruiters' Preferences." Business Education Forum 45:7 (1991): 13–17.

Kleiman, Carol. "Let the Interviewer Know You're Eager." Sun-Sentinel 18 Nov. 1996: B12.

Mannix, Margaret. "Writing a Computer-Friendly Resume: The Old Rules of Presenting Yourself Might Now Hurt." U.S. News and World Report 26 Oct. 1992: 90–93.

"Quick Guide to Resume Writing." <Kristin_Guenov@career.umeadm.maine.edu>

Racal-Datacom. "Software Engineers Job Posting." <http://www.careermosaic.com> 9 June 1997.

"Writers and Editors." Occupational Outlook Handbook. <http://stats.bls.gov/ocos089.html> 12 May 1997.

CHAPTER 13

Bowman, Joel P., and Bernadine P. Branchaw. How to Write Proposals That Produce. Phoenix, AZ: Oryx, 1992.

Vaughan, David K. "Abstracts and Summaries: Some Clarifying Distinctions." The Technical Writing Teacher 18.2 (1991): 132–41.

CHAPTER 14

Barden, Daniel F., Reference Service Trainer for Broward County Library/Main. Personal interview. 6 Aug. 1997.

Harnack, Andrew, and Eugene Kleppinger. Online! A Reference Guide to Using Internet Sources. New York: St. Martin's Press, 1997.

Kehoe, Brendan P. Zen and the Art of the Internet. Upper Saddle River, NJ: Prentice Hall, 1996.

Lester, James D. Writing Research Papers: A Complete Guide. 8th ed. New York: HarperCollins College Publishers, 1996.

Mann, Thomas. A Guide to Library Research Methods. New York: Oxford University Press, 1987.

Mann, Thomas. Library Research Models: A Guide to Classifications, Cataloging, and Computers. New York: Oxford University Press, 1993.

Tritt, Merrill, Computer Consultant. Personal interview. 15 July 1997.

CHAPTER 15

Pearsall, Tom, Cynthia Chapman, and Melinda Kreth. "Interchange: What Are the Differences Between 'Scientific' and 'Technical' Writing?" ATTW Bulletin 7:1 (Fall 1996): 6–7.

Sobel, Dava. "A Reporter At Large: Among Planets." The New Yorker 9 Dec. 1996: 47–52.

CHAPTER 16

American Psychological Association. Publication Manual of the American Psychological Association. 4th ed. Washington, DC: American Psychological Association, 1994.

Council of Biology Editors. The CBE Manual for Authors, Editors, and Publishers. Bethesda, MD: Council of Biology Editors, Inc., 1994.

Gibaldi, Joseph. MLA Handbook for Writers of Research Papers. 4th ed. New York: Modern Language Association of America, 1995.

Grossman, John, ed. Chicago Manual of Style. 14th ed. Chicago: University of Chicago Press, 1993.

Harnack, Andrew, and Eugene Kleppinger. Online! A Reference Guide to Using Internet Sources. New York: St. Martin's Press, 1997.

Lester, James D. Writing Research Papers: A Complete Guide. 8th ed. New York: HarperCollins College Publishers, 1996.

Moscowitz, John. Professor of English, Broward Community College, North. Personal interview. 5 Aug. 1997.

Riger, Stephanie. "Epistemological Debates, Feminist Voices—Science, Social Values, and the Study of Women." American Psychologist 57 (June 1992): 78–92.

CHAPTER 17

Alvear, Jose. "Meeting and Idea-sharing Online." Internet World August 1997: 33–34.

Greengard, Samuel. "Picture Perfect." U.S. Airways Magazine August 1997: 8–15.

CHAPTER 18

Greengard, Samuel. "Picture Perfect." U.S. Airways Magazine August 1997: 12–14.

Grice, Roger A., and Lenore S. Ridgway. "Presenting Technical Information in Hypermedia. Format: Benefits and Pitfalls." Technical Communication Quarterly 4:1 (1995): 35–36.

Maddison, Gordon, Instructor of English, Broward Community College. Interviews. 1993.

INDEX